MATHEMATICS
AN INFORMAL APPROACH

MATHEMATICS
AN INFORMAL APPROACH

Albert B. Bennett, Jr.

University of New Hampshire

Leonard T. Nelson

Portland State University

Allyn and Bacon, Inc. Boston, London, Sydney, Toronto

Production Editor: Lorraine Perrotta

Designer: Elizabeth Griffiths

Library of Congress Cataloging in Publication Data

Bennett, Albert B
  Mathematics, an informal approach.

  Companion vol.: Mathematics, an activity approach.
  Includes index
  1. Mathematics—1961–   I. Nelson, Leonard T.
joint author. II. Title.
QA39.2.B475   512'.1   78–27590
ISBN 0–205–06519–8

Printed in the United States of America.

ACKNOWLEDGMENTS

Cover adapted from photo by Andy Keech.

Graphic art drawn and supplied by Albert B. Bennett, Jr., and Deborah Schneck.

Unless otherwise stated, the following credits refer to photographs.

CHAPTER 1 CREDITS

pg. 2 (top) Reprinted by courtesy of Hale Observatories. (bottom left) Courtesy of Carl Zeiss, Inc., New York. (bottom right) Reproduced from *Art Forms in Nature,* by Ernst Haeckel (New York: Dover Publications, Inc., 1959). **pg. 3** (top right) Courtesy of Garry Bennett. (middle right) Reproduced from *Patterns In Nature,* by Peter S. Stevens, Atlantic Monthly/Little Brown, 1974. (bottom right) Photo by B.M. Shaub. **pg. 4** (bottom left and right) Reproduced from *Signs, Symbols, and Ornaments,* by Rene Smeets. Reprinted by courtesy of Van Nostrand Rheinhold Company. **pg. 5** (bottom) Reproduced by permission of the publisher from *Aesthetic Measure,* by George D. Birkhoff (Cambridge, MA: Harvard University Press). © 1933 by the President and Fellows of Harvard College; 1961 by Garrett Birkhoff. **pg. 9** (middle) B.C. by permission of Johnny Hart and Field Enterprises, Inc. **pg. 14** (middle) © Lorenz 1970—Reprinted from *Look* Magazine. **pg. 15** (middle) Courtesy Italian Government Travel Office. **pg. 16** (bottom) From *Mathematical Snapshots, Third American Edition, Revised and Enlarged,* by H. Steinhaus. Copyright © 1969 by Oxford University Press, Inc. Reproduced by permission. **pg. 20** (exercise 2a) © 1975 by The New York Times Company. Reprinted by permission. (exercise 2b) Reprinted by permission of United Press International. **pg. 26** (top) B.C. by permission of Johnny Hart and Field Enterprises, Inc. **pg. 27** (bottom right) Reprinted from *How To Take a Chance,* by Darrell Huff. Illustrated by Irving Geis., by permission of W.W. Norton & Company, Inc. Copyright © 1959 by W.W. Norton & Company, Inc. **pg. 31** (bottom right) Photo by Talbot Lovering. **pg. 32** (middle right) Photo by Albert B. Bennett, Jr. **pg. 33** (top) B.C. by permission of Johnny Hart and Field Enterprises, Inc. **pg. 37** Reprinted by permission of the publisher from *Math Puzzles* of Sam Loyd, editor Martin Gardner (New York: Dover Publications, Inc., 1959).

CHAPTER 2 CREDITS

pg. 39 (right) Courtesy of Dr. Jean de Heinzelin.
pg. 40 (bottom right) Courtesy of Musée de l'Homme.

pg. 42 (right) Copyright © 1975. Reprinted by permission of *Saturday Review* and Robert D. Ross. **pg. 47** Reprinted by courtesy of Hale Observatories. **pg. 51** (middle) Reproduced from 1977 Encyclopedia Britannica Book of the Year. Reprinted by courtesy of Encyclopedia Britannica, Inc. **pg. 55** (top) Courtesy of the Trustees of the British Museum. **pg. 56** Drawing by Mal; © 1973 The New Yorker Magazine, Inc. **pg. 59** (top right) Drawing by Kovarsky; © 1960 The New Yorker Magazine, Inc. **pg. 63** (top) Photo by Albert B. Bennett, Jr. (middle) *Chip Trading Activities,* © 1975 Davidson, Galton, Fair. Courtesy of Scott Resources, Inc., P.O. Box 2121, Fort Collins, Colorado. **pg. 65** (top right) Reprinted from *Margarita Philosophica* of Gregor Reisch, 1503. **pg. 68** (top right) Reprinted from *Summa de Arithmetica,* by Luca Pacioli. **pg. 70** (top) Reprinted from *Saturday Review.* Courtesy of James Estes. **pg. 76** Courtesy of Leo D. Geoffrion, Department of Education, University of New Hampshire. **pg. 77** (top) Graphs reprinted by permission of Jared J. Wolf and Bolt Beranek and Newman Inc., Cambridge, Massachusetts. (bottom right) Reproduced from *Personal Dynamic Media,* by the Learning Research Group. Reprinted by courtesy of Xerox Corporation. **pg. 78** Courtesy of Walker Art Center, Minneapolis. Photo by Eric Sutherland.

CHAPTER 3 CREDITS

pg. 84 (top) B.C. by permission of Johnny Hart and Field Enterprises, Inc. **pg. 91** (top right) Reprinted by permission from *More About Computers.* © 1974 by International Business Machines Corporation. (bottom right) Drawing by Dana Fradon; © 1976 The New Yorker Magazine, Inc. **pg. 96** (middle) B.C. by permission of Johnny Hart and Field Enterprises, Inc. **pg. 105** Courtesy of N.Y. State Office of General Services. **pg. 112** (middle right) Courtesy of Deutsches Museum, Munich. **pg. 115** From the exhibition "Mathematica: a World of Numbers & Beyond", made by the Office of Charles and Ray Eames for IBM Corporation. **pg. 123** Courtesy of TEREX Division, General Motors. **pg. 127** (bottom right) Drawing by W. Miller; © 1974 The New Yorker Magazine, Inc. **pg. 129** (top right) Reprinted by courtesy of Hale Observatories. **pg. 135** (exercise 15) Krypto, produced by M.P.H. Company, Inc., South Bend, Indiana. **pg. 140–41** The name Cuisenaire ® and the color sequence of the rods, squares and cubes, are trademarks of Cuisenaire Company of America, Inc. **pg. 142** (middle) Reprinted with permission from Sidney Harris. **pg. 146** Reprinted from *Aftermath IV,* Dale Seymour et al. Courtesy of Creative Publications. **pg. 147** Courtesy of U.S. Bureau of Census. **pg. 148** (middle right) reprinted by courtesy of Agencia J.B., Rio de Janeiro. **pg. 153** Otis Elevator illustration reprinted from *Architectural Record,* March 1970. © 1970 by McGraw-Hill, Inc. with all rights reserved.

CHAPTER 4 CREDITS

pg. 160 Photo by B.M. Shaub. **pg. 161** (top right) Photo by B.M. Shaub. (middle) B.C. by permission of Johnny Hart and Field Enterprises, Inc. **pg. 163** (top right) Hirshhorn Museum and Sculpture Garden, Smithsonian Institution. **pg. 164** (middle right cartoon) Reprinted with permission from Sidney Harris. **pg. 168** (top right) Reproduced from *Handbook of Gem Identification,* by Richard T. Liddicoat, Jr. Reprinted by permission of the Gemological Institute of America. (middle right) Photo by B.M. Shaub. **pg. 171** (top right) Copyright © 1974 by Charles F. Linn. Reprinted by permission of Doubleday & Company, Inc. **pg. 173** (top) Courtesy of General Motors Research Laboratories. (bottom right) Photo by B.M. Shaub. **pg. 177** (top) Courtesy of MAS, Barcelona, Spain. **pg. 179** (exercise 1) Reproduced from *Art Forms in Nature,* by Ernst Haeckel (New York: Dover Publications, Inc., 1959). **pg. 182** (exercise 14) Photo by Talbot Lovering. **pg. 183** "Cubic Space Division," by M.C. Escher. Courtesy of the Escher Foundation, Haags Gemeentemuseum, The Hague. **pg. 184** [figures (a) and (c)] Photos by B.M. Shaub. [figure (b)] Courtesy British Museum (Natural History). **pg. 185** (middle right) Reproduced from Minerology, by Ivan Kostov, 1968. Courtesy of the author. (top) Photo by Talbot Lovering. **pg. 186** (top) Reproduced from *Polyhedron Models for the Classroom,* by Magnus J. Wenninger. Courtesy of the National Council of Teachers of Mathematics. (middle right) Reproduced from *The Public Buildings of Williamsburg,* by Marcus Whiffen, published by the Colonial Williamsburg Foundation and distributed by Holt, Rinehart and Winston. **pg. 187** (bottom) Photo by B.M. Shaub. **pg. 188** (bottom right) Courtesy of National Aeronautics and Space Administration. **pg. 189** (middle) Drawings reprinted with permission from *Encyclopaedia Britannica,* 14th edition, © 1972 by Encyclopaedia Britannica, Inc. **pg. 190** Reprinted with permission from *Collier's Encyclopedia.* © 1976 Macmillan Educational Corporation. **pg. 191** (exercise 1a) "Waterfall," by M.C. Escher. Courtesy of the Escher Foundation, Haags Gemeentemuseum, The Hague. **pg. 195** (exercise 11) Courtesy of Babson College, Wellesley, Massachusetts. (exercise 14) Courtesy of National Aeronautics and Space Administration. **pg. 196** Courtesy of Government of India Tourist Office, New York. **pg. 197** (bottom) Courtesy University of New Hampshire Media Services. **pg. 199** (middle) B.C. by permission of Johnny Hart and Field Enterprises, Inc. (bottom left and right) Reproduced from *Snow Crystals,* by W.A. Bentley and W.J. Humphreys. Reprinted by courtesy of McGraw-Hill Book Company, Inc. **pg. 200** (top right) Photograph reprinted by courtesy of The Magazine *Antiques.* (bottom left and right) Courtesy of Entomology Department, University of New Hampshire. **pg. 201** (top excerpt) *The Unexpected Hanging,* by Martin Gardner, © 1969, Simon and Schuster, Inc. (top left and right) Courtesy of Lothrop's Ethan Allen, Dover, New Hampshire. (top middle) Courtesy of Jamaica Lamp Company, Queensville, New York. (bottom right) "Sphere with Fish," by M.C. Escher. Courtesy of the Escher Foundation, Haags Gemeentemuseum, The Hague. **pg. 202** (top right) Courtesy of Lothrop's Ethan Allen, Dover, New Hampshire. **pg. 203** Courtesy of Spanish National Tourist Office, New York. **pg. 204** (exercise 2) Courtesy of The American Numismatic Society, New York. (exercises 4a, b, c) Reproduced from *Art Forms in Nature,* by Ernst Haeckel (New York: Dover Publications, Inc., 1959). **pg. 205** (top right) photo by Albert B. Bennett, Jr. **pg. 207** (exercise 13, left) Courtesy of Erie Glass Company, Parkridge, Illinois. (exercise 13, middle and right) Courtesy of Lothrop's Ethan Allen, Dover, New Hampshire. **pg. 208** (exercises 16a, b, c) Reprinted from *Early American Design Motifs,* by Suzanne E. Chapman (New York: Dover Publications, Inc. 1974). (exercises 17a, b, c, d) Reprinted from *Symbols, Signs and Signets,* by Ernst Lehner (New York: Dover Publications, Inc. 1950).

CHAPTER 5 CREDITS

pg. 211 (top) Courtesy of British Tourist Authority, New York. **pg. 212** (bottom right) From *The Book of Knowledge,* © 1960, by permission of Grolier Incorporated. **pg. 216** Reprinted from *Popular Science* with permission. © 1975 Times Mirror Magazines, Inc. **pg. 218** (top right) Courtesy of O.L. Miller and Barbara A. Hamkalo, Oak Ridge National Laboratory. **pg. 220** (exercise 1 cartoon) Reprinted with permission from Sidney Harris. **pg. 221** (exercise 4) Photo by Albert B. Bennett, Jr. **pg. 223** (exercise 11) Courtesy of Science Museum, London. **pg. 224** (exercise 12) Reprinted with permission from The Associated Press. **pg. 226** Courtesy of the Federal Reserve Bank of Minneapolis. **pg. 228** (bottom) Photos by Albert B. Bennett, Jr. **pg. 229** (middle) Photo by Talbot Lovering. **pg. 231** (top right) Based on drawing from *Mathematics and Living Things,* Student Text, School Mathematics Study Group, 1965. Reprinted by permission of Leland Stanford Junior University. **pg. 232** (top right) Reproduced from Encyclopaedia Britannica (1974), Volume VIII of the Micropaedia. Reprinted by permission of Encyclopaedia Britannica, Inc. (middle right) By permission *Saturday Review* © 1976 & V. Gene Myers. **pg. 233** (top right) Courtesy of Gemological Institute of America. (middle and bottom right) Photos by Albert B. Bennett, Jr. **pg. 234** (top right) Photo by Albert B.

Bennett, Jr. **pg. 237** (exercise 1) Courtesy of Gunnar Birkerts and Associates, Architects. (exercise 2) By John A. Ruge, reprinted from the *Saturday Review.* **pg. 242** (exercise 12) Reprinted by permission of Peachtree Plaza. (exercise 14) Photo by Albert B. Bennett, Jr. **pg. 244** (exercise 17) Courtesy of Dr. William Webber, University of New Hampshire. (exercise 19) Photo by Talbot Lovering. **pg. 245** Courtesy of General Dynamics, Quincy Shipbuilding Division. **pg. 248** (top right) Courtesy of Renaissance Center. **pg. 249** Reprinted by permission of Transamerica Corporation. **pg. 250** (middle right) From *Crystals* by Raymond Wohlrabe. Copyright © 1962 by Raymond A. Wohlrabe. Reprinted by permission of J.B. Lippincott Company. **pg. 251** (top right) Courtesy of National Aeronautics and Space Administration. **pg. 252** (top left and right) Photos by Talbot Lovering. **pg. 254** (exercise 1) Courtesy of John P. Adams, University of New Hampshire. (exercise 3) B.C. by permission of Johnny Hart and Field Enterprises, Inc. **pg. 256** (exercise 8) Collection: Mrs. Harry Lynde Bradley, Andre Emmerich Gallery. **pg. 257** (exercise 9) Courtesy of General Dynamics, Quincy Shipbuilding Division. (exercise 11) Photo by Herb Moyer, Exeter, New Hampshire. Reprinted by permission from Rodney Sanderson.

CHAPTER 6 CREDITS

pg. 262 Courtesy Edward C. Topple, N.Y.S.E. Photographer. **pg. 263** Reproduced by courtesy of the Trustees of the British Museum. **pg. 264** (bottom right) Official U.S. Navy photograph. **pg. 265** (top left) From *The Book of Knowledge,* © 1960, by permission of Grolier Incorporated. (top right) Redrawn from The Encyclopedia Americana, © 1978 The Americana Corporation. **pg. 266** (bottom) Fraction Bar model from the Fraction Bar materials by Albert B. Bennett, Jr. and Patricia S. Davidson, © 1973, by permission of Scott Resources, Inc., Fort Collins, Colorado **pg. 272** (exercise 1) Courtesy of Minolta Corporation. **pg. 278** (top, and bottom right) Courtesy of Rockwell International. **pg. 279** (top left, middle, and right) Illustrations from *Webster's New International Dictionary,* Second Edition. © 1959 used by permission of G. & C. Merriam Co., Publishers of the Merriam-Webster Dictionaries. **pg. 288** Courtesy of American Stock Exchange. **pg. 290** (exercise 4) Illustration from *Wesbster's New International Dictionary,* Second Edition © 1959 used by permission of G. & C. Merriam Co., Publishers of the Merriam-Webster Dictionaries. **pg. 292** (exercise 11) Drawing by Dana Fradon; © 1973 The New Yorker Magazine, Inc. **pg. 294** (exercise 14) Photo by courtesy of the Trustees of the British Museum. **pg. 295** (top) B.C. by permission of Johnny Hart and Field Enterprises, Inc. **pg. 296** Redrawn from *Sports Illustrated.* © 1976 Time Inc. **pg. 297** (middle right) Courtesy of National Aeronautics and Space Administration. (bottom) Adapted from *The Book of Popular Science,* Courtesy of Grolier In-

corporated. **pg. 306** (exercise 1) Official U.S. Navy Photograph. **pg. 307** (exercise 3) Redrawn from a photo by courtesy of National Aeronautics and Space Administration. **pg. 309** (exercise 7, table) Adapted from *1977 Britannica Book of the Year,* Copyright © 1977 by Encyclopaedia Britannica, Inc.

CHAPTER 7 CREDITS

pg. 314 Courtesy of Professor Erwin W. Mueller, The Pennsylvania State University. **pg. 316** Photo by Albert B. Bennett, Jr. **pg. 319** (top right) Redrawn by permission from Fenga and Freyer, Inc. **pg. 324** (exercise 1) Courtesy of the National Bureau of Standards. **pg. 325** (exercise 3) Redrawn by permission from D. Reidel Publishing Company. **pg. 328** (top) Courtesy of Bulova Watch Co. Inc. **pg. 337** (exercise 1) Reprinted from *U.S. News & World Report,* September 6, 1976 issue. Copyright © 1976 U.S. News & World Report, Inc. **pg. 344** Courtesy of AT & T. **pg. 345** (table) Adapted by permission from *1977 Britannica Book of the Year,* Copyright © 1977 by Encyclopaedia Britannica, Inc. **pg. 346** (bottom right) Photo by Herb Moyer, Exeter, New Hampshire. **pg. 349** Courtesy of the Stanford Research Institute, J. Grippo (Project Manager). **pg. 350** (bottom right) Redrawn from data by courtesy of the Population Reference Bureau, Inc. **pg. 354** (exercise 1) Courtesy of Rockwell International. (exercise 2) Courtesy U.S. Department of Transportation and The Advertising Council. **pg. 356** (exercise 9) Adapted with permission from *U.S. News & World Report.* Copyright © 1976 U.S. News & World Report, Inc. **pg. 357** (exercise 11) Reprinted by permission of the Pillsbury Company. **pg. 358** (exercise 12) Reprinted by permission of the National Association of Home Builders. (exercise 13) Courtesy of National Aeronautics and Space Administration. **pg. 362** (exercise 20) Redrawn by permission of Union Oil Company of California. **pg. 363** (top) Reprinted with permission from Sidney Harris. **pg. 365** (top right) Courtesy of Yale University. **pg. 366** (middle) Courtesy of Lehnert and Landrock, Cairo.

CHAPTER 8 CREDITS

pg. 381 Courtesy U.S. Parachute Association. **pg. 384** (top) Drawing by Chas. Addams; © 1974 The New Yorker Magazine, Inc. **pg. 385** (top right) Courtesy of M.I.T. News Office. **pg. 386** (top right) Courtesy of Daytona International Speedway. **pg. 390** (exercise 1) Courtesy of the National Center for Atmospheric Research (NCAR), Boulder, Colorado. **pg. 391** (exercise 2c) Courtesy of Dr. Richard A. Petrie. **pg. 397** (exercise 14, left) Official U.S. Navy Photograph. (exercise 14, right) Redrawn with permission from Volume 21, Collier's Encyclopedia. © 1977 Macmillan Educational Corporation. **pg. 398** Reproduced from *The Earth and Its Satellite,* John E. Guest, Editor. Reprinted by courtesy of David McKay Company, New York. **pg. 399** (top right) Fountain at Swirbul Library, Adelphi University. Photo courtesy of Madeleine Lane, sculptor. **pg. 405** (exercise 1) Courtesy of General Dynamics. **pg. 410** (top) Courtesy of the American Institute of Steel Construction. (middle left) Courtesy of Calvin Campbell, M.I.T. (middle right) Courtesy of N.Y. State Office of General Services. **pg. 411** (top right, middle left and right) Redrawn from *Mathematics and Living Things,* Teacher's Commentary, School Mathematics Study Group, © 1965, by permission of Leland Stanford Junior University. **pg. 412** (bottom right) Reproduced from *Patterns in Nature,* by Peter S. Stevens, Atlantic Monthly/Little Brown, 1974. Reprinted by courtesy of Peter S. Stevens. **pg. 413** (middle right) Photo by Garry Bennett. **pg. 414** (top right) Redrawn from *The Book of Popular Science,* by permission of Grolier Incorporated. **pg. 416** (exercise 1) Reproduced from *Mathematical Carnival,* by Martin Gardner, © 1975. Reprinted by permission of Alfred A. Knopf, Inc. **pg. 418** (exercise 6) Redrawn from *Mathematics and Living Things,* Teacher's Commentary, School Mathematics Study Group, © 1965, by permission of Leland Stanford Junior University. **pg. 420** (exercise 9) Redrawn from *All About Sound and Ultrasonics,* by Ira M. Freeman, illustrated by Irving Geis, © 1961. Courtesy of Random House, Inc. **pg. 422** (exercise 16) Adapted from *The Limits to Growth: A Report for The Club of Rome's Project on the Predicament of Mankind,* by Donella H. Meadows, Dennis L. Meadows, Jørgen Randers, William W. Behrens III. A Potomac Associates book published by Universe Books, New York, 1972. Graphics by Potomac Associates. Permission from Universe Books. **pg. 423** (exercise 17) Reproduced from *An Introduction to Color,* by Ralph M. Evans, © 1948 John Wiley & Sons, Inc. Reprinted by permission of John Wiley & Sons, Inc.

CHAPTER 9 CREDITS

pg. 425 (top) Reprinted by permission Saturday Review, © 1976 & V. Gene Myers. **pg. 426** (bottom right) Courtesy of U.S. Geological Survey. **pg. 427** (bottom) Courtesy of U.S. Department of State. **pg. 428** (bottom right) Courtesy of Travel Marketing, Inc., Seattle. **pg. 429** (top right) "Swans," by M.C. Escher. Courtesy of the Escher Foundation, Haags Gemeentemuseum, The Hague. **pg. 430** (top) Alhambra drawings by M.C. Escher. Escher Foundation, Haags Gemeentemuseum, The Hague. **pg. 431** (bottom right) Courtesy of Rival Manufacturing Company. **pg. 432** (top right and bottom) Reproduced from the book, *Let's Play Math,* by Michael Holt and Zoltan Dienes. Copyright © 1973 by Michael Holt and Zoltan Dienes used by permission of publisher, Walker and Company. **pg. 435** (top) Courtesy of John P. Adams, University of New Hampshire. **pg. 436** (exercise 3) Courtesy University of New Hampshire Media Services. **pg. 441** (bottom) Courtesy of J.E. Cermak, Fluid Dynamics and Diffusion Laboratory, Colorado State University. **pg. 442** (top right)

Courtesy of Rockwell International Corporation. **pg. 446,** (top right) Reproduced by permission of the National Ocean Survey (NOAA), U.S. Department of Commerce. **pg. 448** (middle right) Photo by Wayne Goddard, Eugene, Oregon. Description of knife reprinted by permission of the Register Guard, Eugene, Oregon. **pg. 450** (top right) Courtesy of Sally Ann Sweeney. **pg. 452** (exercise 1) Courtesy of Boeing. **pg. 457** (exercise 13) Courtesy of Sally Ann Sweeney. **pg. 458** (exercise 14) Courtesy of Sidney Rogers Chair Company, Georgetown, Massachusetts. **pg. 460** (top) Photo. Science Museum, London. (bottom right) British Crown Copyright. Science Museum, London. **pg. 461** (top right, and middle) Photos by Talbot Lovering. **pg. 462** (middle) Photos by Albert B. Bennett, Jr. **pg. 464** (right) "Moebius Strip II," by M.C. Escher. Courtesy of the Escher Foundation, Haags Gemeentemuseum, The Hague. (left) Drawing from patent 3267406, filed by R.L. Davis. Reprinted by courtesy of Sandia Laboratories, Albuquerque, New Mexico. **pg. 471** (exercise 11) Redrawn from *Puzzles and Graphs,* by John Fujii. Published by NCTM.

CHAPTER 10 CREDITS

pg. 476 (bottom right) Courtesy of Joseph Zeis, cartoonist. **pg. 478** (top right) Photo by Albert B. Bennett, Jr. (middle right) Reprinted by permission of New Hampshire Sweepstakes Commission. **pg. 479** (exercise 2) B.C. by permission of Johnny Hart and Field Enterprises, Inc. **pg. 482** (exercise 9) Mortality Table from *Principles of Insurance,* by Robert Mehr and Emerson Cammack, © 1966. Reprinted by courtesy of R.D. Irwin, Inc. **pg. 483** (exercise 13) Reprinted by permission of New Hampshire Sweepstakes Commission. **pg. 484** (exercise 14) Reprinted by permission of Publishers Clearing House. **pg. 485** (top) Courtesy of National Aeronautics and Space Administration. **pg. 491** (exercise 4) Reprinted from *Ladies Home Journal,* February 1976, by courtesy of Henry R. Martin, cartoonist. **pg. 493** (exercise 12) Photo courtesy of The Edward Petritz Family, Butte, Montana. **pg. 495** (top) Drawing by Ziegler; © 1974 The New Yorker Magazine, Inc. **pg. 496** Reproduced from "Bills of Mortality" (p. 83) in *Devils, Drugs, and Doctors,* by Howard M. Haggard, M.D. Reprinted by permission of Harper & Row, Publishers, Inc. **pg. 497** (Miss America table) Reprinted by permission of *The World Almanac and Book of Facts 1977.* Copyright © 1976 by Newspaper Enterprise Association, Inc., New York. **pg. 498** (gestation table) Reprinted by permission of *The World Almanac and Book of Facts 1977.* Copyright © 1976 by Newspaper Enterprise Association, Inc., New York. **pg. 499** (earthquake table) Reprinted by permission of *The World Almanac and Book of Facts 1977.* Copyright © 1976 by Newspaper Enterprise Association, Inc., New York. **pg. 503** Drawing by Webber; © 1975 The New Yorker Magazine, Inc. **pg. 504** (Exercise 1, per capita income table) Reprinted by permission of *The*

World Almanac and Book of Facts 1977. Copyright © 2976 by Newspaper Enterprise Association, New York. **pg. 505** (exercise 3, state gas tax table) Reprinted by permission of *The World Almanac and Book of Facts 1977.* Copyright © 1976 by Newspaper Enterprise Association, New York. **pg. 506** (exercise 5) From *The Peoples Almanac,* 1975 Edition by David Wallechinsky and Irving Wallace. Copyright © 1975 by David Wallechinsky and Irving Wallace. Reprinted by permission of Doubleday & Company, Inc. **pg. 507** (exercise 7, home run table) Reprinted by permission of *The World Almanac and Book of Facts 1977.* Copyright © 1976 by Newspaper Enterprise Association, Inc., New York. **pg. 509** (exercise 13) Reprinted by permission of The Arbitron Company. **pg. 510** Reproduced from the collection of the Library of Congress. **pg. 511** Reproduced from *A Guide to the Unknown,* by Frederick Mosteller et al., 1972. Reprinted by courtesy of Holden-Day, Inc. **pg. 515** (bottom) Redrawn from the Differential Aptitude Tests. Copyright © 1972, 1973 by The Psychological Corporation. Reproduced by special permission of the publisher. **pg. 517** (exercise 1) Reprinted by courtesy of *Washington Post.* **pg. 520** (exercise 11) Redrawn from the Stanford Achievement Test. Copyright © 1973 by Harcourt Brace Jovanovich, Inc. Reproduced by special permission of the publisher. **pg. 521** (exercise 13) From the exhibition "Mathematica: a World of Numbers & Beyond," made by the Office of Charles and Ray Eames for IBM Corporation. **pg. 522** (exercise 14) Copyright 1956. Reprinted by permission of *Saturday Review* and Ed Fisher.

CONTENTS

PREFACE

It is easier to believe what you see than what you hear: but if you both see and hear, then you can understand more readily and retain more lastingly; I wish, therefore, so to arrange my work that everything may be understood as easily as possible.

Albrecht Dürer

TO STUDENTS

If you are among the many people who feel they dislike mathematics, this book may change your attitude. It was written to help you relate mathematical ideas to your life. You will see many illustrations of the ways in which mathematics occurs in nature and in a wide variety of applications. Just as Albrecht Dürer stated in the above quote, we have attempted to arrange the ideas so that everything may be understood as easily as possible.

There are sections throughout the text on the use of calculators. You will find it helpful to have a calculator for these exercises. The starred (★) exercises are either completely or partially answered for your convenience. A complete answer guide may be purchased. The exercises contain puzzles, tricks, and other recreational mathematics which we think you will enjoy.

You will discover at least three facts by using this book. First, mathematics is more than a collection of computational skills—it is a way of thinking. Second, mathematics occurs around you and in your life in many surprising and interesting ways. Third, mathematics can be enjoyable. You may also discover, as have many students who used this book, that you like mathematics.

TO THE INSTRUCTOR

This text together with the mathematical activities in *Mathematics: An Activity Approach* represents the content and style of the introductory mathematics courses we teach. Because we believe mathematics is learned by doing, we compiled detailed problem and exercise sets and inductive activity sets for our classes. No current texts were compatible with our approach, or sufficiently readable and interesting to our students—so this book is the result.

The ten chapters of this book contain most of the topics from number systems, geometry, probability, and statistics which are usually taught in introductory mathematics courses. The topics are presented in an informal and intuitive manner with a minimum of symbolism and terminology. The approach is basically inductive rather than deductive. If the text is motivating, covers the essential topics, provides applications, exercises, and problems, then the instructor can add the degree of rigor he/she desires.

Special Features

Metric system—The metric system is used throughout the text. In particular, the chapter entitled *Measurement* presents concepts and applications of the metric units for length, area, volume, weight, and temperature.

Calculators—There are instructions for using a calculator in the sections on numeration, whole numbers, fractions, integers, decimals, and functions. There are calculator exercises which follow each of these sections.

History—Many of the topics in this book are introduced by historical highlights. As Aristotle has said, "Here and elsewhere we shall not obtain the best insight into things until we actually see them growing from the beginning. . .".

Pedagogy—The sections on numeration, whole numbers, fractions, integers, and decimals contain illustrations of models for understanding and teaching the basic operations with these numbers.

Exercises—"I *hear* and I *forget*. I *see* and I *remember*. I *do* and I *understand*." This ancient Chinese proverb shows that getting students actively involved in learning is not a new idea. The exercise sets are extensive, occupying over 40 percent of the book. They contain a wide range of applications.

Readings—Each of the 36 sections of the book has a list of selected readings from books and journals. These sources were chosen as appropriate reading assignments for students.

Activity Book

Mathematics: An Activity Approach is an activity book which can be used to supplement the text. It contains 36 activity sets, one for each section of the text. Each activity set is a sequence of inductive activities and experiments which enables the student to build up mathematical ideas through the use of models and the discovery of patterns. The activity sets extend the ideas presented in the corresponding sections of the text. Many of the activity sets are followed by *Just for Fun* enrichment activities. There is also an appendix of 40 material cards for use with the activities.

Organization and Format

The ten chapters of this book together with the activity book provide sufficient material for a two-semester course. Because variety is an important course ingredient, chapters on geometry are interspersed with those on number systems. Moreover, this enables an early introduction to measurement and the metric system. Each chapter contains from three to six sections, and each section is followed by an exercise set. Answers are provided for approximately 40 percent of the questions. A star (★) beside the exercise indicates that the question is either completely or partially answered in the answer section. Complete answer guides are available for the activity book as well as the text. The text and activity book can be used together in several ways:

Lecture course with assignments from the exercise sets. The activity book can be used as a supplement to the text.

Combination lecture and lab course in which lectures are followed by a lab class on the related activity set, either during the same class period or in two consecutive class meetings.

Lab course based on the activity sets. The text readings and exercise sets can be assigned as a supplement.

Sequences of Topics

Here are four possibilities for selections of topics for one-semester courses. Recommendations III and IV contain combinations of topics from number systems, geometry, probability, and statistics.

I. Number Systems

    Chapters: 1—Nature of Mathematics
                  2—Counting, Sets, and Numeration
                  3—Whole Numbers and Their Operations
                  6—Fractions and Integers
                  7—Decimals: Rational and Irrational Numbers

II. Geometry

    Chapters: 1—Nature of Mathematics
                  4—Geometric Figures
                  5—Measurement
                  8—Geometry with Coordinates
                  9—Motions in Geometry

III. Chapters: 1—Nature of Mathematics
                  2—Counting, Sets, and Numeration
                  3—Whole Numbers and Their Operations
                  4—Geometric Figures
                  5—Measurement
               10—Probability and Statistics

IV. Chapters: 1—Nature of Mathematics
                  2—Counting, Sets, and Numeration
                    (Sections 2.1 and 2.2)
                  3—Whole Numbers and Their Operations
                    (Sections 3.1, 3.2, 3.3, and 3.4)
                  4—Geometric Figures
                    (Sections 4.1 and 4.2)
                  5—Measurement
                    (Sections 5.1 and 5.2)
                  7—Decimals: Rational and Irrational Numbers
                    (Sections 7.1 and 7.2)

9—Motions in Geometry
(Sections 9.1 and 9.2)
10—Probability and Statistics
(Sections 10.1 and 10.2)

ACKNOWLEDGMENTS

We wish to express our appreciation to some of the many people who provided assistance and encouragement during the development of this book. We are grateful to: S. Jeanne Gamlen of Western Washington University, Richard Balomenos and Carol Findell of the University of New Hampshire, and Eugene Maier of the Oregon Mathematics Education Council, for using preliminary versions of this text in their courses; Richard Shumway of Ohio State University, Joe Dan Austin of Emory University, Ruth Wong of the University of Hawaii, Patricia Davidson of Boston State College, Richard Kratzer of the University of Maine, Portland-Gorham, Phares O'Daffer of Illinois State University, and S. Jeanne Gamlen for their reviews of this book and valuable suggestions; Peter Warren, Pamela Bradley, and Salley Sweeney for editing the manuscript at various stages and contributing ideas to improve it; Carmela Minicucci for skillful typing of the preliminary manuscript; Warren Celli for instruction in graphic arts; reference librarians Hugh Pritchard, Diane Tebbetts, Jane Russell, Reina Hart, and especially Dick Pantano who always did more than was expected; Ron Bergeron and Sylvia Stimpson at media services; Margarie Bickford at printing service; Mary Ann Morris, Graphics Etc., for typesetting the final manuscript; Carl Lindholm, the series editor, Lorraine Perrotta, and Libby Griffiths for guiding the manuscript through publication; and our many students whose enthusiasm and criticism inspired numerous revisions.

Finally, we wish to express our love and gratitude to our wives, Jane and Roxanne, and our children Al, Andrea, Garry, Greg, Kris, Marcy, and Nicky for their understanding and patience during the many hours that we were unable to be with them.

Albert B. Bennett, Jr.
Leonard T. Nelson

MATHEMATICS
AN INFORMAL APPROACH

CHAPTER ONE

NATURE OF MATHEMATICS

All Nature is but Art, unknown to thee;
All Chance, Direction, which thou canst not see;
All Discord, Harmony not understood. . . .

Alexander Pope

When Plato was asked what God does, he is said to have answered,
"God eternally geometrizes."

The great spiral galaxy in Andromeda

1.1 MATHEMATICAL PATTERNS

Patterns in Nature   From the smallest of microscopic plants and organisms to the great galaxies
of the universe, Nature has shown no lack of imagination in creating mathematical patterns. The variety of shapes shown below
on the left are sea plants, so small that a
thimbleful contains millions of them.
Which shapes have mathematical names
you are familiar with?

Diatoms, minute unicellular algae

Microscopic sea urchin

The spiral is one of the most intricate of growth patterns. Many kinds of seashells spiral outward as they develop. The sundial seashell has a fairly flat top with a diameter of about 5 centimeters. Its spirals are alternating dashes of light and dark colors. Other examples of spirals occur in the heads of sunflowers and daisies, in pine cones, and in the horns of some animals. The spider spins a spiral web, and many galaxies, including our own, are whirling spirals of stars.

Web of an Epurid Spider
Courtesy of
The American Museum of Natural History.

Pine Cone

Gems, crystals, and minerals contain a rich source of geometric figures. The cluster of cubes pictured on the right is a group of galena crystals. Their faces are usually rough but always at right angles to each other.

Intersecting cubes of galena crystals

Ornamental Patterns   Surrounded by geometric
forms, our primitive ancestors gradually
copied these shapes in their art and architecture. The earliest known ornaments
tended to reflect the world of nature and
were probably body decorations. Later
ornaments were more abstract and
geometric.

Mycenean jar, 1200–1125 B.C.
Courtesy The Metropolitan Museum of Art.

Athenian vase, eighth century B.C.
Courtesy Museum of Fine Arts, Boston.

Polynesian woven mat

Japanese bamboo mat

Ornamentation, which means the creation of repetitive figures, has been found in every civilization. This container from Attica, Greece, and the Athenian vase and Polynesian mat shown on the previous page, each have several different repeating patterns.

Container with cover, Attica, Greece, eighth century B.C. Courtesy Museum of Fine Arts, Boston.

Many examples of patterns from fabrics, utensils, and buildings have survived to this day. A few of these are shown next.*

CHINESE ORNAMENT PAINTED ON PORCELAIN

FRENCH RENAISSANCE ORNAMENT FROM CASKET

MASONRY FRET, TEMPLE AT MITLA, MEXICO

INDIAN PAINTED LACQUER WORK

EGYPTIAN MAT DESIGN (in two colors)

PERSIAN ILLUMINATED MANUSCRIPT

PAINTING ON INDIAN (HINDU) VASE

AMERICAN INDIAN BASKETRY

JAPANESE WROUGHT IRON VASE

GERMAN RENAISSANCE EMBROIDERY (BAVARIA)

*For a complete set of mathematically different one- and two-dimensional ornaments, see G.D. Birkhoff, *Aesthetic Measure* (Cambridge, Mass.: Harvard University Press, 1933), between pp. 56–57.

Growth Patterns   Patterns occur in plants and trees in a variety of ways. Many of these patterns are related to a famous sequence of numbers called *Fibonacci numbers.* After the first two numbers of this sequence, which are 1 and 1, each successive number may be obtained by adding the two previous numbers.

$$1 \quad 1 \quad 2 \quad 3 \quad 5 \quad 8 \quad 13 \quad 21 \quad 34 \quad 55 \quad \cdots$$

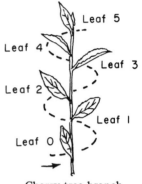

Cherry tree branch

If you select a particular bud or leaf from a branch and then move in a spiral motion around the branch until reaching the bud or leaf directly above your selection, the number of leaves will be a Fibonacci number. This Fibonacci number is governed by the type of plant. For the cherry tree this number is 5. After leaf 0, there are five leaves, up to and including leaf 5, which is directly above. Notice that the beginning leaf is not counted. The number of complete turns in passing around the stem from one leaf to a leaf that is directly above is also a Fibonacci number. Following the dotted path around this stem, you will see that the number of complete turns for the cherry tree is two. (Further examples of this type can be found in Exercise Set 6.1, Exercise 20.)

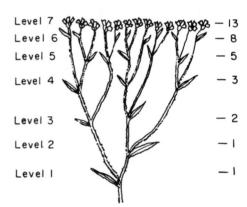

Sneezewort plant

A different occurrence of Fibonacci numbers is found in the number of branch points on the stem of a plant as it develops. The beginning of each new branch in this sketch of the sneezewort plant is marked by a leaf. The number of new branches that develop at each level is a Fibonacci number. At each of the first two levels there is only one new branch. At the third level there are two new branches, and so on. The total number of branches at each level, as well as the number of blossoms at the top of the plant, is a Fibonacci number.

Fibonacci numbers were discovered by the Italian mathematician Leonardo Fibonacci (ca. 1175-1250). He used these numbers in studying the birth patterns of rabbits. Suppose that in the first month a pair of rabbits are too young to produce more rabbits, but in the second month and every month thereafter they produce a new pair of rabbits. Each new pair of rabbits will follow the same pattern. The increasing numbers of pairs of rabbits are Fibonacci numbers.

The realization that Fibonacci numbers could be applied to the science of plants occurred long after they had been developed to solve the rabbit birthrate problem.

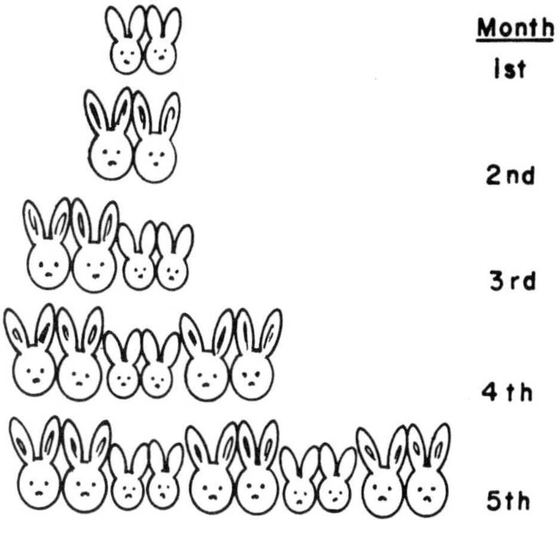

	Month
	1st
	2nd
	3rd
	4th
	5th

Number Patterns and Sequences Number patterns have fascinated people since the beginning of recorded history. One of the earliest patterns was the distinction between even numbers (0, 2, 4, 6, 8, ...) and odd numbers (1, 3, 5, 7, 9, ...). The game of "Even and Odd" has been played for generations. To play this game, one person picks up some stones and the other person guesses "odd" or "even." If the guess is correct, that person wins.

Pascal's Triangle—One of the most familiar number patterns is Pascal's triangle. It has been of interest to mathematicians for hundreds of years, appearing in China as early as 1303. This triangle is named after the French mathematician Blaise Pascal (1623-1662) who wrote a book on some of its uses. There are many patterns in the rows and diagonals. One pattern is used to obtain each row of the triangle from the previous row. In the sixth row, for example, each of the numbers 5, 10, 10, and 5 can be obtained by adding the two numbers in the row above it. What numbers are in the seventh row of Pascal's triangle?

```
        1
      1   1
    1   2   1
  1   3   3   1
1   4   6   4   1
1  5  10  10  5  1
```

Number Sequences—Sequences of numbers are often generated by patterns. In the sequences shown to the right, each number is obtained from the previous number in the sequence by adding the same number throughout.

7, 11, 15, 19, 23, ...
10, 20, 30, 40, 50, ...
7, 4, 1, ⁻2. ⁻5, ...

This number is called the *common difference*. Such a sequence is called an *arithmetic sequence* or *arithmetic progression*. The first arithmetic sequence shown here has a common difference of 4. The common differences for the second and third sequences are 10 and ⁻3.

In the next type of sequence each number is obtained by multiplying the previous number by the same number. This number is called the *common ratio,* and the resulting sequence is called a *geometric sequence* or *geometric progression.* The common ratio in the first sequence shown here is 2, in the second sequence it is 5, and in the third sequence it is 1/2. Find the next number in each of these sequences.

3, 6, 12, 24, 48, ...
1, 5, 25, 125, 625, ...
8, 4, 2, 1, $\frac{1}{2}$, ...

Triangular and Square Numbers—The following sequence of numbers is neither arithmetic nor geometric. These numbers are called *triangular numbers* because of the arrangement of dots that is associated with each number. What is the next triangular number?

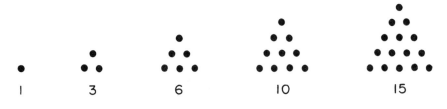

The numbers of dots in square arrays are called *square numbers* or *perfect squares.* The first few of these numbers are shown on the right.

There are other types of numbers that receive their names from the numbers of dots in geometric figures (see Exercise Set 1.1). Such numbers are called *polygonal* or *geometric numbers* and represent one kind of link between geometry and arithmetic.

Finite Differences—Some sequences of numbers have patterns that can be found by computing the differences between the successive numbers. This is called the method of *finite differences.* The method is shown here to determine the next number in the sequence 1, 1, 4, 12, 27, 51, ...

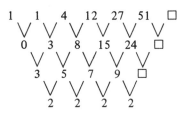

Each number in the second row, 0, 3, 8, 15, and 24, is obtained by subtracting the two numbers just above it. The numbers in the third and fourth rows are also obtained by subtracting the numbers above them. The process is stopped when all the numbers in a row are equal. We can now work our way from the bottom row to the top row, filling in the squares at the ends of the three rows. The next number after 9 is 9 + 2. The next number after 24 is 24 + 11, and the number following 51 is 51 + 35.

SUPPLEMENT (*Activity Book*)

Activity Set 1.1 Geometric Number Patterns (Geometric patterns on arrays of dots and their corresponding number patterns)

Just for Fun: Fibonacci Numbers in Nature

Additional Sources

Carman, R.A. and M.J. Carman. "Number patterns." *The Arithmetic Teacher,* **17** No. 8 (December 1970), 637–39.

Edmonds, G.F. "An Intuitive Approach to Square Numbers." *The Mathematics Teacher,* **63** No. 2 (February 1970), 113–17.

Gardner, M. "The Multiple Fascinations of the Fibonacci Sequence." *Scientific American,* **220** No. 3 (March 1969), 116–20.

Hervey, M.A. and B.H. Litwiller. "Polygonal numbers: a study of patterns." *The*

Arithmetic Teacher, **17** No. 1 (Jan. 1970), 33–38.

Hoffman, N. "Pascal's triangle." *The Arithmetic Teacher,* **21** No. 3 (March 1974), 190–98 (17 number patterns).

Seymour, D. and M. Shedd. *Finite Differences.* Palo Alto: Creative Publications, 1973.

Stevens, P.S. *Patterns in Nature.* Boston: Little, Brown and Company, 1974.

EXERCISE SET 1.1

1. There are many patterns and number relationships on the calendar that can be easily discovered. Here are a few.

   ★ a. The sum of the three circled dates on this calendar is 45. For any sum of three consecutive numbers (from the rows), there is a quick method for determining the numbers. Explain how this can be done. Try your method to find the three consecutive numbers whose sum is 54.

   b. If you are told the sum of any three adjacent dates from a column, it is possible to determine the three numbers. Explain how this can be done and use your method to find the numbers whose sum is 48.

   ★ c. The sum of the 3 by 3 array of numbers outlined on the calendar is 99. There is a shortcut method for using this sum to find the 3 by 3 array of numbers. Explain how this can be done. Try your method to find the 3 by 3 array whose sum is 198.

2. Each digit in the top row of this table has been multiplied by itself, and the units digit of the product is recorded in the second row. For example, $8 \times 8 = 64$ and there is a 4 under the 8.

Digit	1	2	3	4	5	6	7	8	9
2nd power	1	4	9	6	5	6	9	4	1
3rd power									
4th power									
5th power									

★   a. Compute the third power of each digit and record the units digit in the third row of the table.

★   b. Compute the fourth and fifth powers of each digit in the top row and record the units digits in the fourth and fifth rows of the table.

     c. You should have obtained nine different digits in the row for the third power. What is the next power for which this will happen?

     d. Look for a pattern in this table. Without doing any more computing, you should be able to write the numbers that occur in the row for the tenth power.

★ 3. Here are the first few Fibonacci numbers: 1, 1, 2, 3, 5, 8, 13, 21, 34, 55. Compute the following sums and compare the answers with the Fibonacci numbers. Find a pattern and explain how this pattern can be used to find the sums of consecutive Fibonacci numbers.

$$1 + 1 + 2 =$$
$$1 + 1 + 2 + 3 =$$
$$1 + 1 + 2 + 3 + 5 =$$
$$1 + 1 + 2 + 3 + 5 + 8 =$$
$$1 + 1 + 2 + 3 + 5 + 8 + 13 =$$
$$1 + 1 + 2 + 3 + 5 + 8 + 13 + 21 =$$

4. The numbers in the sums to the right were obtained by using every other Fibonacci number (circled).

①   1   ②   3   ⑤   8   ⑬   21   ㉞   55

Compute these sums. There is a surprising relationship between these sums and the Fibonacci numbers. What is this relationship?

$$1 + 2 =$$
$$1 + 2 + 5 =$$
$$1 + 2 + 5 + 13 =$$
$$1 + 2 + 5 + 13 + 34 =$$

★ 5. The sums of the squares of consecutive Fibonacci numbers form a pattern when written as a product of two numbers.

★   a. Complete the missing sums and find the pattern.

★   b. Use this pattern to explain how the sum of the squares of the first n Fibonacci numbers can be found.

$$1^2 + 1^2 = 1 \times 2$$
$$1^2 + 1^2 + 2^2 = 2 \times 3$$
$$1^2 + 1^2 + 2^2 + 3^2 = 3 \times 5$$
$$1^2 + 1^2 + 2^2 + 3^2 + 5^2 =$$
$$1^2 + 1^2 + 2^2 + 3^2 + 5^2 + 8^2 =$$
$$1^2 + 1^2 + 2^2 + 3^2 + 5^2 + 8^2 + 13^2 =$$

6. A Fibonacci-type sequence can be started with any two numbers. Each number after 3 and 4 in this sequence, 3, 4, 7, 11, 18, 29, etc., was obtained by adding the previous two numbers. Find the missing numbers in the first ten numbers of these Fibonacci-type sequences.

★ a. 10,_____, 24, _____,_____, 100,_____,_____,_____, 686

b. 2,_____,_____, 16, 25,_____,_____,_____,_____, 280

c. 1,_____,_____, 11,_____,_____,_____,_____, 118,_____

d. The sum of the first ten numbers in the sequence in part **a** is equal to 11 times the seventh number, 162. What is this sum?

e. Can the sums of the first ten numbers in the sequences in parts **b** and **c** be obtained by multiplying the seventh numbers by 11?

f. Is the sum of the first ten numbers in the Fibonacci sequence equal to 11 times the seventh number in that sequence?

7. The products of 1089 and the first few digits produce some interesting number patterns. Describe one of these patterns. Will this pattern continue if 1089 is multiplied by 5, 6, 7, 8, and 9?

$$1 \times 1089 = 1089$$
$$2 \times 1089 = 2178$$
$$3 \times 1089 = 3267$$
$$4 \times 1089 = 4356$$
$$5 \times 1089 =$$

8. There are many patterns in Pascal's triangle. Add the first few numbers in the 2nd diagonal, starting from the top. This sum will be another number from the triangle. Will this be true for the sums of the first few numbers in the other diagonals?

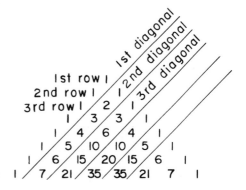

★ 9. Compute the sums of the numbers in the first few rows of Pascal's triangle.

★   a. What kind of sequence (arithmetic or geometric) do these sums form?

b. What will be the sum of the numbers in the 12th row of this triangle?

10. In the familiar song "The Twelve Days of Christmas," the number of gifts that were received each day are triangular numbers. On the 1st day there was one gift, on the 2nd day there were three gifts, on the 3rd day, six gifts, etc., until the 12th day of Christmas. How many gifts were received on the 12th day? What is the total number of gifts received for all 12 days?

11. Use the method of finite differences to find the next number in each of the following sequences.

★   a. 1, 2, 7, 22, 53, 106,   b. 1, 3, 11, 25, 45, 71,

★ 12. Normally, when we ask someone to "find the pattern" in a sequence of numbers, what we really mean is to find the pattern that we have in mind. In the following sequences there are many "next numbers." Select a different "next number" for each sequence and explain what the pattern is in each case.

★   a. 1, 2, 4, 8,   ★   b. 1, 2, 4, 8,

13. Identify each of the following sequences as being arithmetic or geometric. Write the next three terms in each sequence.

★   a. 4.5, 9, 13.5, 18,   b. 15, 30, 60, 120,
★   c. 24, 20, 16, 12,   d. 729, 243, 81, 27,

14. The method of finite differences will sometimes enable you to find the next number in a sequence, but not always.

   a. Write out the first eight numbers of a geometric sequence and try the method of finite differences to find the ninth number. Will this method work?

   b. Answer part **a** for an arithmetic sequence.

15. Square arrays of dots show that each square number can be derived from the previous square number by adding an L-shaped border. This L-shape is called a *gnomon* (nō′ mon) and was used by the ancient Greeks as a sundial. If such a figure is turned to the east in the morning and to the west in the afternoon, the hours can be read from the shadow on the horizontal arm. The Greeks recognized that each odd number, 3, 5, 7, etc., corresponded to the gnomon of a square. For example, $3^2 + 7 = 4^2$, as shown in the following diagram.

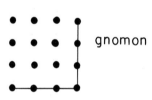 gnomon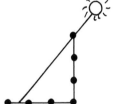

   a. Write the number in each of the following blanks to complete the equations. Look for a pattern. Does this pattern hold for any two consecutive square numbers?

$$4^2 + \underline{\hspace{1cm}} = 5^2 \qquad 5^2 + \underline{\hspace{1cm}} = 6^2 \qquad 6^2 + \underline{\hspace{1cm}} = 7^2$$

★   b. Use the method of finite differences to write the first row of numbers under the following sequence of square numbers. What kind of numbers do you obtain?

        1   4   9   16   25   36   49   64   81   100   121

   c. Do the numbers you obtain in part **b** form an arithmetic sequence or geometric sequence?

16. The Greeks (500–200 B.C.) were deeply interested in numbers associated with patterns of dots in the shapes of geometric figures. Write the next three numbers in each sequence in parts **a, b,** and **c.**

a. Triangular numbers

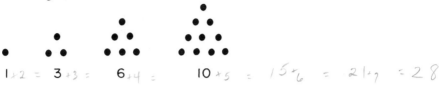

$1+2 = 3+3 = 6+4 = 10+5 = 15+6 = 21+7 = 28$

b. Square numbers

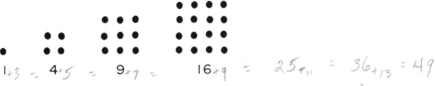

$1+3 = 4+5 = 9+7 = 16+9 = 25+11 = 36+13 = 49$

★ c. Pentagonal numbers

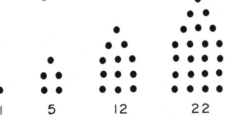

1   5   12   22

d. This array of dots has been partitioned to show that 16 is the sum of two triangular numbers. Partition a 5 by 5 array of dots to find two triangular numbers whose sum is 25.

16 = 10 + 6

e. The pentagonal arrays of dots show that every pentagonal number is the sum of a triangular number and a square number. Write the pentagonal number 70 as the sum of a triangular number and a square number.

17. *Calendar Trick:* Here is a trick you can have some fun with. Give a friend a calendar and ask him or her to secretly circle one number in each row and add these numbers. Without seeing the results, you then ask how many Sundays were circled. For the circled dates shown here the answer is 2. Then ask how many Mondays were circled, etc., repeating the question for each day of the week. After receiving this information you will be able to quickly tell your friend the sum of the circled numbers. The following questions will help you understand how this trick works.

Nov. **1978**

Sun	Mon	Tue	Wed	Thu	Fri	Sat
			1	2	③	4
⑤	6	7	8	9	10	11
12	13	⑭	15	16	17	18
⑲	20	21	22	23	24	25
26	27	28	29	㉚		

★   a. What would be the sum if each of the five dates under Wednesday were circled? (This number is the key to the trick.)

★   b. What is the sum of any three dates under Wednesday and one date under both Tuesday and Thursday? (Remember, select only one date from each row.) Does it matter which of the three Wednesday dates are chosen?

     c. What is the sum of any three dates under Wednesday and one each under the Thursday and Friday columns?

     d. Explain how this trick works.

1.2 MATHEMATICAL REASONING

Inductive Reasoning   The process of forming conclusions on the basis of observations or experiments is called *inductive reasoning.* Everybody uses this type of reasoning, and when our inferences are based on too few examples we hear the phrase, "Don't jump to conclusions." Although inductive reasoning is impor-

tant in mathematics, it sometimes leads to incorrect results. Consider, for example, the number of regions that can be obtained in a circle by connecting points on the circumference of the circle. Connecting 2 points gives 2 regions; connecting 3 points gives 4 regions; and so on. Each time a new point on the circle is used, the number of regions appears to double.

2 points   3 points   4 points   5 points   6 points

2 regions   4 regions   8 regions   16 regions

These numbers of regions are the beginning of the geometric sequence 1, 2, 4, 8, 16, ..., and it is tempting to conclude that 6 points will produce 32 regions. However, no matter how the 6 points are located on the circle, there will not be more than 31 regions.

In spite of the uncertainty of inductive reasoning, its contributions to mathematics and science have been enormous. Galileo Galilei (1564–1643), Italian astronomer and physicist, made a discovery concerning swinging pendulums and it became one of the most famous applications of inductive reasoning. While he was attending a church service, his mind was distracted by a large bronze lamp that oscillated back and forth. Using his pulse to keep time he discovered that the time required for the lamp to complete one swing through an arc was the same, whether the arc was small or large!

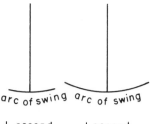

arc of swing   arc of swing

I second   I second

Counterexample—Conclusions that are based on inductive reasoning are sometimes later shown to be false. An example which shows that a statement is false is called a *counterexample.* One famous counterexample by Galileo disproved a statement that had been accepted as true for 2000 years. Aristotle (384–322 B.C.), ancient Greek philosopher and scientist, had reasoned that heavy objects fall faster than lighter ones. Galileo demonstrated this to be false by dropping two pieces of metal from the Leaning Tower of Pisa. In spite of the fact that one was 10 times heavier than the other, both hit the ground at the same moment.

George Polya, the author of *How To Solve It,* offers the following advice: "If you have a general statement, you can at least test it to see if it is true for a few special cases. You may be able to find a counterexample." Consider the statement,

Leaning Tower of Pisa, Pisa, Italy

"The sum of two whole numbers is divisible by 2." This is a true statement for the following pairs of numbers: 4 and 10; 7 and 11; 12 and 16; and 5 and 17. It is not true for 8 and 9 because 17 is not divisible by 2. This is a counterexample which shows that the statement is false.

Analogy and Intuition   Reasoning by *analogy* is the forming of conclusions based on similar situations. For example, we know that for a given perimeter, a circle encloses a greater area than a square or any other plane figure. Reasoning by analogy suggests that a sphere encloses more volume than any other figure having the same surface area. In this case, the conclusion is true.

Reasoning by analogy influences much of our thinking, from trivial conclusions to creative activities. However, there are pitfalls with this type of reasoning, as illustrated by the following example. For 12- and 20-sided regular polygons which are inscribed in circles of the same size, the 20-sided polygon has the larger perimeter and the larger area. Reasoning by analogy suggests that if a regular polyhedron with 12 faces and a regular polyhedron with 20 faces are inscribed in spheres of the same size, then the polyhedron with 20 faces will have the greater surface *area* and the greater volume. However, both assumptions are false! The surface area and volume of the 20-sided polyhedron are significantly smaller.

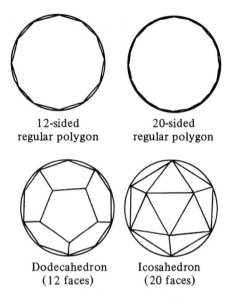

12-sided
regular polygon

20-sided
regular polygon

Dodecahedron
(12 faces)

Icosahedron
(20 faces)

Closely associated with reasoning by analogy is what is known as *intuition*. While the nature of intuition is admittedly vague, it does influence our reasoning and discoveries. There are many well-known examples of false conclusions that were based on intuition. The theories that the earth is a sphere and that it moves were once widely rejected because they were contrary to intuition.

In his paper, "The Crisis in Intuition," Hans Hahn discusses some of the outstanding mathematical discoveries which contradicted ideas that had been previously accepted as intuitively certain.* One of these crises was the discovery of space-filling curves by the Italian mathematician Giuseppe Peano (1858–1932). The following three squares show the beginning of a process for obtaining such a curve. Each curve is obtained by uniting four similar copies of the previous curve. If this process is continued, it can be proven that these curves approach a definite curve which passes through all the points of the square.

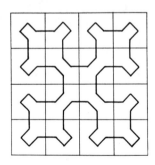 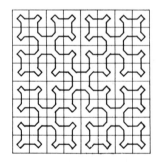 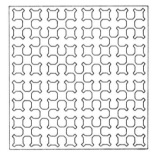

Clearly, the intuitive approach can lead to error, but committing errors can be an important part of the learning process. As the English statesman Francis Bacon (1561–1626) noted, "Truth emerges more readily from error than from confusion."

*See J. R. Newman, *The World of Mathematics,* 3 (New York: Simon and Schuster, 1956), pp. 1956–76.

Deductive Reasoning   *Deductive reasoning* is the process of deriving conclusions from given statements. With inductive reasoning and reasoning by analogy our conclusions are always open to doubt, no matter how fully they may be warranted; whereas, with deductive reasoning we can be sure of the conclusions if the statements from which we are reasoning are true.

In the following illustration of deductive reasoning, the conclusion is deduced from two given statements, called *hypotheses*.

Given Statements
1. In a certain community there are 12 dentists.
2. In one class of 26 students, each student goes to one of these dentists.

Conclusion
At least 3 students go to the same dentist.

By accepting statements 1 and 2 we are compelled to accept the conclusion. This is the essence of deductive reasoning.

We owe the development of deductive reasoning to the ancient Greeks. By insisting on deductive reasoning as a means of mathematical proof, they transformed mathematics from a set of rules with trial and error procedures into a logical deductive system. The crowning mathematical achievement of this period was Euclid's *Elements,* a series of 13 books written about 300 B.C. These books contain over 600 theorems, which were obtained by deductive reasoning from 10 basic assumptions called *axioms* or *postulates.* Euclid's *Elements* stood as the model of deductive reasoning for over 2000 years.

Mathematical Systems—A mathematical system consists of *undefined terms, definitions, axioms,* and *theorems.* There must always be some words that are undefined. *Line* is an example of an undefined term in geometry. We all have an intuitive idea of a line but trying to define it involves more words, such as "straight," "extends indefinitely," "has no thickness," etc. These words would also have to be defined. To avoid this problem certain basic terms, such as point and line, are chosen to be undefined. These words are then used in definitions to define other words. Similarly, there must always be some statements, called axioms, which we assume to be true and do not try to prove. Finally, the axioms, definitions, and undefined terms are used together with deductive reasoning to prove statements called theorems.

Undefined terms
and definitions
↓
Axioms
↓
Theorems

Historically, the axioms for geometry were suggested by the physical world and were considered to be "self-evident truths." For example, Euclid assumed that through a point P not on a line L one and only one line can be drawn parallel to L. This axiom is called the *Parallel Postulate* and was destined to become controversial. There is evidence that Euclid avoided using the Parallel Postulate in proving his first few theorems. It was suspected from the beginning that this axiom could be proven from Euclid's other nine axioms. Attempts at such a proof failed and finally, after 2000 years, lead to the development of non-Euclidean geometry.

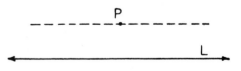

The creation of non-Euclidean geometry has been ranked with the greatest events in the history of thought. As a result of these discoveries there are now hundreds of mathematical systems, each with its own set of undefined terms, definitions, axioms, and theorems. Axioms no longer need to be suggested by the physical world or to appear obviously true. On the contrary, in developing the Theory of Relativity, Albert Einstein (1879-1955) chose an axiom that seems to contradict our intuition. The liberty to choose axioms and create mathematical systems is aptly phrased by the German mathematician Georg Cantor (1845-1918), "The essence of mathematics is its freedom."

Nature of Proof   Sometimes an equation or geometric property is known to be true for thousands of cases, and yet until it is proved deductively for all cases, we cannot be sure it will always be true. Such an unproven statement is called a *conjecture*. One of the best known of these statements is *Goldbach's Conjecture*.

Every even number greater than 4 is the sum of two odd primes.
$$6 = 3 + 3, 8 = 3 + 5, 10 = 3 + 7, 12 = 5 + 7$$

Computers have been used to show that this conjecture is true for all even numbers less than 10,000, but this approach can never be used to show it is true for all the remaining even numbers.

When a statement involves an infinite set of numbers, it is impossible to prove that it is true by trying all numbers. For example, is the sum of any three consecutive whole numbers divisible by 3? Here are two examples for which this is true, but there are an infinite number of such cases to consider.

$$15 + 16 + 17 = 48 \text{ and } 48 \div 3 = 16, \quad 19 + 20 + 21 = 60 \text{ and } 60 \div 3 = 20$$

To show that this is true for any three consecutive whole numbers, we must reason deductively with arbitrary numbers. The first of the three numbers will be represented by $\square$. Symbols like this are called *variables* or *placeholders*. The three consecutive numbers are

$$\square, \square + 1, \square + 2$$

Dividing their sum, $3\square + 3$, by 3 we get the whole number $\square + 1$. This proves not only that the sum of three consecutive whole numbers is divisible by 3, but that dividing by 3 always gives the second of the three numbers.

Number tricks and so-called magic formulas can be analyzed by deductive reasoning. Here is a formula for you to try. Select any number and perform the following operations:

Add 4 to your number; multiply the result by 6; subtract 9; divide by 3; add 13; divide by 2; and then subtract the number you started with.

If you performed these operations correctly your final answer is 9, regardless of the number you started with. Let's denote an arbitrary first number by $\square$ and list the steps in a formula.

1. Select a number . . . . □
2. Add 4 . . . . □ + 4
3. Multiply by 6 . . . . 6□ + 24
4. Subtract 9 . . . . 6□ + 15
5. Divide by 3 . . . . 2□ + 5
6. Add 13 . . . . 2□ + 18
7. Divide by 2 . . . . □ + 9
8. Subtract the number □ 9

This proof shows that it doesn't matter what number is written into the placeholder □. In the final step it gets subtracted and the end result is always 9.

Deductive reasoning is the hallmark of mathematics. Without this type of reasoning no statement is accepted. Nevertheless, with all its advantages the logic of deductive reasoning does not supersede inductive reasoning and experience. This viewpoint is expressed by Leonardo da Vinci:

> "I shall begin by making some experiments before I proceed any further; for it is my intention first to consult experience and then show by reasoning why that experience was bound to turn out as it did. This, in fact, is the true rule by which the student of natural effects must proceed; although nature starts from reason and ends with experience, it is necessary for us to proceed the other way around, that is, as I said above, begin with experience and with its help seek the reason."

SUPPLEMENT (*Activity Book*)

Activity Set 1.2 Games of Reasoning ("Patterns," "Pica-Centro," and "Pick-a-Word")

Just for Fun: Game of "Hex"

Additional Sources

Gardner, M. *The Second Scientific American Book of Mathematical Puzzles and Diversions.* Simon and Schuster, 1961. "Recreational Logic," pp. 119–29 and "Eleusis: The Induction Game," pp. 165–73.

Graening, J. "Induction: Fallible but Valuable." *The Mathematics Teacher,* **64** No. 2 (February 1971), 127–31.

Kline, M. *Mathematics in Western Culture.* New York: Oxford University Press, 1953. pp. 410–31 (non-Euclidean geometries).

Newman, J.R. and E. Kasner. *Mathematics and the Imagination.* New York: Simon and Schuster, 1940. "Pastimes of Past and Present Times," pp. 156–92.

Phillips, H. *My Best Puzzles in Logic and Reasoning.* New York: Dover, 1961.

Rawling, R. and M. Levine. "The Parallel Postulate." *The Mathematics Teacher,* **62** No. 8 (December 1969), 665–69 (a historical sketch).

Willoughby, S.S. "Revolution, Rigor, and Rigor Mortis." *The Mathematics Teacher,* **60** No. 2 (February 1967), 105–8.

EXERCISE SET 1.2

1. Find a pattern in each of the following sets of equations and use inductive reasoning to predict the next equation. Evaluate both sides of your equation.

★ a. $1^2 + 2^2 + 2^2 = 3^2$
$2^2 + 3^2 + 6^2 = 7^2$
$3^2 + 4^2 + 12^2 = 13^2$

b. $1^3 + 2^3 = 3^2$
$1^3 + 2^3 + 3^3 = 6^2$
$1^3 + 2^3 + 3^3 + 4^3 = 10^2$

c. $1 + 2 = 3$
$4 + 5 + 6 = 7 + 8$
$9 + 10 + 11 + 12 = 13 + 14 + 15$

2. What kind of reasoning is used to arrive at the conclusions in these two articles?

a.
Vitamin C student finds a little is best

By NANCY HICKS
New York Times News Service

NEW YORK — A Canadian researcher has reported finding therapeutic value in using Vitamin C to treat symptoms of the common cold in much lower doses than had been previously recommended.

Dr. Terence W. Anderson, an epidemiologist at the University of Toronto, reported a 30 per cent reduction in the severity of cold symptoms in persons who took only a small amount of Vitamin C — less than 250 milligrams a day regularly, and one gram a day when symptoms of a cold began.

The amounts represent a fraction of those recommended by Dr. Linus Pauling, the Nobel Prize-winning chemist who three years ago popularized the Vitamin C regimen in his book "Vitamin C and the Common Cold."

Anderson's conclusion was based on a study of 600 volunteers.

b.
Operating room work may have health hazards

WASHINGTON (UPI) — There is an increase in cancer and other disease rates among hospital operating room personnel and a report Monday said regular exposure to anesthetic gases appears to be the most likely cause.

A survey of 49,585 operating room personnel indicated that female anesthetists and nurses are the most vulnerable, particularly if they are pregnant.

"The results of the survey strongly suggest that working in the operating room and, presumably, exposure to trace concentrations of anesthetic agents entails a variety of health hazards for operating personnel and their offspring."

★ 3. If we begin with the number 6, then double it to get 12, and then place the 12 and 6 side by side, the result is 126. This number is divisible by 7. Try this procedure for some other numbers. Find a counterexample which shows that the result is not always divisible by 7.

4. One line will divide the interior of a circle into 2 regions; 2 lines will divide it into 4 regions; 3 lines will produce 7 regions; and for 4 lines there are 11 regions.

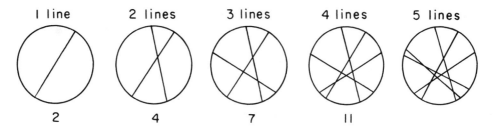

I line   2 lines   3 lines   4 lines   5 lines

2   4   7   11

★   a. Examine the first four circles and then predict the maximum number of regions for 5 lines. Check your conclusion by counting the regions.

b. What type of reasoning did you use in part **a**, if your conclusion was based on your observations for 1, 2, 3, and 4 lines?

c. What is the maximum number of regions for 10 lines?

5. Galileo discovered that the amount of time for one swing of a pendulum depends on the length of the pendulum. Find a pattern in this table and determine the length of a pendulum that will take 5 seconds to complete one swing.

Time of swing (in seconds)	Units of length of pendulum
1	1
2	4
3	9
4	16
5	

★ 6. Using whole numbers there is just one combination of two numbers whose sum is 2 and one combination whose sum is 3. For the numbers 4 and 5 there are two different ways of writing the sum using two numbers. Predict the number of different sums for the numbers from 5 to 10 and then show that your conjecture is either true or false. In general, how many sums of different pairs of numbers are there for a given number?

$$2 = 1 + 1$$
$$3 = 1 + 2$$
$$4 = 1 + 3 = 2 + 2$$
$$5 = 1 + 4 = 2 + 3$$
$$6 =$$
$$7 =$$
$$8 =$$
$$9 =$$
$$10 =$$

7. Here is a chance to use your intuition. If the heads-up dime is rolled along the top half of the circumference of the tails-up dime until it has been moved from left to right, will the head be right-side up or upside-down? Check your intuition by experimenting with coins.

8. Our intuition is often influenced by optical illusions. Are the lines in Figures **a** and **b** straight or bent? Are the lengths of line segments $\overline{AB}$ and $\overline{CD}$ in Figures **c** and **d** equal or unequal?

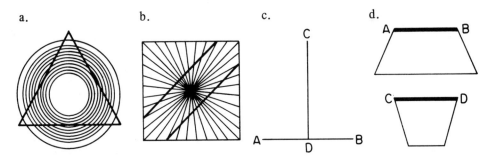

a. b. c. d.

9. Add any five consecutive whole numbers. Is the sum always divisible by 5?

★   a. Let △ represent an arbitrary whole number and write the next four whole numbers.

△, _____ , _____ , _____ , _____

★   b. Add the terms in part **a** and divide the sum by 5. What does this prove?

c. Find a counterexample to show that the sum of four consecutive whole numbers is not divisible by 4.

10. Find a counterexample for each of the following statements.

a. The product of any two whole numbers is divisible by 2.

b. Every whole number greater than 5 is the sum of either two or three consecutive whole numbers. For example, 11 is equal to 5 + 6, and 18 is equal to 5 + 6 + 7.

11. For the three consecutive numbers 8, 9, and 10, the product of the first and last, 8 × 10, differs by 1 from the middle number times itself. Will this be true for any three consecutive whole numbers? Try a few and make a conjecture.

8   9   10

8 × 10 = 80
9 × 9 = 81

Let △, △ + 1, and △ + 2 be three arbitrary consecutive whole numbers and prove or disprove your conjecture.

12. The following questions are from a section called "Intuition Versus Proof" in the text *An Introduction to the History of Mathematics.** Give an intuitive answer to each question and then compute or reason out the answer.

★   a. A car travels 120 miles at 40 miles per hour and then makes the return trip of 120 miles at 60 miles per hour. What is the car's average speed?

b. Is a salary of 1 cent for the first half-month, 2 cents for the second half-month, 4 cents for the third half-month, 8 cents for the fourth half-month, and so on until the year is used up, a good or a poor salary for the year?

*H. W. Eves, *An Introduction to the History of Mathematics*, 3rd ed. (New York: Holt, Rinehart and Winston, 1969), p. 138.

★   c. A clock strikes six in 5 seconds. At midnight how long will it take to strike 12 times?

d. A bottle and a cork together cost $1.10. If the bottle costs a dollar more than the cork, how much does the cork cost?

★   e. Suppose that in one glass there is a certain quantity of liquid A, and in a second glass an equal quantity of another liquid B. A spoonful of liquid A is taken from the first glass and put into the second glass; then a spoonful of the mixture from the second glass is put back into the first glass. Is there now more or less liquid A in the second glass than there is liquid B in the first glass?

13. *Non-Euclidean Geometry:* The German mathematician Bernhard Riemann (1826–1866) invented the non-Euclidean geometry whose axioms and theorems can be illustrated on the sphere. In this geometry the "lines" are the great circles of the sphere. A *great circle* is a circle on the surface of the sphere whose center is the center of the sphere. The equator and the meridians through the North and South Poles are some of the great circles. In this figure, *X* is a great circle and *Y* is not. Which of the following axioms and theorems from Euclidean geometry are true in Riemann's geometry, if we interpret "line" to be "great circle" and "point" to be a "point on the surface of the sphere"?

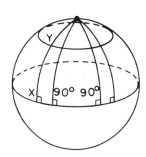

★   a. Two points determine one and only one line.

b. Through a point *P* not on a line *L*, one and only one line can be drawn perpendicular to *L*. [Great circles through the North and South Poles form 90-degree angles (that is, are perpendicular) with the equator, as shown in the above figure.]

★   c. For every line there is a point not on the line.

d. There are 180 degrees in every triangle.

★   e. Two lines intersect in one and only one point.

f. If points *A*, *B*, and *C* are on the same line and *B* is between *A* and *C*, then *C* is not between *A* and *B*.

g. Through a point *P* not on a line *L*, there is another line through *P* which is parallel to *L* (the Parallel Postulate).

★ 14. *Birthday Trick:* To guess a person's birthday have him/her write down the number of the month of his/her birth (1 for January, 2 for February, etc.) and then carry out the steps in the following formula: Multiply the number by 5; add 6 to the result; multiply by 4; add 9; multiply by 5; and finally have him/her add the number of the day he/she was born on.

Next ask for the result of the computation: The number you get after subtracting 165 will tell the month and day he/she was born on. For example, if the number is 622, it is June 22. If the number is 1104, it is November 4.

Using □ as a placeholder for month and △ as a placeholder for day, finish writing down the steps for the Birthday Formula.

```
Number of month . . . . . . . . . . . . . □
Multiply by 5 . . . . . . . . . . . . . . . . 5□
Add 6 . . . . . . . . . . . . . . . . . . . . . 5□ + 6
Multiply by 4 . . . . . . . . . . . . . . . .
Add 9 . . . . . . . . . . . . . . . . . . . . .
Multiply by 5 . . . . . . . . . . . . . . . .
Add the number of the day . . . . . . .
Subtract 165 . . . . . . . . . . . . . . . . .
```

Explain why this formula works. Why is 165 subtracted in the last step of the formula?

★ 15. After the first term the top sequence shown here is a geometric sequence. Write the next two numbers in this sequence. Add 4 to each number in the top sequence and divide the results by 10 to complete the lower sequence.

0 3 6 12 24 __ __
.4 .7 __ __ __ __ __

★ a. This famous sequence of numbers is known as Bode's Law and gives an amazingly close approximation of the distances from the first seven planets to the sun in astronomical units. (An astronomical unit of 1 is the earth's distance from the sun.) Using this sequence of numbers and inductive reasoning, astronomers predicted that there would be a planet between Mars and Jupiter. What was the predicted distance of this planet from the sun? (Asteroids were eventually found between Mars and Jupiter, the biggest of which is Ceres, about 800 kilometers in diameter.)

Planet	Distance from Sun in Astronomical Units	Bode's Law
Mercury	0.4	0.4
Venus	0.7	0.7
Earth	1.0	1.0
Mars	1.52	1.6
Ceres	2.77	
Jupiter	5.2	5.2
Saturn	9.5	10.0
Uranus	19.2	
Neptune	30.1	38.8
Pluto	39.5	77.2

b. In the 1770s when Bode's Law was discovered, only the first five planets in the table had been discovered. Using Bode's Law astronomers found Uranus. What should its distance have been to satisfy Bode's Law?

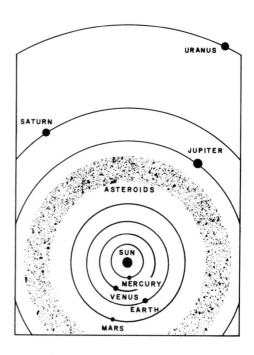

16. *Deductive Reasoning Challenge:* There are three containers that are labelled *incorrectly,* as shown. One contains two nickels, another two pennies, and another a nickel and a penny. Explain how it is possible to determine the correct labellings for all three containers by selecting just one coin. (*Hint:* Each can is labelled incorrectly.)

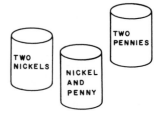

★ 17. *Logic about Logic Prize:* Smith, Brown, Jones, and Robinson are told that they have each won one of the four prizes for mathematics, English, French, and logic, but none of them know which. They are speculating about it. Smith thinks that Robinson has won the logic prize. Brown thinks that Jones has won the English prize. Jones feels confident that Smith has not won the mathematics prize, and Robinson is of the opinion that Brown has won the French prize. It turns out that the winners of the mathematics and logic prizes were correct in their speculations, but the other two were wrong. Who won which prize? (*Hint:* There are only a few possibilities for combinations of right and wrong opinions. Try them all to see which combination has no contradiction.)

1.3 PROBLEMS AND PROBLEM SOLVING

It is a widely held opinion that learning to solve problems is the principal reason for studying mathematics. There are two parts to every problem: (1) The given information; and (2) The question. Sometimes the main difficulty is in deciding which parts of the given information are needed or in determining what question is to be answered.

George Polya, the author of *How to Solve It,* points out that a person becomes a good problem solver by solving a variety of problems. This philosophy is similar to the often-heard phrase, "practice makes perfect," and implies that there is a certain amount of skill required for problem solving. Polya identifies the following four aspects of problem solving: Understanding the problem; devising a plan; carrying out the plan; and looking back (that is, checking solutions, generalizing results, or varying the conditions to obtain new problems).*

While it is not possible to give a checklist of strategies for understanding a problem or devising a plan for the solution, there are several approaches that are frequently useful. These will be discussed and illustrated in the following paragraphs.

Drawing Pictures   One of the most helpful suggestions for understanding a problem and obtaining ideas for a solution is to draw pictures and diagrams. There is a widely known story about a mathematician who got stuck in the middle of a proof while lecturing before a class. He went over to the side of the board, drew a few pictures, erased them, and then was able to continue his lecture.** In the following problem a picture provides some information that you might otherwise have overlooked. It will help you to think through the solution.

A well is 20 meters deep. A snail at the bottom climbs up 4 meters each day and slips back 2 meters each night. How long will it take the snail to reach the top of the well?

You may be surprised to learn that the answer is not 10 days. The accompanying sketch of a well will help in plotting the day-by-day progress of the snail. On the first day the snail gets up to 4 meters but finishes the day at the 2-meter mark. What is the highest level the snail reaches on the second day and what level does the snail end at? Tracing out the snail's daily locations shows that the snail takes 9 days to get out of the well.

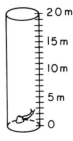

*G. Polya, *How to Solve It* (New York: Doubleday, 1957), pp. 5–19.
**Related by M. Kline, *Why Johnny Can't Add* (New York: St. Martin's, 1973), p. 166.

The problem in the above cartoon contains several bits of information. Let's see how Peppermint Patty could have gotten a better understanding of the problem by drawing pictures. First, the towns, *A*, *B*, *C*, and *D*, are one after the other, so they can be represented by points on a line.

A   B   C   D

Next, the relative distances between towns can be used to sketch more approximate locations. It is farther from *A* to *B* than from *B* to *C* and farther from *B* to *C* than from *C* to *D*.

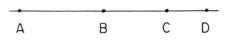

Finally, the distance from *A* to *D* is given as 390 miles.

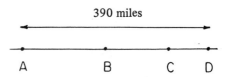

These sketches of lines present the given information and can be helpful in checking the reasonableness of the solution, which will be pursued in the following paragraphs under *Guessing*.

Guessing   Sometimes it doesn't pay to guess, as illustrated by the cartoon on the right. On the other hand, many problems can be solved by guessing or by trial and error procedures. If your first guess is off, it may lead to a better guess. Even if guessing doesn't produce the correct answer, you may increase your understanding of the problem and strike upon an idea for solving it. As Polya once said, "No idea is really bad unless we are uncritical. What is really bad is to have no idea at all."

Let's try guessing to solve Peppermint Patty's problem. It is 10 miles farther from A to B than from B to C, and it is 10 miles farther from B to C than from C to D. From A to D it is 390 miles.

Guess 1 If the distance from A to B is 150 miles ($AB = 150$ miles), then $BC = 140$ miles and $CD = 130$ miles. But these distances are too great because their total is 420 miles.

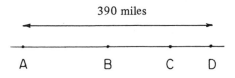

Guess 2 If the distance from A to B is 145, then $BC = 135$ miles and $CD = 125$ miles. The sum of these distances is 405 miles, which is still too great.

We might now notice how much closer Guess 2 is than Guess 1 and use this information to make a third guess. This process will eventually produce the correct answer. While this problem can be solved with simple algebra, the guessing approach puts it within the reach of Peppermint Patty and other pre-algebra students.

Here is another problem that can be solved fairly easily with a little guessing.

How can single-digit numbers be placed in these boxes so that whenever a number is placed in a box in the lower row, the number above it must occur that many times in the lower boxes?

0	1	2	3	4	5	6	7	8	9

For example, if a 2 is placed in the box below the 4, then two 4s must be placed somewhere in the lower boxes.

The surprising thing about this problem is that out of the 10,000,000,000 (or 10^{10}) different ways in which digits can be put in these boxes, there is one and only one solution.

A common first guess is to put a 9 under the 0 and nine 0s in the other boxes. However, this is not a solution because a 9 has been used and there is no 1 in the lower boxes under the 9.

0	1	2	3	4	5	6	7	8	9
9	0	0	0	0	0	0	0	0	0

The next logical guess is to put 8 under the zero box, 2 under the 1, 1 under 2, and 1 under 8. Why doesn't this solution work? You may wish to continue guessing and refining your guesses to solve this problem. (For a solution see the answer to **6b** of Exercise Set 1.3.)

Many people are content to solve a problem and are not anxious to create any more. Others, with a more inquisitive mind, will ask questions of the form "What will happen if . . . ?", in which they change some of the given conditions or look for a more general result. Once, after the previous problem had been presented to a class of students, one of them asked if there would be a solution for

0	1	2	3	4	5	6	7	8

the digits from 0 to 8. They quickly solved this problem and a multitude of similar questions ensued. Would it work for the digits from 0 to 7? or 0 to 12? or 0 to any whole number? They found a pattern in the answers, which helped them to find other solutions. This is an example of how Polya's advice to "look back" at a problem can lead to variations and new discoveries.

Simplifying   Another approach to problem solving is to simplify the problem and solve it for some special cases. Solutions to special cases sometimes lead to ideas for solving the original problem. This approach will be illustrated in solving a counterfeit coin problem.

There are 8 coins and a balance scale. The coins are alike in appearance but one of them is counterfeit and lighter than the other seven. Find the counterfeit coin using two weighings on the balance scale.

If 4 coins are placed in each pan, the counterfeit coin can be narrowed down to 4 coins. However, using this approach it will take three weighings to discover the counterfeit coin. Perhaps solving this problem for fewer coins will suggest a solution. For 2 or 3 coins, only one weighing is needed to discover the counterfeit coin. How many weighings are needed for 4, 5, or 6 coins? Hopefully, thinking through these special cases will suggest a solution to the 8-coin problem. Try it.

Other types of problems that are well suited to simplifying are those stated for the general case or for large numbers. For example, for 20 given points how many line segments have these points as end points? It is natural to find answers for 2 points, 3 points, 4 points, etc., to get a feeling for what the problem involves and then look for a pattern.

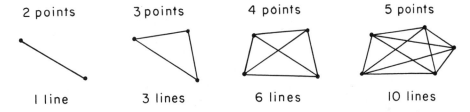

For 2, 3, 4, and 5 points, there are 1, 3, 6, and 10 lines, respectively. This is the beginning of a familiar pattern of numbers, the triangular numbers. At this point we could guess that there will be 15 line segments for 6 points. However, there is a deeper insight which can be discovered from the first few cases. For example, with 4 points, such as *A, B, C,* and *D,* there are 6 lines. Now when a 5th point *E* is brought in, there are 4 new lines, one from each of the points *A, B, C,* and *D* to the new point *E*. Similarly, in going from 5 to 6 points there will be 5 new lines. We have now perceived a way of going from any number of points to the next number of points, and it seems reasonable to conjecture that the pattern of triangular numbers will continue. Solving the problem for 20 points is now just a matter of listing the first 19 triangular numbers.

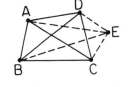

Forming Tables   A problem can sometimes be solved, or at least better understood, by listing the given information and related facts. Here is an example.

John and Harry earned the same amount of money, although one worked 6 days more than the other. If John earns $12 a day and Harry earns $20 a day, how many days did each work?

The following table shows the cumulative amounts each person earned day by day. It took John 5 days to earn as much as Harry made in 3 days. Then after 10 days John had earned as much as Harry earned in 6 days. Finally, John's pay of $180 required 6 more days than Harry's pay of $180.

Number of Days	John's Pay	Harry's Pay
1	12	20
2	24	40
3	36	60
4	48	80
5	60	100
6	72	120
7	84	140
8	96	160
9	108	180
10	120	200
11	132	220
12	144	240
13	156	260
14	168	280
15	180	300

Working Backwards   When something of value has been lost or misplaced it is natural to backtrack your steps in hopes of discovering where the loss might have occurred. This is similar to a common technique for solving problems in mathematics. The solution to the following problem illustrates the *working backwards* strategy.

A certain gambler took his week's paycheck to a casino. Aside from a $2 entrance fee and a $1 tip to the hatcheck person upon leaving, there were no other expenses. Bad luck plagued the gambler. The first day he lost one-half the money he had left after paying the entrance fee, and the same thing happened on the second, third, and fourth days. At the end of the fourth day he had $5 remaining. What was his weekly paycheck?

Think of retracing the gambler's steps back into the casino during the fourth day. After obtaining $1 for the hatcheck he would have $6. Doubling this, because of his losses, gives $12. Receiving the entrance fee brings his total to $14 for the beginning of the fourth day. Continuing backwards through each day shows that the original paycheck was $140.

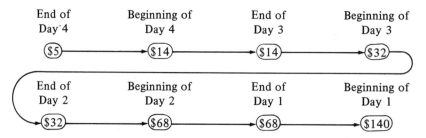

This solution should now be checked by beginning with $140 and going through the expenditures for the 4 days to see if $5 remains.

Unsolved Problems   It comes as a surprise to some people that there are unsolved problems in mathematics. In fact, many of these problems can be described in such nontechnical language that they can be understood by young school children. Yet, their solutions have resisted the world's best mathematicians for years.

Some of the unsolved problems from number theory are the easiest to state and understand. Goldbach's Conjecture (page 18) is one example. A related unproven conjecture by A. de Polignac states that "Every even number is the difference of two consecutive primes in an infinite number of ways." For example, here are four pairs of consecutive primes whose difference is 6, and the conjecture says that there are an infinite number of pairs of such primes.

$$29 - 23 = 6, \quad 37 - 31 = 6, \quad 59 - 53 = 6, \quad 67 - 61 = 6$$

Another of Goldbach's conjectures is that "every odd number greater than 5 is the sum of three primes." Here are four examples.

$$7 = 2 + 2 + 3, \quad 9 = 2 + 2 + 5, \quad 11 = 2 + 2 + 7, \quad 13 = 3 + 3 + 7$$

No one has ever found an odd number for which this condition does not hold, but the conjecture has not been proven.

Many of the unsolved problems are in geometry. Consider the problem of fitting a square onto a curve. On the simple closed curve (one that does not cross itself) shown here, there are 4 points that are the vertices of a square. Does every simple closed curve contain 4 points that are the vertices of a square? (The 4 points must be on the curve, but the square does not have to lie within the curve.) This is an unsolved problem.

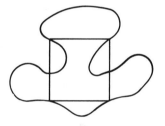

Another unsolved problem involves surfaces, which as yet have never been characterized by equations. These surfaces are formed by dipping wire frameworks into a soap solution. The wire cube pictured here has 12 film surfaces that connect its 12 edges to a small cube-shaped bubble at the center. Each of these 12 surfaces is nearly a trapezoid. Surprising surfaces are also obtained for other types of wire frameworks. Courant and Robbins in *What Is Mathematics* call it a challenging unsolved problem to describe just what the surfaces of such figures are.* The soapy surfaces connecting the edges of the cube have a smaller area than the six square faces of the cube. Wire models were first suggested by the Belgium physicist Joseph Plateau (1801–1883) in order to discover the minimal surfaces of figures. Though totally blind by middle age, he conceived of a set of physical experiments that were carried out by assistants to study this problem.

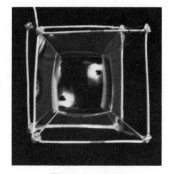

Wire cube with
soap film surfaces

*R. Courant and H. Robbins, *What Is Mathematics* (New York: Oxford University Press, 1951), pp. 354–61, 385–97.

One of the most famous of all unsolved problems, the *Four-Color Problem*, was solved in 1976 by Kenneth Appel and Wolfgang Hanken of the University of Illinois. The problem is simple to state: Can every map (on a plane surface) be colored with four or fewer colors so that any two countries with a common border have different colors? This problem was first proposed in 1852 and since that time has challenged some of the best mathematicians. Some maps, such as Map A, can be colored with fewer than four colors, and others, such as Map B, require four colors. It is impossible to draw a map that requires five colors, since it has been proven that four colors are all that will ever be needed.

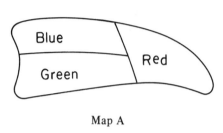

Map A

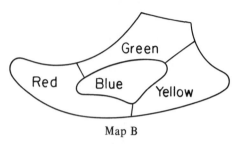

Map B

If you play chess you may want to try this unsolved problem.

Can the eight major black pieces (no pawns) be placed on the chessboard in such a way that every space, including the occupied ones, is challenged by at least one piece?

It is not difficult to challenge 63 squares in this manner, but no one has ever succeeded in challenging all 64 squares. The eight pieces on the board pictured here are challenging all but 3 of the 64 squares. One of these squares is occupied by the king. What are the two remaining unchallenged squares?

SUPPLEMENT (*Activity Book*)

Activity Set 1.3 Tower of Brahma

Just For Fun: Instant Insanity

Additional Sources

Almgren, F.J., Jr. and J.E. Taylor. "The Geometry of Soap Films and Soap Bubbles." *Scientific American,* **234** No. 1 (July 1976), 82–93.

Brown, S.I. and M. Walter. "What if Not?" *Mathematics Teaching,* **46** (1969), 38–45 (procedures for generating new problems).

Ehrmann, R.M. "Minimal Surfaces Rediscovered." *The Mathematics Teacher,* **69** No. 2 (February 1976), 146–52 (soap film surfaces).

Emmett, E.R. *Puzzles for Pleasure.* New York: Barnes and Noble, 1972.

Gardner, M. *aha! Insight.* New York: W.H. Freeman, 1978.

Oglivy, C.S. *Tomorrow's math* (2nd ed). New York: Oxford University Press, 1972 (unsolved problems, easily understood).

Schuh, F. *Master Book of Mathematical Puzzles and Recreations.* New York: Dover, 1969.

IF A BIRD FLIES AT AN AVERAGE OF 5 MILES PER HOUR, AND IS HEADING INTO A 50 MILE PER HOUR WIND....

THE ANSWERING SERVICE

HOW LONG WILL IT TAKE HIM TO FLY ONE HUNDRED MILES?

LESS THAN TWO HOURS IF HE TURNS AROUND.

THE ANSWERING SERVICE.

EXERCISE SET 1.3

After each of the problems from 1 through 9 there is a suggested problem-solving technique. There are usually many ways to solve a problem and you may find other approaches more helpful.

★ 1. How can boiling vegetables be timed for 15 minutes with an 11-minute hourglass and a 7-minute hourglass? (Guess and work backwards.)

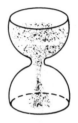

2. A well is 30 meters deep. A snail at the bottom climbs up 3 meters each day and slips back 2 meters each night. How long will it take the snail to reach the top of the well? (Draw pictures.)

★ 3. A collection entitled *Problems for the Quickening of the Mind,* dating from the Dark Ages, contains the following question.

If 100 bushels of corn are distributed among 100 people in such a manner that each man receives 3 bushels, each woman 2 bushels, and each child 1/2 a bushel, how many men, women, and children are there? (*Hint:* Try guessing. There are several solutions in which there are more children than women and more women than men.)

4. How can you bring exactly 6 liters of water from the river when you have only two containers, a 4-liter pail and a 9-liter pail? Is it possible to bring any whole number of liters from 1 to 13 using these two containers? (Guess and work backwards.)*

★ 5. If there are 15 people in a room and everybody shakes hands once with each of the other people, how many handshakes will there be? (Simplify.)

*For a general solution to this type of problem, see G. Goff, "There is a lot of mathematics in two buckets," *The Mathematics Teacher,* **70** No. 6 (October 1977), 611–12.

6. Place single-digit numbers in the lower boxes in such a way that whenever a number is placed in a box in the lower row, the number above it must occur that many times in the lower boxes. (Guess and look for a pattern in your answers.)

a.

★ b.

0	1	2	3	4	5	6	7	8	9

c.

0	1	2	3	4	5	6	7	8	9	10	11	12

7. Given 4 pieces of chain that are 3 links each, explain how all 12 links can be joined into a single circular chain by cutting and rejoining only 3 links. (Work backwards by beginning with a circular chain.)

★ 8. In driving from town A to town D, you pass first through town B and then through town C. It is ten times further from A to B than from B to C, and ten times further from B to C than from C to D. If it is 1332 miles from A to D, how far is it from A to B? (Guess and draw pictures.)

9. There are 5 identical coins and a balance scale. One of these coins is counterfeit and either heavier or lighter than the other 4. Explain how the counterfeit coin can be identified and whether it is lighter or heavier than the others with only 3 weighings on the balance scale. (Simplify.)

★ 10. *Perceptual Blocks:* We are often prevented from finding a solution to a problem because of unnecessary constraints or perceptual blocks. As examples, in solving part **a**, it is not necessary that the 4 lines be confined to the smallest rectangle containing the dots; in part **b** the 6 matchsticks do not have to be on a flat surface; and in part **c** the cut does not have to be a straight line. Use these hints to solve the following problems.

★ a. Draw 4 lines, without lifting your pencil from the paper, which cross through all 9 points.

• • •

• • •

• • •

★ b. Form 4 triangles using 6 matchsticks so that they touch only at their end points (do not cross).

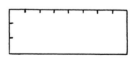

★ c. Cut a 3 by 8 rectangle into two pieces with one cut, and form a 2 by 12 rectangle.

11. Trick problems such as the following ones can help improve problem-solving ability. In such problems it is necessary to think carefully about the given information and the question. They also can be lots of fun.

★ a. If you went to bed at 8 o'clock and set the alarm to get up at 9 o'clock in the morning, how many hours of sleep would you get?

   b. Take 2 apples from 3 apples and what do you have?

   c. A farmer had 17 sheep. All but 9 died. How many did he have left?

★ d. I have two U.S. coins that total 30 cents. One is not a nickel. What are the two coins?

★ 12. *Train Puzzle:* Use the following facts to determine the name of the engineer. On a train Smith, Robinson, and Jones are the fireman, brakeman, and engineer, but NOT respectively. Also aboard the train are three businessmen who have the same names: Mr. Smith, Mr. Robinson, and Mr. Jones.

   1. Mr. Robinson lives in Detroit.

   2. The brakeman lives exactly halfway between Chicago and Detroit.

   3. Mr. Jones earns exactly $20,000 per year.

   4. The brakeman's nearest neighbor, one of the passengers, earns exactly three times as much as the brakeman.

   5. Smith beats the fireman at billiards.

   6. The passenger whose name is the same as the brakeman's lives in Chicago.

13. There are many unsolved problems involving dissections. The square shown here is divided into 7 smaller squares of two different sizes. It is possible to cut a square into 24 smaller squares *all of different sizes*. However, it is not known whether or not a square can be cut into fewer than 24 different-sized squares, with each square occurring only once. The smallest number of squares that a rectangle can be divided into is 9. Assemble the squares on the following page into a 33 by 32 rectangle. The given numbers are the dimensions of the squares. (*Hint:* Trace the squares, cut them out, and fit them into a 33 by 32 rectangle.)

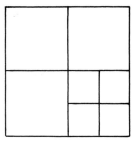

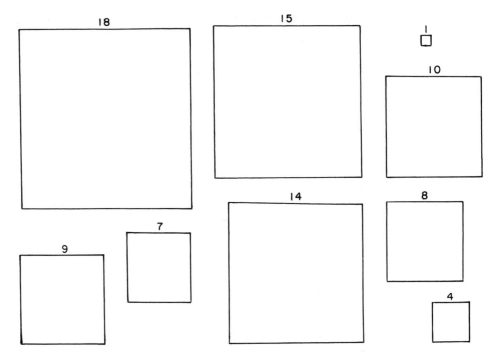

14. *Five-House Puzzle:* There are five houses in a row, which are each occupied by a person of different nationality. The Spaniard owns a dog. The Englishman lives in the red house. Coffee is drunk in the green house. The Ukrainian drinks tea. The green house is immediately to the right of the ivory house. The Old Gold smoker owns snails. Kools are smoked in the yellow house. Milk is drunk in the middle house. The Norwegian lives in the first house. The man who smokes Chesterfields lives next door to the man with the fox. Kools are smoked in the house next to the house where horses are kept. The Lucky Strike smoker drinks orange juice. The Japanese smokes Parliaments. The Norwegian lives next door to the blue house. Each man has one home, one pet, one type of smoke, a different nationality, and a different drink. Who drinks brandy? Who owns a skunk? (Draw pictures to illustrate the given information.)

15. *Balance Pan Puzzle:* This puzzle was posed by one of America's greatest puzzle experts, Sam Loyd (1841-1911).* The following questions show how guessing can lead to a solution.

Mathematical Puzzles of Sam Loyd, ed. Martin Gardner (New York: Dover, 1959), p. 101.

a. As a first guess, let's suppose each block weighs 2 marbles. What would be the weight of the top, according to the first picture of the balance?

b. Explain why your answer in part **a** contradicts the information given by the picture of the second balance.

c. As a second guess, you might assume that each block weighs: 1 marble; 3 marbles; or 4 marbles. Which is the only guess that agrees with the given information in the first two balances?

d. Describe another method of solving this problem without guessing.

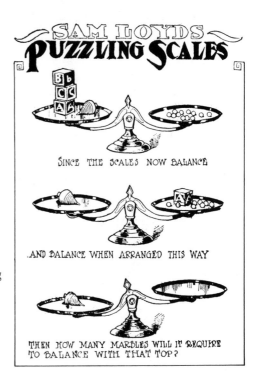

SAM LOYDS PUZZLING SCALES

SINCE THE SCALES NOW BALANCE

...AND BALANCE WHEN ARRANGED THIS WAY

THEN HOW MANY MARBLES WILL IT REQUIRE TO BALANCE WITH THAT TOP?

CHAPTER TWO

COUNTING, SETS, AND NUMERATION

Our minds are finite, and even in these circumstances of finitude we are surrounded by possibilities that are infinite; and the purpose of human life is to grasp as much as we can out of that infinitude.

Alfred North Whitehead

2.1 COUNTING AND SETS

Long before numbers were invented numerical records were kept by means of tallies. These were maintained by making collections of pebbles, cutting notches in wood, or making marks in dirt or stone. The oldest example of the use of a tally stick dates back 30,000 years and was found in 1937 in Czechoslovakia. It is a 7-inch wolf bone engraved with 55 notches occurring in 11 sets of 5.

A more recent example of tallies is found on the 8000-year-old Ishango bone, which was discovered on the shore of Lake Edward in the Congo. The marks on this bone occur in several groups that are arranged in three distinct columns (see Exercise Set 2.1, Exercise 2). The tip of this bone is provided with a quartz flake, which, according to geologist-archaeologist Dr. de Heinzelin, is very likely a writing tool.*

At one time, tally sticks were common for keeping records of debts and payments. The tally stick would be split lengthwise across the notches, with the debtor receiving one piece and the creditor the other. Such tally sticks were used by the British Royal Treasury in the thirteenth century and continued in use until the nineteenth century.

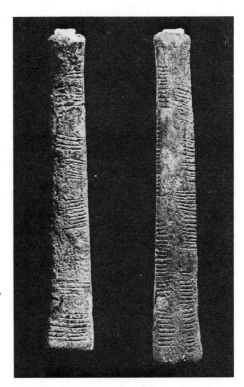

Two views of the Ishango bone, found on the shores of Lake Edward in the Congo

*Jean de Heinzelin, "Ishango," *Scientific American,* **206,** June 1962, pp. 113–14.

One-to-One Correspondence and Counting

It is not difficult to imagine our ancient ancestors keeping track of their flocks by matching each animal with a pebble, a mark in the dirt, or a notch on a stick. Such methods of tally keeping are examples of one-to-one correspondence.

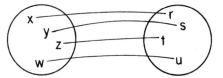

Two sets in one-to-one correspondence

Definition: Two sets or collections of objects are said to be in *one-to-one correspondence* if for every element in the first set there is just one element in the second set and for every element in the second set there is just one element in the first set.

Counting is an extension of the idea of one-to-one correspondence. The numbers 0, 1, 2, 3, 4, etc., are called *whole numbers*. To *count* the elements of a set we match these elements with the whole numbers that are greater than zero: 1, 2, 3, 4, and so forth. To count the number of letters in the word "straight" (one of the longest one-syllable words) we match these letters with the numbers 1 through 8.

$$s \leftrightarrow 1$$
$$t \leftrightarrow 2$$
$$r \leftrightarrow 3$$
$$a \leftrightarrow 4$$
$$i \leftrightarrow 5$$
$$g \leftrightarrow 6$$
$$h \leftrightarrow 7$$
$$t \leftrightarrow 8$$

Whole Numbers and Their Uses

Fifty thousand years ago, people were living in caves and using fire. Implements from the remains of this period suggest the existence of barter and the need for numbers. During the Old Stone Age (10,000–15,000 B.C.) figures of people, animals, and abstract symbols were painted in caves in Spain and France. The symbols were composed of many geometric forms: straight lines, spirals, circles, ovals, and dots. The rows of dots and rectangular figures in this photograph are from El Castillo caves, Spain (12,000 B.C.). It is conjectured by some scholars that this was a system for recording the days of the year. It seems likely that numbers were in existence by this time. The earliest dated event in history is the Egyptian calendar, 4236 B.C. This calendar predates the earliest known writing by over 500 years.

At first it may only have been necessary to distinguish between one object, two objects, and many objects. The first words for numbers were probably associated with specific things. This influence can be seen in the expressions we have for two, such as

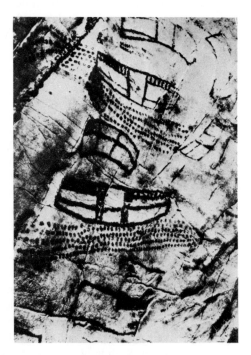

Art symbols (12,000 B.C.)
El Castillo caves, Spain

a "couple" of people, a "brace" of hens, and a "pair" of shoes. Eventually the concepts of twoness, threeness, etc., were separated from physical objects, and the abstract notion of "number" developed.

Definition: **Two sets of objects are said to have the *same number* of elements if they can be put into one-to-one correspondence.**

Look around you and find a set of objects that can be put into one-to-one correspondence with the fingers of your hand. We express this relationship by saying that the set has "five" elements. The number five is an abstract idea expressing a common property of sets. Similarly, the numbers zero, one, two, three, etc., are each abstract ideas associated with sets.

Some of the earliest symbols for numbers were sets of marks that could be put into one-to-one correspondence with the set of objects being counted. For example, "2" was often represented by two marks, $||$; "3" by three marks, $|||$; etc. Such symbols are easy to remember because they suggest the meanings of the numbers.

It is usually assumed that numbers arose in answer to the practical needs of people, such as counting possessions, days, years, and so on. However, there is some evidence that numbers developed in connection with religious rituals and that they may have been used first to denote the rank or order of people in a group. We now recognize three different uses of whole numbers: cardinal, ordinal, and naming.

Uses of Whole Numbers

1. *Cardinal Use*—If a number gives information about quantity and answers questions such as, "How many?" or "How much?", this is the *cardinal use* of numbers. The following are examples of cardinal uses of whole numbers: There are 50 states; he is a 21-year-old lieutenant; and the House voted by 162 to 52 to shrink itself by one-third.

2. *Ordinal Use*—If a number gives information about the order or location of objects, this is an *ordinal use.* Here are a few examples: She is in row 27; the fifth ranked player beat the second ranked; and we meet on the third Wednesday of every month.

3. *Naming Use*—Social Security numbers, license plate numbers, and checking account numbers are tags to name and identify objects. This is a *naming use* of numbers. For example: Flight 972 is bound for Kennedy Airport; she flies a Sikorsky S-61 helicopter; and number 82 is a guard.

In the following sentence you will recognize all three of the uses of whole numbers.

Nineteen cars have arrived for the 8th annual stock car derby, but car 37 has withdrawn.

Historical Background of Sets   The notion of sets or collections of objects is not only fundamental to counting and numbers, but it permeates every branch of mathematics. The importance of sets was first recognized during the nineteenth century development of logic. George Boole (1815–1864) was one of the first mathematicians to systematically use sets in logic. He developed an algebra of sets which is now called *Boolean algebra.* John Venn (1834–1923), another pioneer in the early development of logic, published the book *Symbolic Logic* in which he uses circles to represent sets. There is a circle for each set and the elements of the set are represented by points inside the circle. These circular diagrams are called *Venn diagrams.* Here is an example of how these diagrams are used to reason deductively.

All salamanders are amphibians, so the circle for the salamanders is placed inside the circle for the amphibians. All animals that develop an amnion are not amphibians. This set is represented by drawing a circle outside of the amphibian circle. From this diagram we can conclude that salamanders do not develop an amnion.

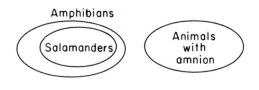

The person most often associated with sets is Georg Cantor (1845–1918), the Russian-born mathematician who moved to Germany at the age of eleven. He is referred to as "the father of set theory" because of his extensive development of sets. In 1874, he published a controversial paper on set theory, and the resulting criticism by other mathematicians led to a series of mental breakdowns, which continued to his death. In recent years Cantor has won widespread recognition for his theory of sets, which has contributed to every area of mathematics.

Sets and Their Elements   There are many words for sets of objects: a *flock* of birds, a *herd* of cattle, a *collection* of paintings, a *bunch* of grapes, a *group* of people, and so forth. The word "set" is so basic in mathematics it is one of the undefined terms. Intuitively, it is described as a collection of objects called elements, which can be either concrete, such as the set of seats in a theater, or abstract, as a set of ideas.

There are two conditions that a set of objects must satisfy. For the first condition it must be possible to determine if a given object is in the set. We refer to this condition by saying that the set is *well-defined*. This just means that sets must be described carefully enough to avoid confusion about which elements belong to the set. For example, "the set of all armed forces personnel in a particular country" is a well-defined set. Given any person, it can be determined whether or not he/she is in this set. On the other hand, "the set of all beautiful paintings" is not well-defined because what is beautiful is a matter of opinion. The second condition is that a set must not contain an element more than once. A set of numbers, for example, should not have the same number listed twice.

"We understand you tore the little tag off your mattress."

There are two common methods of specifying a set. One is by *describing* the elements of the set, such as, "The capitals of the six New England states." The second method is by *listing* each element in the set. When this is done the members of the set are written between braces. Here is the set of capitals in New England.

$$\{\text{Augusta, Concord, Boston, Hartford, Providence, Montpelier}\}$$

If the set of elements is large, we sometimes begin the list and then use three dots to mean, "and so forth." The following set contains the multiples of 10 from 10 to 500.

$$T = \{10, \ 20, \ 30, \ 40, \ 50, \ 60, \ 70, \ ..., \ 490, \ 500\}$$

Sometimes a set containing no elements will be described. The set of all prime numbers greater than 31 and less than 37 is a set with no elements. This is called the *empty set* or *null set* and is denoted by the set braces with no elements between them, $\{\ \}$, or by the Greek letter phi, ϕ. This is a very useful set and occurs quite frequently.

It is customary to denote sets by uppercase letters and the elements of sets by lowercase letters. If k is an *element of* a set S we write $k \in S$ and if it is *not an element of* S, this is written $k \notin S$. Using the previous set T as an example, $60 \in T$ and $55 \notin T$.

Attribute Pieces—Attribute pieces are geometric models of various shapes, sizes, and colors for illustrating sets. Those shown here have three different shapes [triangular, (t), rectangular (r), and hexagonal (h)], two sizes [large (l) and small (s)], and two colors [black (b) and white (w)].

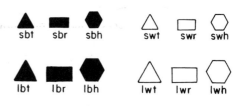

These objects can be classified into *sets* in many different ways. Here are a few: S is the set of all small attribute pieces; L is the set of all large attribute pieces; T is the set of all triangles; H is the set of all hexagons; W is the set of all white attribute pieces; LB is the set of all large black attribute pieces; etc. These attribute pieces will be used in the following paragraphs to provide examples of relationships between sets and operations on sets.

Relationships between Sets   There are relationships between sets, just as there are relationships between numbers. For any two numbers, either one is less than the other or they are equal. Two sets may have no elements in common, some elements in common, or all elements in common. The following sets of attribute pieces illustrate two of these situations. On the left, set H (hexagons), has some attribute pieces in common with set S (small pieces). On the right, every attribute piece in BT (black triangles) is also in T (triangles).

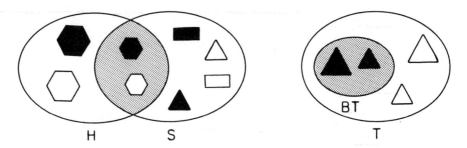

Subsets—The relationship between BT and T is described by saying BT is a *subset* of T, because every attribute piece in BT is also in T.

Definition: If every element of set A is also an element of set B, then A is a *subset* of B, written $A \subset B$.

In the preceding example, $BT \subset T$. The sets H and S have some elements in common but neither is a subset of the other. This can be indicated by writing $H \not\subset S$ and $S \not\subset H$. According to the definition of subset, every set is a subset of itself. For example, $BT \subset BT$, $T \subset T$, $H \subset H$, etc., because every element in the first set is also an element of the second set.

If we know $A \subset B$ and that one or more elements in B are not in A, then A is sometimes called a *proper subset* of B. As examples, BT is a proper subset of T, but H is not a proper subset of itself. The null set, $\{\ \}$, is a proper subset of every non-empty set. Perhaps you can see why this is true. The only way one set can fail to be a subset of another is if it has elements that are not in the other. Thus, the null set cannot fail to be a subset of every set.

Equal Sets—Sets that contain the same elements are *equal*. Sometimes two equal sets may look different or have different descriptions. The set of all solutions to the equation $x^2 - 5x + 6 = 0$ and the set $\{2,3\}$ are examples of equal sets.

Definition: If A is a subset of B, and B is a subset of A, then both sets have exactly the same elements and they are *equal,* written $A = B$. In this case, A and B are just different letters for the same set.

The set of attribute pieces with less than four sides is equal to the set of triangular attribute pieces, as shown by the following two sets. Notice that equality of sets does not depend on the order of the elements.

$$\{lbt, \ sbt, \ lwt, \ swt\} = \{lwt, \ lbt, \ swt, \ sbt\}$$

Equivalent Sets—The set of small black attribute pieces can be put into one-to-one correspondence with the set of large white attribute pieces. We refer to this fact by saying the two sets are *equivalent.* In other words, they have the same number of elements.

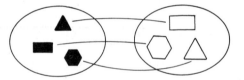

Definition: Two sets are *equivalent* if their elements can be placed in one-to-one correspondence.

Note that two equal sets are also equivalent, but it is possible for two sets to be equivalent without being equal. The sets in the previous diagram are equivalent but they are not equal.

Disjoint Sets—The two sets of attribute pieces shown here have no elements in common. We describe this fact by saying the two sets are *disjoint.*

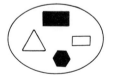
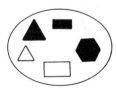

Definition: Two sets are *disjoint* if they have no elements in common.

Operations on Sets   There are operations that replace two sets by a third set just as there are operations on numbers which replace two numbers by a third number. An operation that assigns each pair of elements to another element is called a *binary operation*. Addition and multiplication are two examples of binary operations on whole numbers. Intersection and union are binary operations on sets.

Intersection of Sets—The set of small attribute pieces and the set of black attribute pieces have three elements in common. If we form a third set containing these common elements, it is called the *intersection* of the two sets.

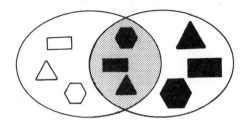

Definition: The *intersection* of two sets *A* and *B* is the set of all elements that are in both *A* and *B*. This operation is written as *A* ∩ B.

The intersection of the above two sets is the set of attribute pieces that are small and black. This is illustrated by shading the common region inside the two curves.

$$\{swr,\ swh,\ swt,\ sbt,\ sbr,\ sbh\} \cap \{sbt,\ sbr,\ sbh,\ lbt,\ lbr,\ lbh\} = \{sbt,\ sbr,\ sbh\}$$

If two sets are disjoint, such as the set of large attribute pieces (*L*) and the set of small attribute pieces (*S*), their intersection is the null set. This can be briefly written as $L \cap S = \{\ \}$, or $L \cap S = \phi$.

Union of Sets—The set of small attribute pieces and the set of black attribute pieces have nine different elements. A third set containing all of these elements is called the *union* of the two sets.

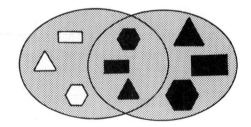

Definition: The *union* of two sets *A* and *B* is the set of all elements that are either in *A*, or in *B*, or in both *A* and *B*. This operation is written as *A* ∪ *B*.

The union of the two sets pictured above is the set of all attribute pieces that are small or black. This is illustrated by shading the total region inside the two curves.

$$\{swr,\ swh,\ swt,\ sbt,\ sbr,\ sbh\} \cup \{sbt,\ sbr,\ sbh,\ lbt,\ lbr,\ lbh\} =$$
$$\{swr,\ swh,\ swt,\ sbt,\ sbr,\ sbh,\ lbt,\ lbr,\ lbh\}$$

Notice that in the preceding example, the union of the set of small attribute pieces and the set of black attribute pieces contains each attribute piece only once, even though three of these elements are contained in both sets. The union of the two disjoint sets shown here contains the

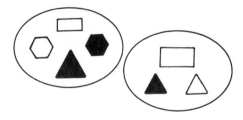

four attribute pieces from one set and three from the other. The union of disjoint sets can be written by listing the elements in the first set followed by those in the second.

$$\{lbt,\ swr,\ swh,\ sbh\}\ \cup\ \{lwr,\ sbt,\ swt\}\ =\ \{lbt,\ swr,\ swh,\ sbh,\ lwr,\ sbt,\ swt\}$$

Complement of a Set—Frequently, we wish to compare a set that has a certain property with another set that does not have this property. Consider the set of small black attribute pieces and the remaining ones without this property. In the following diagram the small black pieces are inside the circle and the others are outside. These two sets are called *complements* of each other.

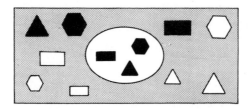

Definition: For any given set, if two subsets A and B are disjoint and their union is the whole set, then A is the *complement* of B, written $A = B'$, and B is the complement of A, written $B = A'$.

The complement of *SB*, the set of small black attribute pieces, is the set of attribute pieces not contained in *SB*.

$$SB' = \{lbt,\ lbh,\ lbr,\ lwt,\ lwh,\ lwr,\ swt,\ swr,\ swh\}$$

The "given set" in the previous definition is sometimes called the *universal set*. We have been using a universal set of 12 attribute pieces. In solving problems with whole numbers, the universal set is often the set of all whole numbers. If the universal set is all whole numbers, the set of odd numbers and the set of even numbers are complements of each other. To give another example, the complement of the set of whole numbers less than 10 is the set of whole numbers greater than or equal to 10.

The universal set can be any set, but once it is established, each subset has a unique (one and only one) complement. In other words, *complement* is an operation that assigns each set to another set.

Infinite Sets   Are time and space infinite quantities? What do we mean by "infinitely large" and "infinitely small"? The problems of the infinite have challenged the human mind as no other single problem in the history of thought.

First of all, we must realize that "very big" and "infinite" are entirely different concepts. There are about 100 billion stars in our galaxy, The Milky Way, and about one billion other galaxies. The number of electrons and protons in all of these stars and in interspace has been estimated to be less than 10^{80} (1 followed by 80 zeros). While this number of objects is very large, it is still not infinite.

In the sense that "infinite" means "without end" or "without bound," we have an intuitive notion of its meaning. These vague terms, however, were not precise enough to be useful to mathematicians, and in 1874 Cantor developed the following definition of infinite.

Galaxy Messier 81,
10 million light-years away

Definition: A set is *infinite* if it has a proper subset with which it can be put into one-to-one correspondence.

Let's apply this definition to the set of whole numbers to show it is infinite. The set of even numbers is a proper subset of the set of whole numbers, because every even number is a whole number and the set of whole numbers contains numbers that are not even, namely, the odd numbers. By matching every whole number with the even number that is twice as big, the numbers of both sets can be placed in one-to-one correspondence (see the following diagram). Intuitively, it may seem that we should "run out" of even numbers before whole numbers, but this does not happen. For every whole number n there corresponds an even number $2n$, and for every even number there is a whole number that is half as big.

$$
\begin{array}{ccccccccccc}
0 & 1 & 2 & 3 & 4 & 5 & 6 & 7 & \cdots & n & \cdots \\
\updownarrow & \updownarrow & \updownarrow & \updownarrow & \updownarrow & \updownarrow & \updownarrow & \updownarrow & \updownarrow\updownarrow\updownarrow & \updownarrow & \cdots \\
0 & 2 & 4 & 6 & 8 & 10 & 12 & 14 & \cdots & 2n & \cdots
\end{array}
$$

Notice that this definition distinguishes between finite and infinite sets. Try as we may, there is no way to form a one-to-one correspondence between the finite set $\{1, 2, 3, 4\}$ and any of its proper subsets.

We have shown that the set of whole numbers is infinite because it can be put into one-to-one correspondence with the set of even numbers. Remember that earlier we defined two sets to have the *same number* of elements if they can be put into one-to-one correspondence. This means that the set of whole numbers and the set of even numbers have the same number of elements.

Georg Cantor was especially interested in infinite sets and was the first to recognize that there are many different types of these sets. That is, there is a hierarchy of infinite sets, in which each set is so much bigger than the previous set they cannot be put into one-to-one correspondence. The first or smallest of the infinite sets is the type that can be put into one-to-one correspondence with the set of whole numbers. These sets are said to be *countably infinite.* (For another example of a countably infinite set, see Exercise 14 in the following exercise set.)

SUPPLEMENT (*Activity Book*)

Activity Set 2.1 Sorting and Classifying (Operations on sets of attribute pieces)
Just for Fun: Games with Attribute Pieces

Additional Sources

Association of Teachers of Mathematics. *Notes on Mathematics in Primary Schools.* London: Cambridge University Press, 1969. pp. 301–5 (comments on sets).

Dantzig, T. *Number The Language Of Science* (4th ed.). New York: Macmillan, 1954. pp. 57–65 (quote from Archimedes' *The Sand Reckoner* and discussion of infinity).

Eves, H.W. *Mathematical Circles Revisited.* Boston: Prindle, Weber and Schmidt, 1971. pp. 41–45 (tally sticks).

Glenn, W.H. and D.A. Johnson. *Exploring Mathematics on Your Own.* New York:

Doubleday, 1961. "Sets, Sentences, and Operations," pp. 139–96.

Newman, J.R. *The World Of Mathematics,* **3.** New York: Simon and Schuster, 1956. "Infinity," pp. 1593–611.

Newman, J.R. and E. Kasner. *Mathematics and the Imagination.* New York: Simon and Schuster, 1940. "Beyond the Googol," pp. 27–64.

Smeltzer, D. *Man And Number.* London: A. and C. Black, 1970. pp. 1–26 (early number sense and finger counting).

© 1965 United Features Syndicate, Inc.

EXERCISE SET 2.1

1. The notches in the 30,000-year-old Czechoslovakian wolf bone are arranged in two groups. There are 25 notches in one group and 30 in the other. Within each series the notches are in groups of 5. Could this recording system have been accomplished without: number names? number symbols?

2. Both sides of the 8000-year-old Ishango bone are shown next. There is one column of marks on one side of the bone and there are two columns of marks on the other side. Anthropologists have questioned the significance of the numbers of these marks: Are they arbitrary records of game killed or belongings, or are they intended to show relationships between numbers?* For each group of marks in these columns, write the number of marks.

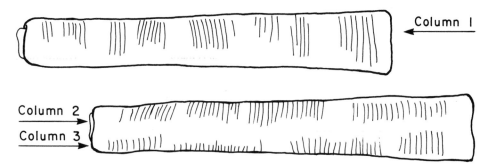

★ a. These sets of numbers form definite number patterns. Which of the three columns of notches represent prime numbers?

b. Which of these columns suggests a knowledge of multiplication by 2?

★ c. Explain how the numbers in the remaining column are related.

d. Do you think these marks were intended to show relationships between numbers?

3. For each number in the following sentences determine whether it is a cardinal, ordinal, or naming use of numbers.

★ a. About 25 percent of the output is harvested by hand.

★ b. This report continues on page 742.

c. Loch Ness, more than 900-feet deep in places, is part of the Great Glen, a fault line bisecting Scotland.

d. The Datsun 810 has a tilt steering wheel and maintenance warning system.

e. He is seeded number 5 in the tennis tournament.

4. Which of these sets are well-defined?

★ a. The number of letters on this page.

b. All difficult mathematics problems.

★ c. All useful books.

d. All Ford trucks built in 1980.

*For a discussion of these marks, by de Heinzelin, see A. Marshack, *The Roots of Civilization* (New York: McGraw-Hill, 1972), pp. 21–26.

5. The 12 attribute pieces shown here have 3 shapes, 2 sizes, and 2 colors. Use the following sets to answer parts **a** through **f**. *W:* white attribute pieces; *H:* hexagonal attribute pieces; *SW:* small white pieces; *SB:* small black pieces; *L:* large pieces.

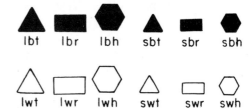

★   a. Which pairs of sets, if any, are equivalent?

     b. Which pairs of sets, if any, are equal?

★   c. Which pairs of sets, if any, are disjoint?

     d. Which set is a subset of another?

★   e. Which attribute pieces are in the intersection of *W* and *L*?

     f. Which attribute pieces are in the union of *W* and *L*?

     g. The union of which two sets is equal to the complement of *SB*?

6. Sometimes the best way to find the elements which have a given property is to find those elements which do not have this property. That is, we find one set and then take its complement. There are nine attribute pieces that are not(large and black), and these can be found by taking the complement of the set of large black attribute pieces. Use complements to find the elements in the following sets.

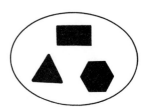

★   a. All attribute pieces that are not(hexagonal and small).

     b. All attribute pieces that are not(white and triangular).

7. The attribute pieces shown here are hexagonal or small (some pieces have both properties). The elements in the complement of this set are the remaining four attribute pieces that are not(hexagonal or small). Use complements to find the elements in the following sets.

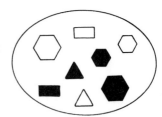

★   a. All attribute pieces that are not(small or white).

     b. All attribute pieces that are not(triangular or large).

8. The shaded region in the accompanying diagram represents $(A \cap B)'$, the complement of $A \cap B$. Shade in the regions on the following page for $(A \cup B)'$, $A' \cup B'$, and $A' \cap B'$. (*Hint:* Two of the following four diagrams should be shaded exactly like the remaining two.)

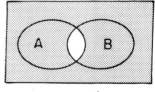

$(A \cap B)'$

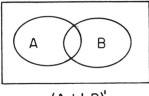
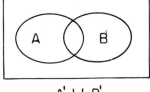
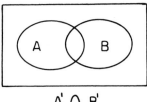

$(A \cup B)'$ $A' \cup B'$ $A' \cap B'$

Only two of the following equations are true for all sets A and B. These two equations are called De Morgan's Laws. Use the diagrams to determine which two equations hold and which two do not.

★   a. $(A \cap B)' = A' \cap B'$ b. $(A \cap B)' = A' \cup B'$

★   c. $(A \cup B)' = A' \cap B'$ d. $(A \cup B)' = A' \cup B'$

9. This Venn diagram of human populations was used in investigations
correlating the presence or absence of
B27+ (a human antigen), RF+ (an
antibody protein), spondylitis (an
inflammation of the vertebrae), and
arthritis (an inflammation of the
joints), with the incidence of various
rheumatic diseases.

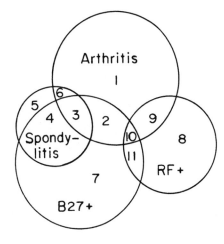

★   a. What is the number of the region
that corresponds to spondylitis and
B27+ but not to arthritis?

    b. What is the number of the region
that corresponds to arthritis and
B27+ but not to RF+ or to
spondylitis?

10. Venn diagrams are sometimes helpful in representing relationships between sets for
the purpose of forming conclusions.

★   a. Use the statements I, II, and III and Venn diagrams to show that all porpoises and
dolphins are mammals.

     I. All porposies are whales.

     II. All dolphins are whales.

     III. All whales are mammals.

   b. Use statements IV and V and Venn diagrams to show that we cannot conclude
from the given information that mosses and ferns do not have chlorophyll.

     IV. All seed plants have chlorophyll.

     V. Mosses and ferns are not seed plants.

11. The Venn diagram shown here satisfies the following conditions for sets Y, B, and G.

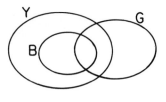

$$B \subset Y, B \neq Y, G \cap B \neq \{\}, G \not\subset Y, B \not\subset G$$

Draw diagrams for sets Y, B, and G which satisfy the following conditions.

★   a. $Y \subset B, Y \neq B, G \subset Y, G \neq Y$

     b. $B \cap G \neq \{\}, B \not\subset G, G \cap Y \neq \{\}, Y \not\subset G, Y \cap B = \{\}$

12. Here are the results of a survey of 120 people which show the television networks they watched in a given evening.

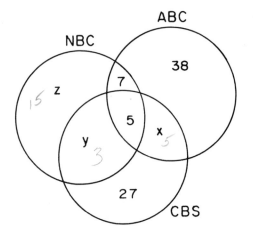

Networks	Numbers of People
ABC	55
NBC	30
CBS	40
ABC and CBS	10
ABC and NBC	12
NBC and CBS	8
NBC and CBS and ABC	5

★   a. Write the numbers of people for the three remaining regions of the circles. (*Hint:* Find x and y by using the 5 in the intersection of all three sets. Then find z.)

     b. How many people watched ABC or CBS, but not NBC?

     c. How many people watched CBS but not(ABC or NBC)?

★   d. How many people watched NBC and ABC but not CBS?

     e. How many people did not watch any of these three networks?

13. In a survey of 6500 people, 5100 had a car; 2280 had a pet; 5420 had a television set; 4800 had a TV and a a car; 1500 had a TV and a pet; 1250 had a car and a pet, and 1100 had a TV, a car, and a pet. Begin with the 1100 in the intersection of all three sets and find the numbers for r, s, and t.

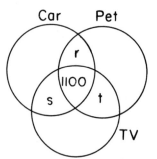

★   a. Write the numbers of people in the 7 regions.

     b. How many of these people have a TV and a pet, but do not have a car?

★   c. How many of these people didn't have a pet, or a TV, or a car?

14. Show that the set of counting numbers and the set of multiples of 10, $\{10, 20, 30, 40, ...\}$, have the same number of elements, by putting them into one-to-one correspondence.

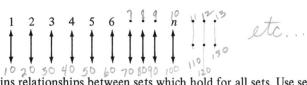

15. Boolean algebra contains relationships between sets which hold for all sets. Use sets A, B, and C to evaluate both sides of the equations in a through e. Which two of these equations do not hold for all sets?

$$A = \{1, 3, 7\} \qquad B = \{3, 5, 7\} \qquad C = \{5, 9\}$$

★ a. $A \cap (B \cup C) = (A \cap B) \cup (A \cap C)$ ★ b. $(A \cup B) \cap C = A \cup (B \cap C)$

   c. $A \cup (B \cup C) = (A \cup B) \cup C$ d. $A \cap (B \cup C) = (A \cap B) \cup C$

   e. $A \cup (B \cap C) = (A \cup B) \cap (A \cup C)$

★ 16. *Rainy Day Puzzle:* During a vacation it rained on 13 days, but when it rained in the morning, the afternoon was fine and every rainy afternoon was preceded by a fine morning. There were 11 fine mornings and 12 fine afternoons. How long was the vacation? (*Hint:* Try some numbers in the 3 regions of this Venn diagram. The intersection of these two sets will be the days with both fine mornings and fine afternnons.)

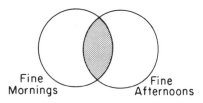

Fine Mornings   Fine Afternoons

17. *Hotel Puzzle:* There once was a certain hotel in never-never land with an infinite number of rooms. On one very busy holiday in which every room was taken, the hotel manager found that a room had not been saved for a very important person who was due to arrive at any minute. Luckily, there was a mathematician staying at the hotel and he proposed the following solution. Have the people in room 1 move to room 2; those in room 2 move to room 4; those in room 3 move to room 6; and in general whatever the number of a room, its occupants moved to a room whose number was twice as big. The important and late arriving guest was then placed in room 1, and in the end, after some grumbling about moving around, everyone was happy, especially the hotel manager.*

a. Explain why this solution was possible.

b. How many new rooms were made available by this solution?

c. Would the mathematician's plan have worked if each person had been moved to a room whose number was 10 times greater than their first room number?

---

*For more problems and solutions involving a hotel with an infinite number of rooms, see N. Y. Vilenkin, *Stories about Sets* (New York: Academic Press, 1968), pp. 4–15.

★18. *Hundred-Men Puzzle:* In a certain town there live 100 men; 85 are married; 70 have a telephone; 75 own a car; and 80 own their own home. What is the least number who are married, own a car, have a phone, and own their own house? (*Hint:* Use complements of sets. What is the greatest number of men who are not married; do not have a phone; etc.)

19. *Game* (2 players or 2 teams): One player selects a number from the sets *S, C,* or *T.* The other player tries to guess this number by asking yes or no type questions. Count the number of questions it takes to guess the chosen number. After both players have had a chance at choosing a number and guessing, the player who requires the least number of questions is the winner. Here are 4 questions and their answers in a game. Determine the number.

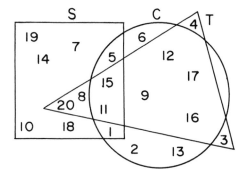

1. Is the number in *C*? No.
2. Is the number in *S*? Yes.
3. Is the number in $S \cap T$? Yes.
4. Is the number 20? No.

Egyptian stone giving an account of the expedition of Amenhotep III in 1450 B.C.

2.2 NUMERATION SYSTEMS

The first uses of numbers, their names, and their symbols have been lost in early history. Number symbols or numerals probably preceded number words, since it is easier to cut notches in a stick than to establish phrases to identify a number.

Early numeration systems appear to have grown from such tallying. In many of these systems one, two, and three were represented by $|$, $||$, and $|||$. By 3400 B.C. the Egyptians had an advanced system of numeration for numbers up to and exceeding one million. Their first few number symbols show the influence of the simple tally strokes.

$|$	$||$	$|||$	$||||$	$|||||$	$||||||$	$|||||||$	$||||||||$	$|||||||||$
1	2	3	4	5	6	7	8	9

Their symbol for "3" can be seen in the third row from the bottom of the stone inscriptions shown above. What other symbols for single-digit numerals can you see on this stone?

Grouping and Number Bases　As soon as it became necessary to count larger numbers of objects, the counting process was extended by groupings. Since the fingers furnished a convenient counting device, grouping by fives was used in some of the oldest methods of counting. Even today, people keep tallies by four straight marks and one drawn across, ⊞. In certain parts of South America and Africa, it is still customary to "count by hands":

one, two, three, four, hand, hand and one, hand and two, hand and three, etc.

The left hand was generally used to keep a record of the number of objects being counted, while the right index finger pointed to the objects. When all five fingers had been used, the same hand would be used again to continue counting.

As soon as people became accustomed to counting by the fingers on one hand, it was natural to use the fingers on both hands to group by tens. In most numeration systems today, grouping is done by tens. The names of our numbers reflect this grouping process. "Eleven" derives from "ein lifon," meaning *one left over,* and "twelve" is from "twe lif," meaning *two over ten.* There are similar derivations for the number names from 13 to 19. "Twenty" is from "twe-tig," meaning *two tens,* and hundred means *ten times* (ten).*

Grouping by twenty originates from the barefoot days when the toes as well as the fingers were accessible aids for counting. There are many traces of the use of twenty. The Mayas of Yucatan and the Aztecs of Mexico had elaborate number systems based on twenty. In Greenland they used the expression "one man" for twenty, "two men" for forty, and so on. A similar system was used in New Guinea. For example, their word for 90 was "4 men and 2 hands." Evidence of grouping by twenty among the ancient Celtics can be seen in the French use of "quatre-vingt" (four-twenty) for 80. In our language the use of "score" suggests past tendencies to count by twenties. Lincoln's familiar Gettysburg Address begins with "Four score and seven years ago." Another example occurs in the childhood nursery rhyme "Four and twenty blackbirds baked in a pie."

"You're probably all wondering why I called you here today."

Number Bases—The number of objects, which is used in the grouping process, is called the *scale* or *base.* In general, a base b is selected and names are assigned to the numbers 1, 2, 3, etc., up to and including b. Names for numbers greater than b are then established by combinations of the number names for the numbers less than or equal to b.

In the example of counting by hands, the base is five. By using names for the first four numbers and "hand" for the name of the base, the numbers up to and including 24 (4 hands and 4) can be named. The next number, 25 or 5^2, requires a new name. This could be called "hand of hands." With this new power of the base, numbers up to 124 (4 hand of hands, 4 hands, and 4) can be named. The next step is to name 125 or 5^3. In this manner, number names can be generated indefinitely by creating new names for powers of the base five.

*H. W. Eves, *An Introduction to the History of Mathematics,* 3rd ed. (New York: Holt, Rinehart and Winston, 1969), pp. 8–9.

Simple Grouping Systems   The ancient Egyptian numeration system had a base of ten. The numerals for the first few powers of 10 were picture symbols called *hieroglyphics.* Here are the Egyptian hieroglyphics and the corresponding powers of 10.

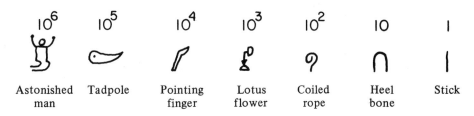

10^6	10^5	10^4	10^3	10^2	10	1
Astonished man	Tadpole	Pointing finger	Lotus flower	Coiled rope	Heel bone	Stick

These symbols were repeated the required number of times to represent a number. Some examples of this numeration can be seen in the stone inscriptions on the first page of this section. Notice the numeral for 743 in the third row from the bottom. The symbols for 3 ones, 4 tens, and 7 hundreds are written from left to right. It was the Egyptian custom to write increasing powers of 10 from left to right, rather than from right to left as we do today. When a power of 10 occurred zero times, as in 14,026, it was expressed by omitting the symbol for that power of 10. In this case, there are 0 hundreds and so the Egyptian numeral for 100 is omitted.

2342

14,026

Roman Numeration—Roman numerals can be found on clock faces, buildings, gravestones, and the preface pages of books. Like the Egyptians, the Romans used a base ten simple grouping system. In addition to the symbol for one and powers of 10, there are symbols for five, fifty, and five hundred. In all, there are seven basic symbols.

I	V	X	L	C	D	M
1	5	10	50	100	500	1000

There is historical evidence that C is from "centum," meaning hundred, and M is from "milli," meaning thousand. The origin of the other symbols is uncertain. The Romans wrote their numerals with the highest powers of 10 on the left and the units on the right.

1000 900   70   9

MDCCCCLXXVIIII   1979

When a Roman numeral is placed to the left of a numeral for a larger number, such as IX for nine, it indicates subtraction. The subtractive principle was recognized by the Romans but they did not make much use of it.* Compare the preceding Roman numeral for 1979 with the following numeral written with the subtractive principle.

*D. E. Smith, *History of Mathematics,* **2** (Lexington, Mass.: Ginn, 1925), p. 60.

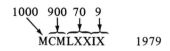

MCMLXXIX   1979

The Romans had relatively little need for large numbers, and so they developed no general system for writing them. One early example of this is the inscription on a monument commemorating the victory over the Carthaginians in 260 B.C. The symbol ⌒, for 100,000 is repeated 23 times to represent 2,300,000.

Multiplicative Grouping Systems   A natural improvement to the simple grouping system is to use numerals to indicate the number of times the powers of the base are to occur. Such a system is called a *multiplicative grouping system.* To give a fictitious example with our digits and the Egyptian symbols, the number 2346 would be written as shown in the first illustration on the right.

Such a system is much more economical than the Egyptian simple grouping system. Imagine the convenience of using this system to write the number 800 if you were chipping these numerals in stone.

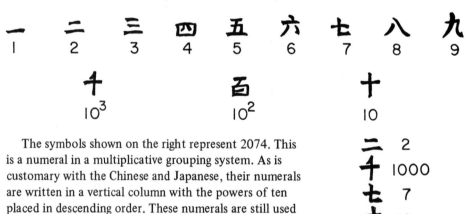

Chinese Numeration–The Chinese numerals form one of the oldest numeration systems. Here are the symbols for the digits and the first few powers of 10.

The symbols shown on the right represent 2074. This is a numeral in a multiplicative grouping system. As is customary with the Chinese and Japanese, their numerals are written in a vertical column with the powers of ten placed in descending order. These numerals are still used today by some businesses and on most checks.

Positional Numeration Systems   The next logical improvement in numeration systems was to omit writing the powers of the base. Now the only number symbols needed are those from zero up to one less than the base. Thus in our base ten system, only the digits 0, 1, 2, 3, 4, 5, 6, 7, 8, and 9 are needed to represent all numbers, but it must be agreed that the positions of these digits represent various powers of ten. Such a system is called a *place value* or *positional numeration system.*

Mayan Numeration—There is archaeological evidence that the Mayas were in Central America before 1000 B.C. During the Classical Period (A.D. 300–900) they had a highly developed knowledge of astronomy and a 365-day calendar with a cycle going back to 3114 B.C. The Mayas were one of the first people to use a positional numeration system with a zero symbol. Their system has a base of 20, requiring basic numerals for the numbers from 0 to 19. These numerals are listed below. Notice that they indicate a grouping by fives within the first 20 numbers.

"No, no, no! *Thirty* days hath September!"

0	〇	5	—	10	=	15	≡
1	•	6	•̄	11	=̇	16	≡̇
2	••	7	••̄	12	=̈	17	≡̈
3	•••	8	•••̄	13	=⋮	18	≡⋮
4	••••	9	••••̄	14	=⋮⋮	19	≡⋮⋮

It is interesting that the Mayan and Chinese numeration systems were developed on opposite sides of the earth at about the same period of time and that both chose to write their numbers vertically. In the example shown here the position of the top numeral, $\underset{=}{\bullet\bullet}$, represents 12 times

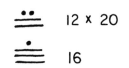

20, and the lower numeral, $\underset{=}{\bullet}$, represents 16. This "two-place" numeral represents $12 \times 20 + 16$, or 256. Whenever the Mayas wrote one numeral over the other, as in this example, the top numeral indicated the number of 20s and the bottom numeral the number of 1s. The following numerals from left to right represent 20, 60, 106, and 158.

Hindu-Arabic Numeration—Our system for writing numbers is another example of positional numeration. It is named for the Hindus, who invented it, and the Arabs, who transmitted it to Europe. Being a base ten system, it is also commonly called the *decimal numeration system.* The word "decimal" comes from the Latin "decem" which means ten.

There are various theories about the origin of the digits 1, 2, 3, 4, 5, 6, 7, 8, and 9. However, it is widely accepted that they originated in India. Notice the resemblance of the Brahmi numerals for 6, 7, 8, and 9 to our numerals. The Brahmi numerals for 1, 2, 4, 6, 7, and 9 were found on stone columns in a cave in Bombay dating from the second or third century B.C.* It seems likely that our numerals "2" and "3" are merely cursive forms of these early symbols for two and three.

Our method of numeration is remarkably efficient. Each digit represents a power of 10 according to its position, and these positions are referred to by name. In 75,063 the 3, 6, 0, 5, and 7 are called the *units digit, tens digit, hundreds digit, thousands digit,* and *ten thousands digit,* respectively. When we write a number as the sum of numbers represented by each digit in the numeral, the number is said to be written in *expanded form.* Here are three ways of writing a number in expanded form.

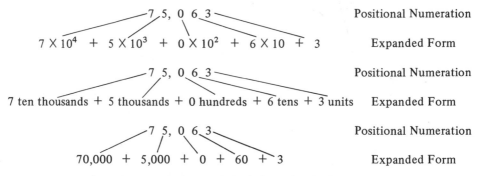

Brahmi numerals

7 5, 0 6 3   Positional Numeration

$7 \times 10^4 + 5 \times 10^3 + 0 \times 10^2 + 6 \times 10 + 3$ Expanded Form

7 5, 0 6 3   Positional Numeration

7 ten thousands + 5 thousands + 0 hundreds + 6 tens + 3 units   Expanded Form

7 5, 0 6 3   Positional Numeration

70,000 + 5,000 + 0 + 60 + 3   Expanded Form

Many years of development and struggle lie behind the Hindu-Arabic numeration system. The oldest dated European manuscript that contains our numerals was written in Spain in A.D. 976. In 1299, merchants in Florence were forbidden to use these numerals. Gradually, over a period of centuries, the Hindu-Arabic numeration system replaced the more cumbersome Roman numeration.

Reading and Writing Numerals—The number names for the whole numbers from one to twenty are all single words. The names from twenty-one to ninety-nine, with the exceptions of thirty, forty, fifty, etc., are compounded number names that are hyphenated. These names are also hyphenated when they occur as parts of other names. For example, we write, "three hundred forty-seven," for 347. Some people write or read this number as "three hundred and forty-seven," but the "and" is not necessary.

*J. R. Newman, *The World of Mathematics,* 1 (New York: Simon and Schuster, 1956), pp. 452–54.

Numerals with more than three digits are read by naming periods for each group of three digits. Within each period the digits are read as we would read any number from 1 to 999, and then the name of the period is recited. Beginning with the second group of three digits, the names of several periods are listed at the right. Here is an example of a number in the trillions.

thousand 000
million $000,000$
billion $000,000,000$
trillion $000,000,000,000$
quadrillion
quintillion
sextillion
.
.
.

$$2\ \ 3,\ 4\ \ 7\ \ 8,\ 5\ \ 0\ \ 6,\ 0\ \ 4\ \ 2,\ 3\ \ 1\ \ 9$$

trillion billion   million thousand

This number is read as, twenty-three trillion, four hundred seventy-eight billion, five hundred six million, forty-two thousand, three hundred nineteen.

Rounding Off   If you were to ask a question, such as, "How many people voted in the 1976 presidential election?" you might be told that in "round numbers" it was about 80 million. Approximations such as this are often as helpful as knowing the exact number, which in this example was 79,973,371.

To *round off* a number to a given place value, locate the place value and then check the digit to its right. If the digit to the right is 5 or greater, then all digits to the right are replaced by 0s and the given place value is increased by 1. If the digit to the right is 4 or less, then all digits to the right of the given place value are replaced by 0s. Rounded off to the nearest million, the number of votes in the 1976 election is 80 million, because the 9 to the right of the millions place increases 79 to 80.

millions place
↓
79,973,371

Here are some more examples of rounding off to place values in 79,973,371.

Rounding to   Rounded off

1. Hundred thousands
   Hundred thousands place
   ↓
   79,973,371   80,000,000

2. Ten thousands
   Ten thousands place
   ↓
   79,973,371   79,970,000

3. Thousands
   Thousands place
   ↓
   79,973,371   79,973,000

4. Hundreds
   Hundreds place
   ↓
   79,973,371   79,973,400

Models for Numeration   Of all the arithmetical concepts, positional numeration and place value lend themselves most readily to the use of physical materials. Here are several models which will be described: multibase pieces, bundles of sticks, bean sticks, the abacus, and chip trading.

Multibase Pieces—The powers of the base in this model are represented by objects called *units, longs, flats,* and *blocks.* In base ten, a *long* equals 10 units, a *flat* equals 10 longs, and a *block* equals 10 flats. These pieces represent the increasing powers of ten, namely, 1, 10, 100, and 1000. This model can be extended to higher powers of the base by using groups of blocks. For example, 10,000 is represented by a column of 10 blocks, and is called a *long-block.* The following set of multibase pieces represents 1347.

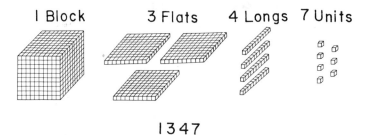

| Block   3 Flats   4 Longs   7 Units

1347

Bundles of Sticks and Bean Sticks—The next two models are similar. In the *bundle of sticks model,* units and tens are represented by single sticks and bundles of 10 sticks. One hundred is represented by a bundle of 10 bundles. In the *bean stick model,* single beans represent units; a stick (or strip of cardboard) with 10 beans glued to it represents 10; and 100 is a collection of 10 bean sticks. Here are two examples of these models.

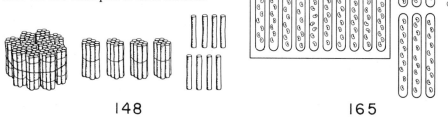

148   165

Abacus—The abacus is considered to be the oldest of all computing devices. Some form of it was used in Egypt at least as early as 500 B.C. There are several types in use in the world today. They are found in the Middle East, Russia, and the Orient. The basic idea of the abacus is to represent numbers by beads on rods. A Japanese abacus, called a soroban, is pictured at the top of the following page. Each bead below the horizontal bar (from right to left) represents a power of 10 $(1, 10, 10^2, 10^3, ...)$, and the corresponding bead above the bar represents five times as much $(5, 50, 500, 5000, ...)$. The two beads in the first column, which have been pushed to the horizontal bar, represent 6 ones. The one bead in the second column represents 10, and the four beads in the third column represent 400. All together, the beads represent 416.

In Western countries today the abacus is used mainly as a model for teaching place value and the basic operations. In the simplified abacus shown here, any number of markers may be placed on a column. For base ten, each column represents increasing powers of 10 from right to left (from 10^0 to 10^6). The 3 markers on the 10^4 column of this abacus represent 3×10^4 or 30,000. The 7 markers on the 10^3 column represent 7000, etc. If there are no markers on a column, this means that particular power of the base occurs zero times. The markers shown represent 37,204.

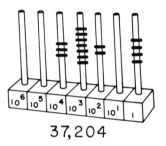

Chip Trading—This model uses colored chips and a mat and is similar to the abacus. There are columns for the chips, and in base ten the columns represent powers of 10. One yellow chip represents a unit, a blue chip stands for ten, green corresponds to hundreds, and each red chip is worth one thousand. Viewed in another way, 10 yellow chips can be traded for 1 blue chip, 10 blue chips can be traded for 1 green chip, etc. The mat keeps the chips in columns as a model for the place value concept of positional numeration. The 5 yellow chips on the mat illustrated represent 5 ones; the 7 blue chips represent 7 tens; etc. All together, these chips represent 4375.

Red	Green	Blue	Yellow

4375

Place Value on Calculators   On most calculators the digits that are entered first appear on the right in the display, then they move left as successive digits are entered. For example, to enter 428 on the calculator, we press $\boxed{4}$, then $\boxed{2}$, then $\boxed{8}$.

$\boxed{4} \quad \boxed{2} \quad \boxed{8}$

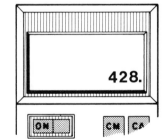

First a "4" appears in the display, then the "4" moves over and "42" appears, and finally the "42" moves and "428" appears. The three stages of the display are thus:

4.	42.	428.

On some calculators the numbers appear in the display from left to right. When 428 is entered, first a "4" appears at the left end of the display, then "42," and finally "428."

4.	42.	428.

The calculator can be used to increase understanding of positional numeration and place value by writing a number in expanded form and computing the sum. If we enter the value indicated by the position of each digit in 48,023 and *add* these numbers, then "48023" will appear in the display. Here is the sum to enter:

$$40000 \quad \boxed{+} \quad 8000 \quad \boxed{+} \quad 0 \quad \boxed{+} \quad 20 \quad \boxed{+} \quad 3$$

This process can be reversed by entering a number into the calculator and then removing one digit at a time from the left in the display. If 93706 is entered, it can be changed to 3706 by subtracting 90000. Next, the 3706 can be changed to 706 by subtracting 3000. Finally, subtracting 700 from 706 leaves the units digit in the display.

93706
3706
706
6

SUPPLEMENT (*Activity Book*)

Activity Set 2.2 Models for Numeration (Base five multibase pieces and the variable base abacus)

Additional Sources

Byrkit, D.R. "Early Mayan mathematics." *The Arithmetic Teacher*, **17** No. 5 (May 1970), 387–90.

Hogben, L. *Mathematics in the Making.* New York: Doubleday, 1960. "Our Hindu Heritage," pp. 26–49.

Menninger, K. *Number Words and Number Symbols.* Cambridge, Massachusetts: The M.I.T. Press, 1969. pp. 305–15 and 349–52 (current use of abacus and its history).

Smeltzer, D. *Man and Number.* London: A. and C. Black, 1970. pp. 29–55 and 79–89 (descriptions of numeration systems).

Smith, D.E. and J. Ginsburg. "Numbers and Numerals," Monograph No. 1. Reston, Virginia: NCTM, 1953.

Willerding, M. *Mathematical Concepts, A Historical Approach.* Boston: Prindle, Weber and Schmidt, 1967. pp. 1–16 (numeration systems).

Zaslavsky, C. "Black African Traditional Mathematics." *The Mathematics Teacher*, **63** No. 4 (April 1970), 345–56.

1. In the fifteenth and sixteenth centuries there were two opposing opinions on the best numeration system and methods of computing. There were the "abacists" who used Roman numerals and computed on the abacus and the "algorists" who used the Hindu-Arabic numerals and place value. The accompanying sixteenth century picture shows an abacist competing against an algorist. The abacist has a reckoning table with four horizontal lines and a vertical line down the middle. Counters or chips placed on lines represented powers of 10, as shown in the next diagram. The thousand's line was marked with a cross to aid the eye in reading numbers. If more lines were needed, every third line was marked with a cross. This was the origin of separating groups of three digits in a numeral by a comma.*

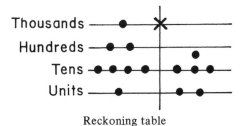

An algorist computing with numerals and an abacist computing with counters

★   a. What number is represented on the left side of this reckoning table?

     b. Each counter in a space between the horizontal lines represents half as much as it would on the line above. What number is represented on the right side of this reckoning table?

★   c. Each line and the spaces between them can be labelled by Roman numerals similar to the labelling of lines and spaces on sheet music. Use the seven Roman numerals (I, V, X, L, C, D, M) to label the lines and spaces on the reckoning table.

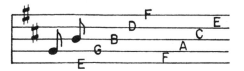

Reckoning table

*D. E. Smith, *History of Mathematics,* **2** (Lexington, Mass: Ginn, 1925), pp. 183–85.

2. Here is the complete counting system of a twentieth century Australian tribe that uses only 4 numbers.

Neecha	Boolla	Boolla Neecha	Boolla Boolla
1	2	3	4

a. If this system were continued, what would be the names of the numbers for 5 and 6?

b. How would even numbers differ from odd numbers in a continuation of this system?

★ 3. Using the base five system of counting by fingers and grouping by hands, the name for 7 is "one hand and two."

★   a. What would be the name for 22 in this system?

★   b. If 25 is called a "hand of hands," how would 37 be represented in this system?

★ 4. The names of the following numbers are some of the literal meanings of number words taken from primitive languages in various parts of the world.* Write in the missing names that follow this pattern.

5, whole hand; 6, one on the other hand; 8, _____;
10, _____; 11, one on the foot; 15, _____;
16, _____; 20, man; 21, one to the hands of the next man;
25, _____; 30, _____;
40, _____.

5. The following numeration systems were used at widely varying times and in different geographical locations. Compare these sets of numerals for the numbers 1 through 10. What similarities can you find? What evidence is there of grouping by fives?

Babylonian Numerals

Mayan Numerals

Roman Numerals

Chinese Numerals

6. Perhaps the greatest achievement in the development of numeration systems was the invention of a symbol for zero. The following ancient numeration systems had no zero symbol. How is 603 written in each of these systems?

a. Egyptian   ★ b. Roman   c. Chinese

*D. Smeltzer, *Man and Number* (London: A. and C. Black, 1970), pp. 14–15.

★ 7. These Greek numerals have been dated about 1200 B.C. Use these symbols and the simple grouping system concept to write 2483.

| 10 | 100 | 1000
| — | ○ | □

8. The Attic-Greek numerals were developed sometime prior to the third century B.C. and came from the first letters of the Greek names for numbers. Use the clues in the following table to find the missing numerals.

Hindu-Arabic	I	4	8		26	32	52	57	206	
Attic-Greek			ΓΙΙΙ	ΔΓΙ		ΔΔΔΙΙ		ΡΓΙΙ	ΗΗΓΙ	ΓΔΙ

a. What base is used in this system?

★ b. Is this system most like the simple grouping, multiplicative grouping, or positional numeration system?

9. In a positional numeration system, any amount of numbers can be written by using only a finite number of different symbols.

★ a. How many different symbols are needed in the Hindu-Arabic numeration system?

b. Can any amount of different numbers be written in the Mayan numeration system with only a finite number of different symbols?

★ c. Answer part **b** for the Chinese numeration system.

d. Answer part **b** for the Roman numeration system.

10. Write the names of the numbers in parts **a** through **d**.

★ a. 5,438,146   b. 31,409

★ c. 816,447,210,361   d. 62,340,782,000,000

           tr. b. M. th.

11. The Japanese abacus on page 63 has 23 columns.

★ a. What is the largest number that can be represented on this abacus?

b. Describe the positions of the beads for the number in part **a**.

c. Write the name of this number.

12. Write each of the following numbers in Mayan numeration.

★ a. 30   b. 177

13. Around 2000 B.C. the Babylonians had a base sixty positional numeration system. How many different number symbols would be needed in a base sixty positional numeration system?

14. Most ancient civilizations developed ways to represent numbers with various positions of their hands and fingers. By the Middle Ages this method had become international for business transactions. In the system shown here the numbers from 1 to 10,000 can be represented.

★ a. Represent 9090 on your hands with this system. On this chart circle the hand positions that you used.

   b. Circle the hand positions for representing 404.

★ c. Which pairs of these columns are similar?

   d. How can this chart be used to explain the following phrase found in Juvenal's tenth satire: "Happy is he indeed who has postponed the hour of his death so long and finally numbers his years upon his right hand."*

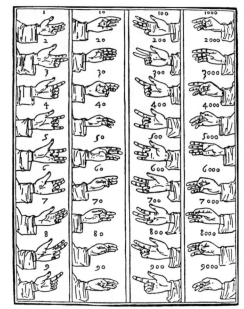

Finger counting for numbers from 1 to 10,000

   e. In medieval times the finger symbol for 30 was called "the tender embrace" and was a symbol for marriage.** Have you ever seen this symbol used as a gesture of something nice?

15. Use a sketch to represent the following numbers as they would be illustrated by the given models.

★ a. 136 by multibase pieces   b. 47 on the bundle of sticks model

   c. 64 on the bean stick model

16. On the following diagrams draw markers or chips to represent the numbers.

★ a. 4908   b. 7052

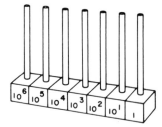

Red	Green	Blue	Yellow

*H. W. Eves, *Mathematical Circles* (Boston: Prindle, Weber and Schmidt, 1969), p. 29.
**H. W. Eves, *Mathematical Circles Squared* (Boston: Prindle, Weber and Schmidt, 1972), p. 4.

17. *Calculator Exercise:* In parts **a** through **c**, explain what you would do to the calculator to change the top display to the one under it without changing the digits that are the same in both displays.

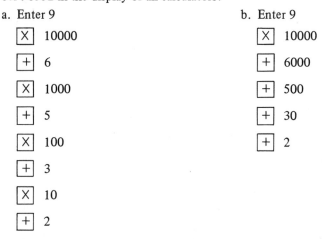

a.

| 1034692. |

| 1834692. |

b.

| 938647. |

| 908047. |

c.

| 40000. |

| 400000. |

18. *Calculator Exercise:* Which of the following sequences of steps will result in the number 96532 in the display of all calculators?

a. Enter 9

  ☒ 10000

  ⊞ 6

  ☒ 1000

  ⊞ 5

  ☒ 100

  ⊞ 3

  ☒ 10

  ⊞ 2

b. Enter 9

  ☒ 10000

  ⊞ 6000

  ⊞ 500

  ⊞ 30

  ⊞ 2

19. *Positional Numeration without Zero:* It is possible to represent all whole numbers using positional numeration and only the digits from 1 through 9 (no zero) by selecting a new symbol for 10. If we choose # as the new symbol for 10, then 20 is represented by 1#, that is, $1 \times 10 + 10$. Similarly, 30 would be written as 2#, which represents $2 \times 10 + 10$. Write each of the following numbers in this system without using a zero.*

★ a. 100   b. 140   ★ c. 300   d. 1000

*Described by C. Brumfiel, "Zero is highly overrated," *The Arithmetic Teacher,* **14** No. 5 (May 1967), 377–78.

"Nope, no chess problems until you've finished your quadratic equations!"

2.3 BINARY SYSTEMS AND COMPUTERS

Binary Sequence   The game of chess has been played for centuries. The time and place of its origin are unknown. However, the general opinion is that it was first played in India as early as the sixth century A.D.

There is a legend that chess was invented for the Indian King Shirham by the Grand Vizier Sissa Ben Dahir. As a reward, Sissa requested that he be given 1 grain of wheat to place on the first square of the chess board, 2 grains for the second square, 4 grains for the third square, then 8 grains, 16 grains, etc., until each square of the

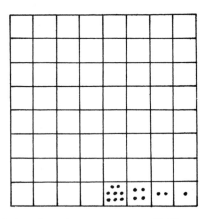

board was accounted for. The King was surprised at such a meager request until Sissa informed him that this was more wheat than existed in the entire kingdom.

The numbers of grains of wheat which Sissa requested on the squares of the chess board are called *binary numbers.* Beginning with 1, each succeeding number is obtained by doubling the previous number. This infinite sequence of numbers is called the *binary sequence.* These numbers are also successive powers of 2.

$$
\begin{array}{cccccccc}
1 & 2 & 4 & 8 & 16 & 32 & 64 & 128 \quad \cdots \\
2^0 & 2^1 & 2^2 & 2^3 & 2^4 & 2^5 & 2^6 & 2^7 \quad \cdots
\end{array}
$$

Binary Numeration   If we had been born with two fingers rather than ten, base two would probably be used today by most countries rather than base ten. Very likely, base two was the first base to be used by primitive people. In recent years, this base has been found among the African pygmies. Their first few number names show the influence of grouping by twos for the numbers greater than three.

a	oa	ua	oa-oa	oa-oa-a	oa-oa-oa
1	2	3	4	5	6

In base two positional numeration, 0 and 1 are the only symbols needed. The positions of 0 and 1 in a base two numeral indicate powers of two just as the positions of the digits 0 through 9 in base ten indicate powers of 10. These powers of 2 increase from right to left.

$$1011_{two} \text{ represents: } (1 \times 2^3) + (0 \times 2^2) + (1 \times 2) + 1$$

Each base two numeral represents a sum of binary numbers. To convert a base two numeral into base ten, just add the binary numbers indicated by the 1s. The number 101101_{two} equals $32 + 0 + 8 + 4 + 0 + 1$. The numeral for this sum in base ten is 45.

$$32 + 8 + 4 + 1$$

Conversely, to express any number in base two numeration, list the first few binary numbers and note the ones whose sum equals the given number. For example, 26 is equal to the sum of the 2nd, 4th, and 5th binary numbers. Therefore, the base two number for 26 is 11010_{two}.

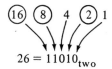

$$26 = 11010_{two}$$

Finding the base two numeral for a given number of objects can be easily done by grouping the objects by powers of 2. Always begin with the greatest possible binary number. The base two numeral for the number of paper clips is 1101_{two}.

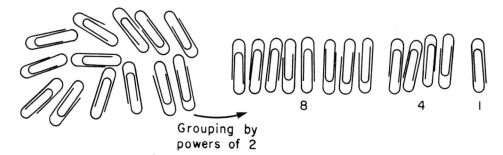

Grouping by powers of 2

Base two is the smallest base for which there can be any grouping. In terms of the number of different basic symbols needed for writing numbers in positional numeration, base two numeration is the simplest. However, this numeration system may seem confusing at first because the grouping process occurs so frequently. For example, in counting there are only two single-digit numbers, 0 and 1, before the first two-digit number, 10_{two}. Here are the first few whole numbers in binary numeration.

$0 = 0$	$4 = 100_{two}$	$8 = 1000_{two}$
$1 = 1$	$5 = 101_{two}$	$9 = 1001_{two}$
$2 = 10_{two}$	$6 = 110_{two}$	$10 = 1010_{two}$
$3 = 11_{two}$	$7 = 111_{two}$	$11 = 1011_{two}$

Models for Binary Numeration   The following models illustrate the grouping process and positional numeration in base two.

Multibase Pieces—Here are the multibase pieces for the first few powers of 2. From right to left these pieces represent 1, 2, 4, 8, and 16.

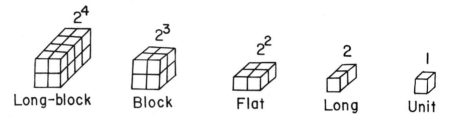

To represent a number using this model, each multibase piece is either used once or not at all. Let's see why this is true. First of all, a number should always be represented by the fewest number of pieces. Furthermore, any two multibase pieces can always be replaced by one piece for the next higher power of 2. Two units can be replaced by 1 long; 2 longs equal 1 flat; etc. Therefore, there is no need to use any particular multibase piece more than once.

To determine the multibase pieces for a given number, write the number as a sum of binary numbers (1, 2, 4, 8, 16, etc.). The number 21, for example, equals $16 + 4 + 1$. These powers of 2 correspond to 1 long-block, 1 flat, and 1 unit.

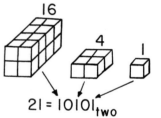

$21 = 10101_{two}$

Abacus for Base Two—Each column of the abacus shown here represents a power of 2. From right to left these numbers are $1, 2, 2^2, 2^3, 2^4, 2^5$, and 2^6. The numbers represented on this abacus will have either 1 marker or no markers on the columns. The markers on this abacus represent 110111_{two}.

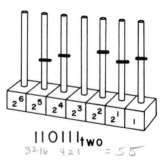

110111_{two}

To determine what 110111_{two} equals in base ten numeration, it is only necessary to add the powers of 2 which are indicated by the markers on the columns. This abacus has markers on the columns for $1, 2, 2^2, 2^4$, and 2^5. The sum of these numbers is $1 + 2 + 4 + 16 + 32$, which equals 55 in base ten numeration.

Early Electronic Computers   The first large-scale automatic computer, the Mark I, was developed by IBM and presented to Harvard University in 1944. This was a 5-ton machine containing more than 3000 electromechanical relays and 500 miles of wire. It was retired in 1959 and some of its parts are on display at the Smithsonian Institution.

The first fully electronic computer, the ENIAC (Electronic Numerical Integrator and Calculator), was developed for the U.S. Army at the University of Pennsylvania. Mechanically moving parts, such as counter wheels to represent digits and numbers, were eliminated in this computer. Instead, numbers were represented by sending electronic pulses to vacuum tubes to switch them on and off. Since electronic pulses can move thousands of times faster than electromechanical devices, the ENIAC was a significant breakthrough in the development of fast calculators. However, its speed of 5000 additions per second is slow by today's standards. In recent years, the replacement of tubes by miniaturized circuits has increased the speed of computers to capabilities of 10 million additions per second.

Computers and Binary Numbers   Computers are able to print messages, compose music, play chess, inventory supplies, compute payroll deductions, supply proofs to theorems, and perform many other tasks. Each letter, note, operation, punctuation mark, and instruction for the computer is coded in base two numerals. The reason for this is that the two binary digits of 0 and 1 can be represented by electric switches which are turned either on or off. In this illustration the three lighted bulbs represent the base two numeral 1101_{two}, which in turn is equal to 13. Each bulb from right to left represents increasing powers of 2.

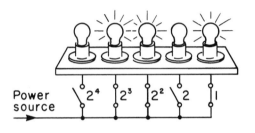

A power of 2 is counted, depending upon whether or not the bulb corresponding to that power is lighted. By lighting various combinations of these five bulbs, the numbers from 0 (no bulbs lighted) to 31 (all bulbs lighted) can be represented. Here are two more sets of lights which represent numbers in base two. The first number is equal to 14, and the second number is equal to 27.

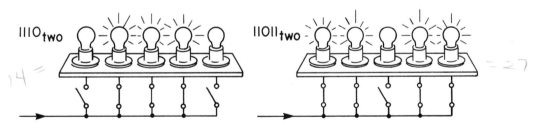

With larger sets of lights the computer can read and store billions of binary numerals.* The code shown on the next page is for the letters of the alphabet, commas, periods, and blanks, and is used in the IBM system/360. The letters from A through I are represented by placing 1100 in front of the base two numerals from 1 to 9. For the letters J through R, the base two numerals for 1 through 9 are used again but this time they are preceded by 1101. Other letters, symbols, and punctuation marks are coded in a similar manner.

*Rather than lights, computers use small two-state devices such as metal rings which can be magnetized in clockwise or counterclockwise directions.

A	1100 0001	J	1101 0001	S	1110 0010
B	1100 0010	K	1101 0010	T	1110 0011
C	1100 0011	L	1101 0011	U	1110 0100
D	1100 0100	M	1101 0100	V	1110 0101
E	1100 0101	N	1101 0101	W	1110 0110
F	1100 0110	O	1101 0110	X	1110 0111
G	1100 0111	P	1101 0111	Y	1110 1000
H	1100 1000	Q	1101 1000	Z	1110 1001
I	1100 1001	R	1101 1001		

Blank 0100 0000   Period 0100 1011

Comma 0110 1011

How Computers Add   To add two numbers in base two numeration the computer only has to be able to add 0s and 1s. There are four combinations of these numbers, as shown in the next table. If the two numbers are both 0, their sum is 0. If the numbers are 1 and 0 or 0 and 1, their sum is 1. In the fourth line of the table both digits are 1 and their sum is recorded as 0. This is because $1 + 1 = 10_{two}$ in base two and so "0" is recorded and provisions are made to carry the "1" in "10" to the next column or place value.

First number	Second number	Sum
0	0	0
1	0	1
0	1	1
1	1	0

Now let's see how this simple example of addition is carried out electronically to add two binary digits. There are four switches: A_0 and A_1 for the first number; and B_0 and B_1 for the second number. If the first number is 0, close A_0, and if it is 1, close A_1. If the second number is 0, close B_0, and if it is 1, close B_1.

First Number $\boxed{A_0}$ $\boxed{A_1}$ (0 or 1)

Second Number $\boxed{B_0}$ $\boxed{B_1}$ (0 or 1)

The electric circuit that will compute these four sums ($0 + 0$, $1 + 0$, $0 + 1$, and $1 + 1$) has two special types of switches, called *two-pole switches*. Switch A_1 and switch A_0 are connected so that if A_1 is closed then A_0 is open, and if A_0 is closed then A_1 is open. Switches B_1 and B_0 work in the same way. Now, to compute $1 + 0$, A_1 and B_0 are closed and this opens A_0 and B_1 as shown in diagram (a). The current will flow through the top wire and the light will be on, indicating a sum of 1. To compute $0 + 1$, A_0 and B_1 are closed, opening switches A_1 and B_0, and so the current will flow through the lower wire as shown in diagram (b).

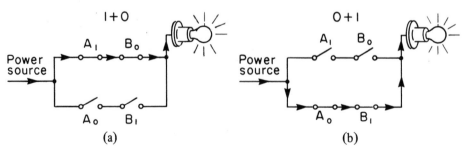

(a)

(b)

For the remaining two sums, $0 + 0$ and $1 + 1$, the bulb should not be lighted in order to indicate a sum of 0. Consider the sum $0 + 0$. To compute this sum, switches A_0 and B_0 are closed, which opens up switches A_1 and B_1. Diagram (c) shows the positions of the switches for this sum, and we see that the current will not flow to the bulb.

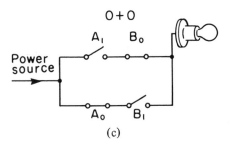

(c)

Computer Input and Output   There are many devices for feeding information (input) into a computer. One of these is punched cards, on which numbers and letters are coded by patterns of holes. These holes are punched by a special typewriter called a keypunch. The computer reads a card by electrical impulses which pass through the holes.

The more recently developed magnetic tapes provide much faster (but more expensive) methods of storing and feeding in computer instructions. These tapes are made by depositing a layer of magnetic material on plastic tape and magnetizing this tape in a code of 1s and 0s. For example, a magnetized spot can represent 1, and another spot which is magnetized in the opposite direction will represent 0. With this method, information can be packed closely together and hundreds of thousands of 1s and 0s can be read per second.

Time-sharing systems enable people to use a computer from a remote terminal. The results from the computer (output) are received immediately on the terminal, typewritten either on paper or on a screen. In some cases, the output from the computer can be in the form of words or music. The terminals can be located anywhere and connected to a central computer installation through telephone lines or cable television. In such a system the computer automatically divides its time among all the users. The computer works so fast, in comparison to the terminal users, that there are seldom any delays at the terminals. Time-sharing is used in a wide variety of ways: by school and college students for problem solving; by scientists and engineers for calculations; by stockbrokers, lawyers, police, and realtors to obtain information from large data banks; and by clerks and store managers for making inventories and ordering new supplies.

There are five parts to a computer system: The *Input Device* gives instructions through punched cards, typewriters, tapes, or discs; the *Compiler* translates the instructions into a binary code; the *Control Unit* performs the operations; the *Decoder* translates the results from the base two code back into base ten numerals and words; and the *Output Device* presents the information by teletype printouts, TV tubes, graph plotters, or speakers.

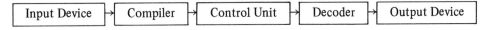

Computer Programs and Language   The set of instructions and the necessary data for performing a task on the computer is called the program. Each program is written in one of several existing computer languages, which require strict adherence to rules of form and grammar. One of the early and widely used languages was FORTRAN (FORmula TRANslation). FORTRAN is based on the language of algebra plus a few rules of grammar imposed by the computer. It was developed primarily for engineers and scientists familiar with mathematics.

When businesses and industries started using computers, FORTRAN wasn't the answer for problems involving inventory control and payroll billing. The language developed for these needs is called COBOL (COmmon Business Oriented Language). COBOL consists of about 250 key words for expressing problems in a language very similar to business English.

One of the easiest languages to learn is BASIC (Beginners All-purpose Symbolic Instruction Code), developed in 1963–1964 at Dartmouth College. The vocabulary of BASIC includes about 20 words which you must know how to assemble into logical or acceptable statements. For example, you can't tell the computer to "add a column of numbers"—the statement has to be more explicit. The following program in BASIC computes the average of three numbers. It contains five BASIC vocabulary words which are printed in capital letters since there are no lowercase letters on the teletype keyboard. Here is the meaning of each of the five statements in this program: Statement 100 tells the computer to let the three numbers in Statement 400 (the data) equal the three variables $X1, X2, X3$; Statement 200 gives the computer the formula for the average of three numbers; Statement 300 tells the computer to compute the average and print the result; Statement 400 lists the numbers to be operated on in the program; Statement 500 tells the computer the program is finished.

```
100 READ X1, X2, X3
200 LET S = (X1 + X2 + X3)/3
300 PRINT S
400 DATA 4472, 1086, 3260
500 END
```

Computer programmers are continually striving to decrease the time and effort it takes to write programs and, in general, to make it easier to communicate with the computer. On some visual display terminals you can roughly sketch a gear or curve with a light-sensitized pen, and the computer will display an accurate sketch, giving the equations and certain related numerical values. The student in this photo is using a light-sensitized pen to select a word from a list of verbs. The dog on the display will then act out this verb.

Perhaps the ultimate in giving a computer instruction is by talking to it. Some computers have been programmed to respond to a limited number of words. However, contrary to the impression one might get from the talking computer in the movie "2001," one cannot interact with a computer by merely speaking in its presence. The task of getting the computer to understand continuous speech (unlike isolated words) is very complex. In continuous speech there is no indication of where one word ends and the next begins; nor are the boundaries between the individual speech sounds (called phonemes) clearly indicated. The following graphs show just two of the speech sounds in "potassium." Graph (a) shows the vowel "a" and graph (b) is for "i." The sound for "a"

lasted about 3/20 of a second and the sound for "i" required about 1/20 of a second. These graphs show how energy levels vary during these letter sounds.*

"a"

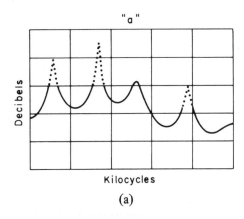

(a)

"i"

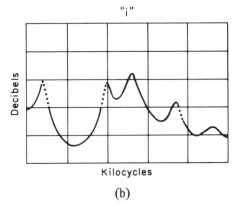

(b)

Although general speech recognition is not presently practical, many imaginative methods of computer input are being used. In some experiments computers have been programmed to listen to the notes of a piano or organ and to print this music on a visual display. Then the printed version of the music can be edited by a light-sensitized pen and the computer can be instructed to play back the edited music. The organ keyboard in this picture is connected electrically to a computer in much the same way that a typewriter keyboard is connected. The editing that is done on the display screen allows for changes in volume, harmonics, and pitch variation.**

Organ keyboard connected to computer

*From research by J.J. Wolf and W.A. Wood, "HWIM Speech Understanding System," Bolt Beranek and Newman, Inc., Cambridge, Mass. (1977).

**From research reported in "Personal Dynamic Media," Xerox Corporation, Palo Alto Research Center, California (March 1976).

SUPPLEMENT (*Activity Book*)

Activity Set 2.3 Applications of Binary Numbers (Mind reading cards, binary card sorting system, and Martingale system of gambling)

Just for Fun: Game of "Nim"

Additional Sources

Baer, R.M. *The Digital Villain.* Reading, Massachusetts: Addison-Wesley, 1972.

Dunkum, W. "Another use for binary numerals." *The Arithmetic Teacher,* **17** No. 3 (March 1970), 225–26.

Dutton, P.E. "Computer Page." *Mathematics in Schools,* **2** No. 6 (November 1973), 18–21.

Franta, W.R. "Computer Arithmetic." *The Mathematics Teacher,* **64** No. 5 (May 1971), 409–14.

Gardner, M. *New Mathematical Diversions from Scientific American.* New York: Simon and Schuster, 1966. "The Binary System," pp. 13–22.

National Council of Teachers of Mathematics. *Instructional Aids in Mathematics.* 34th yearbook. Reston, Virginia: NCTM, 1973. "The Role of Electronic Computers and Calculators," pp. 153–201.

Perisho, C.R. "Applications of binary notation." *The Arithmetic Teacher,* **14** No. 5 (May 1967), 388–90.

Network III, a computer-generated environment, Walker Art Center, Minneapolis

EXERCISE SET 2.3

1. The network of lights in the photograph on the facing page is activated by a computer as people walk on the carpet. The carpet is subdivided into small squares, each containing a pressure-actuated switch that detects the presence of a person on that square. The computer is programmed to determine where people are within the space, which direction they are moving in, how fast they are travelling, etc. The computer uses this information to generate patterns in the network of lights. A person walking into the space might receive a response in the form of a circle of lights on the grid, or perhaps a pair of intersecting lines of light. If a person should pause for some length of time, his/her pattern—say, a circle of light, might begin to expand and contract. The sense of a highly elaborate abstract game is intended to be conveyed, tempting the person into testing the environment to see what effect he/she has on the responses.

    a. The 24-by-24-foot carpet is partitioned into 4-by-4-foot squares. How many of these squares will there be?

    ★   b. The computer will read the array of switches (one for each small square) ten times each second to determine how people are moving about the region. How many readings is this per minute?

2. In the legend of the origin of chess the total number of grains for all 64 squares of the chessboard is the sum of the first 64 binary numbers.

    a. Write the sums of the first few binary numbers. What happens to the total number of grains of wheat each time one more square is used on the board?

    $1 = 1$
    $1 + 2 = 3$
    $1 + 2 + 4 =$
    $1 + 2 + 4 + 8 =$
    $1 + 2 + 4 + 8 + 16 =$
    $1 + 2 + 4 + 8 + 16 + 32 =$

    ★   b. How many grains of wheat will there be on the first 16 squares?

3. Each person has 2 parents, 4 grandparents, 8 great grandparents, and so on.

    ★   a. Assuming that there have been 20 generations before you (since 1492) how many lineal relations would you have had living at the time Columbus sailed for the New World?

    b. What is the total number of lineal relations you would have had since 1492? (Count those who were living in 1492 as well as those who lived after.)

4. Write the base two numerals and the base ten numerals for the numbers that are represented by the models in parts **a** through **d**.

        $22 = 10110_{two}$

    ★   a.   b.

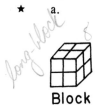

    16

    Block   Flat   Unit   Long-block   Flat   Long

    $13 = 110_{two}$

$87 = 1010111_{two}$

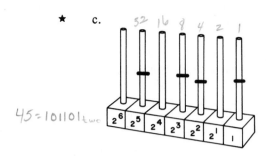

$45 = 101101_{two}$

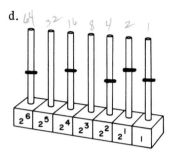

5. Express each of the following numbers as a sum of binary numbers (1, 2, 4, 8, 16, ...). Do not use a binary number more than once. Then write these numbers using base two positional numeration.

★   a. 58   b. 108

6. Write the following numbers using base ten numeration.

★   a. 1001_{two} b. 111011_{two}

7. Close the switches and light the bulbs to show which lights should be lighted to represent the given numbers in base two numeration.

★   a. 23   b. 37

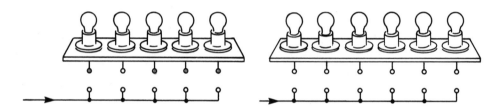

8. Draw in the positions of the switches $(A_1, A_0, B_1,$ and $B_0)$ for computing the sums. Indicate in each example whether or not the bulb should be lighted.

★   a. $1 + 0$ b. $1 + 1$

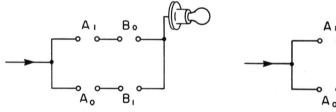

9. The 20 symbols listed next are commonly used in computer programs. Devise a binary code for these symbols using five-place base two numerals.

["   @   # $ %   ¢   &   *   (+   )   ?   ]   =   !   →   <   ←   >

10. Three-place binary numerals can be formed by placing 1s or 0s in three boxes. The smallest number is 000.

   ★ a. What is the largest number that can be represented by a three-place binary numeral? How many different numbers can be represented by three-place binary numerals?

   b. What is the largest number that can be represented by a four-place binary numeral? How many four-place binary numerals are there?

   ★ c. How many numbers can be represented by five-place binary numerals?

11. *Subsets:* Suppose you were going on a trip and you could take any combination of 4 boxes, including taking all of them or none of them. For example,

   Box 1   Box 2   Box 3   Box 4

   you might take boxes 1, 3, and 4. In how many ways could you make your selection? (*Hint:* Place a 1 in a box to indicate it will be taken and a 0 in a box if it will not be taken. In how many ways can 1s and 0s be placed in these boxes?)

12. *Binary Finger System:* Here is a system for representing the numbers from 0 to 31 on your right hand. As you look at the palm of your hand, the thumb, index finger, middle finger, ring finger, and little finger represent the binary numbers 1, 2, 4, 8, and 16, respectively. To represent 3, raise the thumb and index finger. To represent 18 raise the index finger and little finger.

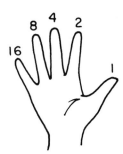

   ★ a. How is 27 represented in this system?

   b. What finger is always raised for odd numbers?

   ★ c. If you also use the left hand and continue assigning the binary numbers 32, 64, 128, 256, and 512 to the little finger, ring finger, middle finger, index finger, and thumb, respectively, then what is the greatest number you can represent with both hands?

13. A computer memory is made up of many groups of binary numerals. Each group of 32 digits has a certain location called an address. A small portion of a memory with 6 addresses is shown here.

Addresses in Computer Memory

1st	1100 1001	0100 0000	1110 0011	1100 1000
2nd	1100 1001	1101 0101	1101 0010	0100 1011
3rd	0100 0000	1110 0011	1100 1000	1100 0101
4th	1101 1001	1100 0101	1100 0110	1101 0110
5th	1101 1001	1100 0101	0100 0000	1100 1001
6th	0100 0000	1100 0001	1101 0100	0100 1011

★   a. Any 32-digit base two number can be represented at an address. If all 32 digits are 0, the number is 0. What number is represented if all 32 digits are 1s? (*Hint:* Use a power of 2 to write your answer.)

b. Write the 32-digit address for the number 253. (*Hint:* Most of the digits will be 0s.)

★   c. The computer can be instructed to output the contents of each address in letters (alphanumeric characters). Using the binary code on page 74, the first eight digits in the first address represent I; the next eight digits stand for a blank; the next group of eight represents T; etc. Complete this translation to decode the sentence in the six addresses.

14. *Balance Pan Puzzle:* It is possible with just 6 different weights to weigh any object weighing from 1 to 63 g on a balance pan scale. (These objects have whole number weights and the weights are placed on just one side of the balance beam.) What are these 6 weights?

15. *Four Weights Puzzle:* Using the balance scale and only the four weights 1 g, 3 g, 9 g, and 27 g, it is possible to weigh any object from 1 to 40 g (whole number weights in grams). Explain how you could show that an object had the following weights.

★   a. 5 g   b. 16 g

CHAPTER THREE

WHOLE NUMBERS AND THEIR OPERATIONS

In learning any kind of subject matter it is necessary to concentrate on the high spots, the most significant things. A good memory should be like a fisherman's net—that is, it should retain all the big fishes but let the little ones escape.

Eramus

3.1 ADDITION

Addition is first encountered with concrete objects. If 4 clams are *put together* with 3 clams, the total number of clams is the *sum* 4 + 3. The close association between the operation of *addition* and the set operation of *union* can be seen in the words that have been used over the years. In the thirteenth through fifteenth centuries the words *aggregation, composition, collection, assemble,* and *join* were all used at various times to mean addition.

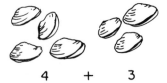

$$4 \qquad + \qquad 3$$

Definition of Addition   The idea of *putting sets together* or *taking their union* is often used as a definition for addition. If set R has r elements and set S has s elements, and if these two sets have no elements in common, then r added to s is defined as the number of elements in the union of sets R and S. This is written symbolically as

$$r + s = n(R \cup S)$$

In the definition of addition the requirement that R and S be disjoint sets is necessary. Suppose, for example, that there are 8 people in a group who play a guitar and 6 who play a piano. We cannot assume that the total number of people in the group who play a guitar or a piano is 14, because there may be some who play both. The case in which there are 2 people who play both instruments is illustrated in the accompanying diagram. The total number of people is 12, as shown by the total number of different points inside the circles. In this example, $n(G) = 8, n(P) = 6$, but $n(G \cup P) \neq 8 + 6$.

not disjoint sets

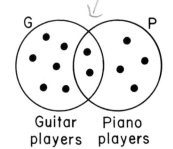

Guitar   Piano
players   players

Number Properties   To illustrate some of the abuses of what has been popularly called "Modern Mathematics" or "New Math," Morris Kline relates the following conversation in his book *Why Johnny Can't Add.*

Teacher: Why is 2 + 3 = 3 + 2?

Student: Because both equal 5.

Teacher: No, the correct answer is because the commutative law of addition holds.

At the beginning of the "new math" era one of the major criticisms of traditional mathematics was the amount of rote learning and unthinking repetition. One attempt to promote understanding in the new math programs was the use of number properties, such as the commutative and associative properties shown in the paragraphs that follow. While it is doubtful that the student in the previous example has acquired a better understanding of mathematics, these number properties, nevertheless, can be helpful in computing and in providing a better understanding of the steps used in computing.

Associative Property for Addition—Sometimes we will want to add two numbers by breaking one of the numbers into a convenient sum of numbers. The sum $147 + 26$ can be easily added in your head by computing $147 + 20$ and then adding 6. The change in the second of the following two equations is permitted by the associative property for addition.

$$147 + 26 = 147 + (20 + 6) = (147 + 20) + 6$$

Associative Property
for Addition

In a sum of three numbers the middle number can be added to (associated with) either of the two end numbers. This is called the *associative property of addition.* In general, for any three numbers that are used for the placeholders $\square$, $\triangle$, and $\diagup\!\diagup$,

$$(\square + \triangle) + \diagup\!\diagup \quad = \quad \square + (\triangle + \diagup\!\diagup)$$

The associative property of addition also plays a roll in arranging numbers to produce sums of ten, called "making tens." As an example,

$$8 + 7 = 8 + (2 + 5) = (8 + 2) + 5 = 10 + 5 = 15$$

Associative Property
for Addition

Commutative Property for Addition—This property cuts in half the number of basic addition facts that must be memorized. It also allows us to select convenient combinations of numbers when adding. The numbers 26, 4, and 37 are arranged more conveniently in the right side of the following equation than on the left because $26 + 4 = 30$ and it is easy to compute $30 + 37$. This rearrangement of numbers is permitted by the commutative property for addition.

$$26 + (37 + 4) = 26 + (4 + 37)$$

Commutative Property
for Addition

In any summand the positions of two numbers can be interchanged or commuted without changing the sum. This is called the *commutative property for addition.* In general, for any two numbers that are used for the placeholders $\square$ and $\triangle$,

$$\square + \triangle = \triangle + \square$$

Taken together, the associative and commutative properties for addition permit us to compute a sum of numbers in any order we wish. In the following sum we might begin by adding the 10s, then the 5s, and then the 17 and 23. Or, you might prefer a different combination. Regardless of the order and groupings, the sum will always be the same.

$$5 + 17 + 10 + 5 + 10 + 23 + 10 + 5 + 10$$

Models for Addition Algorithm   An *algorithm* is a step-by-step routine for computing. The algorithm for addition involves two separate procedures: (1) adding separate digits; and (2) "regrouping" or "carrying" (when necessary) so that the sum is written in positional numeration. Each of the models used on pages 61–63 for numeration is also appropriate for illustrating the four basic operations on whole numbers. Several of these models are presented in the following paragraphs and in Exercise Set 3.1 to illustrate the addition algorithm.

Bundles of Sticks—To illustrate the sum of two numbers, the sticks representing these numbers can be placed below each other, just as the numerals are in the addition algorithm. The sum is the total number of sticks in the bundles plus the total number of individual sticks. To compute 46 + 33 there are 7 bundles of sticks (7 tens) and 9 sticks (9 ones), so the sum is 7 × 10 + 9 or 79.

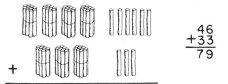

The mathematical justification for counting the total number of individual sticks (adding the units digits), and then counting the total number of bundles of sticks (adding the tens digits), requires the associative and commutative properties for addition. In the first of the following two equations, 46 and 33 are written in expanded form. The sums on the right side of this equation can then be rearranged to obtain the second equation, by using the commutative and associative properties.

$$46 + 33 = (40 + 6) + (30 + 3)$$
$$= (40 + 30) + (6 + 3)$$

Whenever there is a total of 10 or more individual sticks, they can be regrouped into a bundle of 10. In the following example the total number of individual sticks is 14, which groups into 1 bundle of 10 sticks and 4 more. Thus, there is a total of 6 bundles and 4 sticks. In the algorithm a "4" is recorded in the units column and the extra ten is recorded by writing a "1" in the tens column.

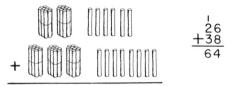

Abacus—The abacus is more abstract (place value is indicated by columns) than the bundles of sticks model and in this respect is closer to the algorithm for addition. Addition is accomplished by pushing the markers on each column together and regrouping or "carrying" if necessary. The expression "to carry" is an old one and probably originates from the time when a counter or chip was actually carried to the next column on an abacus. When the markers for 256 and 183 are pushed together, column by column, on the abacus, regrouping will be needed on the tens column. This is done by removing 10 markers from the tens column and placing another marker on the hundreds column. There will be 3 markers left in the tens column. In the algorithm this step is indicated by recording a "3" for the sum in the tens column and writing a "1" above the hundreds column.

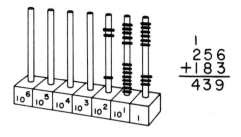

$$\begin{array}{r} 1 \\ 256 \\ +183 \\ \hline 439 \end{array}$$

Addition Algorithms   The primary purpose of an algorithm is to compute in the quickest and most efficient manner. The familiar algorithm for addition does just this. The digits of the numbers to be added are placed under one another, units under units, tens under tens, etc., and regrouping from one column is indicated by writing small numerals above the next column, as shown in the accompanying example.

$$\begin{array}{r} 121 \\ 4375 \\ 893 \\ + 1464 \\ \hline 6732 \end{array}$$

Expanded Forms—There are several ways of providing intermediate steps between the physical models for computing sums (bundles of sticks, abacus, etc.) and the algorithm for addition. One of these is to represent numbers in their *expanded forms*. Here are two examples of this method. The regrouping in Example 2 has been done in two stages.

EXAMPLE 1

$$452 = 4 \text{ hundreds} + 5 \text{ tens} + 2$$
$$+ 243 = 2 \text{ hundreds} + 4 \text{ tens} + 3$$
$$\overline{ 6 \text{ hundreds} + 9 \text{ tens} + 5}$$
$$= 695$$

EXAMPLE 2

$$345 = 3 \text{ hundreds} + 4 \text{ tens} + 5$$
$$+ 278 = 2 \text{ hundreds} + 7 \text{ tens} + 8$$
$$\overline{ 5 \text{ hundreds} + 11 \text{ tens} + 13}$$

Regrouping $\begin{cases} 5 \text{ hundreds} + 12 \text{ tens} + 3 \\ 6 \text{ hundreds} + 2 \text{ tens} + 3 \end{cases}$
$$= 623$$

Partial Sums—Another intermediate step for providing understanding of the addition algorithm is computing *partial sums*. In this method the digits in each column are added, and these partial sums are placed on separate lines. This eliminates carrying (in most examples), because the digits of the sums are placed in the appropriate columns. The "18" in the first line of the partial sums shown here represents 18; the "12" in the second line of the sums represents 120; and the "8" in the third line represents 800. Some people prefer to write in the missing 0s.

$$\begin{array}{r} 238 \\ 521 \\ + 179 \\ \hline \end{array}$$

Partial $\begin{cases} 18 & \text{1st line} \\ 12 & \text{2nd line} \\ 8 & \text{3rd line} \end{cases}$
Sums
$$\overline{938}$$

Scratch Method—Sometimes it is instructive to examine algorithms from the past. While it is now customary to add from right to left, beginning with the units digits, this was not always the practice. The early Hindus, and later the Europeans, added from left to right. For instance, to compute $897 + 537$ from left to right, the first step is to add 8 and 5 in the hundreds column. In the second step the 9 and 3 are added in the tens column, and because carrying is necessary, the "3" in the hundreds column is scratched out and replaced by a "4." In the third step we add the units digits, and again, carrying is necessary. So the "2" in the tens column is scratched out and replaced by a "3." The Europeans called this the *scratch method.*

First step	Second step	Third step
897	897	897
+537	+537	+537
13	1Ʒ2	1ƷƷ4
	4	43

Lattice Method—In another early algorithm for addition, called the *lattice method*, the numbers in the columns were added and their sums placed in cells under the columns. This is similar to the method of partial sums. The final sum is obtained by adding the numbers along the diagonal cells.

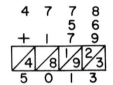

Inequality of Whole Numbers

Inequality of whole numbers can be explained intuitively in terms of the locations of these numbers as they occur in the counting process. For example, 3 is less than 5 because it is named before 5 in the counting sequence. This ordering of numbers can be pictured by a number line with the numbers located at equally spaced intervals. For any two numbers, the one that occurs on the left is less than the other.

0 1 2 3 4 5 6 7 8 9 10 11 12 13 14

Inequality of whole numbers is defined in terms of addition.

Definition: **For any two whole numbers, *m* and *n*, *m* is *less than n* if there is a nonzero whole number *k* such that $m + k = n$. This is written as $m < n$. We also say that *n* is *greater than m* and write $n > m$.**

The symbols $<$ and $>$ were first used by the English surveyor Thomas Harriot in 1631. There is no record of why Harriot chose these symbols, but the following conjecture is logical and will help you to remember their use. The distances between the ends of the bars in the equality symbol (e.g., $3 = 12/4$) are equal, and the number on the left of the sign equals the number on the right. Similary, $3 < 4$ would indicate that 3 is less than 4, because the distance on the left between the bars is less than the distance on the right between the bars. The reasoning is the same whether we write $3 < 4$ or $4 > 3$. These symbols could easily have evolved to our present notation, $<$ and $>$, with the bars completely converging to prevent any misjudgment of these distances.*

*This is one of two conjectures on the origin of the inequality symbols, described by H.W. Eves, *Mathemtical Circles* (Boston: Prindle, Weber and Schmidt, 1969), pp. 111–13.

Addition on Calculators   We have been examining pencil and paper algorithms, that is, sequences of steps for computing sums on paper. Similarly, there are sequences of steps for computing sums on calculators, referred to as calculator algorithms. Most calculators are designed for algebraic (equation type) logic, such as the equation format $475 + 381 + 906 = \square$. On these calculators this sum is computed by the sequence of steps shown next. In Step 4, pushing $\boxed{+}$ gives the total of the first two numbers. The final sum can be obtained by pushing $\boxed{+}$ or $\boxed{=}$ in Step 6.

Calculator Algorithm	Displays at the End of Each Step
Step 1 Enter 475	475.
Step 2 $\boxed{+}$	475.
Step 3 Enter 381	381.
Step 4 $\boxed{+}$	856.
Step 5 Enter 906	906.
Step 6 $\boxed{=}$ or $\boxed{+}$	1762.

Some calculators are equipped to generate an arithmetic sequence by repeatedly pushing a button such as $\boxed{=}$ or $\boxed{+}$. The steps shown below, if continued, will produce the sequence 20, 26, 32, 38, ..., on most calculators

Steps	Displays
1. Enter 20	20.
2. $\boxed{+}$	20.
3. Enter 6	6.
4. $\boxed{=}$	26.
5. $\boxed{=}$	32.
6. $\boxed{=}$	38.

Calculators can be used to strengthen and support our understanding of pencil and paper algorithms. One way is by checking answers to computations. Another is by computing partial sums, as in the steps shown next for determining $384 + 297$. This sum is computed by adding units, then tens, then hundreds. The addition can be done just as easily by beginning with hundreds, then tens, then units. To carry out this process just reverse the steps, performing them from 11 to 1.

Steps		Partial Sums
1. Enter 4	7. Enter 90	384
2. $+$	8. $+$	$+\ 297$
3. Enter 7	9. Enter 300	11
4. $+$	10. $+$	170
5. Enter 80	11. Enter 200	500
6. $+$	12. $=$	681

Large errors from computing on a calculator, such as those produced by hitting an extra button or by worn-out batteries, can sometimes be discovered by rounding off the numbers being added in order to approximate the sum. Suppose, for example, that you wanted to add 417, 683, and 228, but that you entered 2228 on the calculator rather than 228. The sum of the three numbers you intended to add—rounded off to the nearest hundred—is 1300, but the erroneous calculator sum will be 3328. The difference of more than 2000 between the approximate sum and the calculator sum indicates that the computation should be redone.

Sum		Rounded off
417	$\longrightarrow$	400
683	$\longrightarrow$	700
$+\ 228$	$\longrightarrow$	$+\ 200$
		1300

SUPPLEMENT (*Activity Book*)

Activity Set 3.1 Adding with Multibase Pieces and Chip Trading

Additional Sources

Granito, D. "Number patterns and the addition operation." *The Arithmetic Teacher,* **23** No. 6 (October 1976), 432–34.

King, I. "Giving meaning to the addition algorithm." *The Arithmetic Teacher,* **19** No. 5 (May 1972), 345–48.

Madachy, J.S. *Mathematics on Vacation.* New York: Charles Scribner's Sons, 1966. 178–200 (cryptarithms).

School Mathematics Study Group. *Studies in Mathematics,* **13**. Palo Alto: Stanford University, 1966. pp. 113–30.

Smith, D.E. *History of Mathematics,* **2**. Lexington, Mass: Ginn, 1925. pp. 88–94.

Blaise Pascal's arithmetic machine, 1642

EXERCISE SET 3.1

1. This calculating machine was developed by Blaise Pascal for computing sums. The machine is operated by dialing a series of wheels bearing numbers from 0 to 9 around their circumferences. To "carry" a number to the next column, when a sum is greater than 9, Pascal devised a ratchet mechanism that would advance a wheel one digit when the wheel to the right made a complete revolution. The wheels from right to left represent units, tens, hundreds, etc. To compute 854 + 629, first turn the hundreds, tens, and units wheels 8, 5, and 4 turns, respectively. These same wheels are then dialed 6, 2, and 9 more turns. The sum will appear on indicators at the top of the machine.

★   a. Which of these wheels will make more than one revolution for this sum?

      b. Which two wheels will be advanced one digit due to "carrying"?

★   c. Can this sum be computed by turning the hundreds wheel for both hundreds digits, 8 and 6; then turning the tens wheel for 5 and 2; and then turning the units wheel for 4 and 9?

2. Some types of errors and misuses of addition are very common, and others such as the one in this sign, are fairly rare. Describe the type of error occurring in each of the following examples.

★   a.   47
       + 86
       ‾‾‾‾
       123

      b.   16
       + 48
       ‾‾‾‾
       91

★   c.   56
       + 78
       ‾‾‾‾
       1214

      d.   35
       + 46
       ‾‾‾‾
       171

★ 3. Find the number for each of the following expressions $A = \{$John, Mary$\}$

in parts **a** through **d**. In which cases is the number of $B = \{$Sue, Joan$\}$

people in the union of the two sets equal to the sum $C = \{$John, Sue, Al$\}$

of the numbers of people in each set? $D = \{$Bill, Paul, Jane$\}$

★   a. $n(A \cup B)$ ★   b. $n(A \cup C)$ ★   c. $n(B \cup C)$ ★   d. $n(B \cup D)$

4. Look for easy combinations of numbers to compute the following sums in your head. Show the groupings you used.

a. $97 + 68 + 3 + 22$ b. $345 + 207$ c. $4 + 9 + 36 + 18 + 11$

5. Which property, commutativity for addition or associativity for addition, is being used in equations **a** through **d**?

★   a. $(38 + 13) + 17 = 38 + (13 + 17)$

b. $(47 + 62) + 12 = (62 + 47) + 12$

★   c. $2 \times (341 + 19) = 2 \times (19 + 341)$

d. $\dfrac{13 + (107 + 42)}{3 \times (481 + 97)} = \dfrac{(13 + 107) + 42}{3 \times (481 + 97)}$

6. Addition is illustrated on the number line by a series of arrows as shown here.

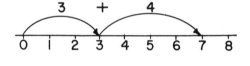

a. Illustrate $2 + 5$ and $5 + 2$ on the same line.

★   b. Devise a way of illustrating 0 plus any number on the number line. Then illustrate $0 + 6$ and $6 + 0$ on the same line.

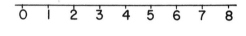

c. Use the number line to show that $(3 + 4) + 1$ and $(4 + 1) + 3$ are equal.

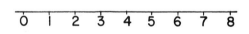

7. Pairs of numbers are represented by the multibase pieces in part **a** and the bean stick model in part **b**. Do the necessary regrouping for each sum and sketch the answers.

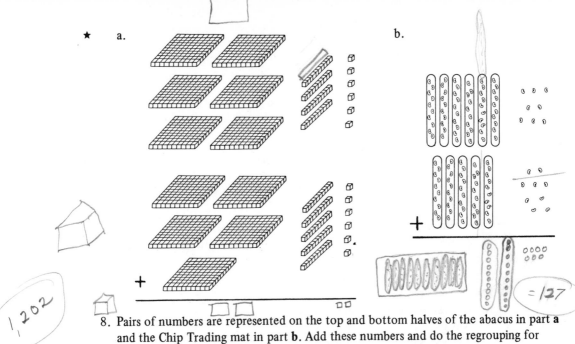

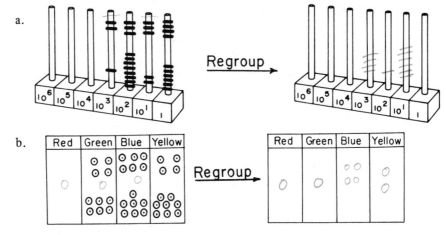

8. Pairs of numbers are represented on the top and bottom halves of the abacus in part **a** and the Chip Trading mat in part **b**. Add these numbers and do the regrouping for each sum. Sketch the answers on the abacus and mat to the right.

★ a.

b.

Red	Green	Blue	Yellow

Red	Green	Blue	Yellow

9. Write the steps to be used on a calculator for obtaining the first five numbers in the arithmetic sequence given. Show the numbers that would appear in the calculator display for each of your steps.

146, 209, 272, ...

10. Whether you use a pencil and paper or a calculator for adding, mistakes can often be detected by rounding off and approximating. Compute the following sums and then round off each number to its largest place value and compute the sums.

Sum		Rounded off
472	→	500
906	→	900
123	→	100
+ 568	→+	600
2069		2100

a.	238	b.	4718	c.	73182
	709		8056		29476
	+ 873		+ 1329		+ 31024

11. A homeowner has the following bills to pay for the month of March, and a monthly salary of $1800. To estimate whether or not these bills can be paid, the homeowner rounds off each bill to its highest place value.

   Electricity $86, Heat $128, Water and sewage $94, Property taxes $163, Life insurance $230, Car insurance $65, House insurance $58, Food $541, Doctor's bills $477, Gas and oil $73, Car payments $148, Home mortgage $225, Dentist's bills $109, and Pleasure $14.

★   a. Round off these numbers and compute their sum.

   b. What can this person conclude about being able to pay these bills for the month of March?

★   c. What is the exact sum of these bills?

12. *Calculator Exercise:* Some calculators will add a constant number to any number that is entered into the calculator. Let's assume that a certain stock has paid off a bonus of $164 to each stockholder and that this amount is to be added to their accounts. The first step on the calculator adds 164 to 4728. To add 164 to the remaining accounts it is necessary only to enter the next number and press $\boxed{=}$. Find the new account balances for these displays.

Stockholder Accounts

$4728
$13491
$6045
$21437
$10418
$9366

Steps		Displays
1. 4728 $\boxed{+}$ 164 $\boxed{=}$		4892.
2. 13491 $\boxed{=}$		
3. 6045 $\boxed{=}$		
4. 21437 $\boxed{=}$		
5. 10418 $\boxed{=}$		
6. 9366 $\boxed{=}$		

13. Compute the sums for the given methods. Name an advantage and disadvantage of each method. Does the lattice method eliminate the need for carrying?

★   a. Scratch Method

   482
   6731
   + 2064

b. Lattice Method

   4 7 6 2
   + 8 7 6

14. Compute the sums using the given methods.

a. Expanded Forms

$$4036$$
$$+\ 784$$

b. Partial Sums

$$736$$
$$+\ 582$$

15. *Number Curiosity:* Choose any four-digit number, reverse its digits, and add the two numbers. In the example shown here, the sum is divisible by 11. Try this for some other numbers. Will this always be true? (Try this exercise on a minicalculator.)

$$7582$$
$$+\ 2857$$
$$10439$$

$$949$$
$$11\overline{)10439}$$

★ 16. *Palindromic Numbers:* Numbers such as 31413 that read the same from either end are called *palindromic numbers.* In the accompanying example the process of reversing the digits and adding the two numbers has been continued until a palindromic number is obtained. Will this process always result in a palin-dromic number by beginning with any two-digit number? (It will be convenient to have a calculator for this exercise.)

$$87$$
$$+\ 78$$
$$165$$
$$+\ 561$$
$$726$$
$$+\ 627$$
$$1353$$
$$+\ 3531$$
$$4884$$

17. *Faded Documents:* Computations in which some of the digits are missing are called faded document puzzles. The idea underlying these puzzles is that ancient documents have been discovered, but so decayed that some of the digits are not recognizable. Supply the missing digits.

★   a.   7 □ 2 □
    + □ 4 □ 2
    1 3 1 9 1

b.   □ 7 □
   + 9 □ 2
   1 9 7 0

18. *Magic Squares:* A square array of numbers in which the sum of numbers in any horizontal row, vertical column, or diagonal is always the same is called a *magic square.* Write the numbers 1 through 9 in the 3 by 3 array to produce a magic square.

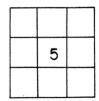

19. *Magic Triangles:* Place each of the numbers 1, 2, 3, 4, 5, and 6 in a circle so that the sum along each side of the given triangle is the same.

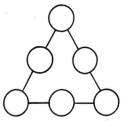

20. *Cryptarithms:* A *cryptarithm* is a puzzle in which a different letter is substituted for each digit (0, 1, 2, 3, 4, 5, 6, 7, 8, 9) in a simple arithmetic problem. In part **a**, a test of each digit shows that B must be 5 and, consequently, A must be 2. Knowing that K is 2, deduce the values of C, G, and F in part **b**.

a.
```
  B 5
  B 5
  B 5
  B 5
+ B 5
  ───
  A B
  2 5
```

b.
```
  C C C 1
  +   K 2
  ──────
  G F F G
  1 0 0 1
```

Here are two more cryptharithms. The first one is easy, but the second one is difficult and intended for the ambitious. Supposedly, it was a message written by a college student to his parents.

★ c.
```
  T H I S
    I S
+ V E R Y
─────────
  E A S Y
```

d.
```
  S E N D
+ M O R E
─────────
M O N E Y
```

★ 21. *Missing Dollar Puzzle:* One day three men went to a hotel and were charged $30 for their room. The desk clerk later realized that he had overcharged them by $5 and sent the refund up with the bellboy. The bellboy knew that it would be difficult to split the $5 three ways. Therefore, he kept a $2 "tip" and gave the men only $3. Each man had originally paid $10 and was given back $1. Thus, it cost each man $9 for the room. This means that they spent $27 for the room plus the $2 "tip." What happened to the other dollar?

3.2 SUBTRACTION

Subtraction is usually explained by *taking away* a subset of objects from a given set. In the preceding cartoon, Peter has had 2 clams taken away from his 5 clams by a series of trades. The accompanying figure shows 2 clams being taken away from 5 clams and illustrates $5 - 2 = 3$. The word "subtract" literally means to *draw away from under,* and

$$5 - 2 = 3$$

the words which have been used for this operation are variations of this meaning. In 1202 the Italian mathematician Fibonacci used words meaning *I take* for the operation of subtraction. Over the years the terms *detract, subduce, extract, diminish, deduct,* and *rebate* have all been used as synonyms for "subtract." We still hear "deduct" being used in place of "subtract": "deduct the tax"; and "deduct what I owe."

Definition of Subtraction   The process of *take away* for subtraction may be thought of as the opposite of the process of *put together* for addition. Because of this dual relationship, subtraction and addition are called *inverse operations.* This relationship is used to define subtraction in terms of addition. For any two whole numbers, r and s,

$$r - s = \square \quad \text{if and only if} \quad r = s + \square$$

The preceding definition says we can compute the difference $17 - 5$ by determining what number must be added to 5 to give 17. We use this approach when making change. Rather than subtract 83 cents from \$1.00 to determine the difference, we pay back the change by counting up from 83 to 100.

In the definition of subtraction there is no need of restricting r to be greater than or equal to s. However, in the early school grades, before negative integers are introduced, the examples involve only subtracting a smaller number from a larger one. Historically, subtracting was limited to this case, as shown in the following explanation by the English author Digges (1572):

To subduce or subtray any sūme, is wittily to pull a lesse frō a bigger nūber.*

Models for Subtraction Algorithm   There are two types of examples to consider in explaining the steps for finding the difference between two multi-digit numbers: those examples where borrowing or regrouping is not needed and those where it is needed. Both of these cases will be illustrated by models in the following paragraphs. The amounts to be taken away in each example will be encircled.

Bundles of Sticks—To find the difference between 48 and 23, 4 bundles of sticks and 8 sticks are used to represent 48. Then, 2 bundles of sticks and 3 single sticks are taken away so that the remaining bundles and sticks represent the difference between 48 and 23.

$$\begin{array}{r} 48 \\ -23 \\ \hline 25 \end{array}$$

Sometimes regrouping is necessary before subtracting. The usual method of computing $53 - 29$ is to borrow a 10 from the tens column so that there will be 13 ones in the units column. This is accomplished with the sticks by regrouping 1 bundle of sticks so that there are 13 single sticks. Once this is done, the subtraction can be completed by removing 2 bundles of sticks and 9 single sticks.

Regroup

$$\begin{array}{r} 4 \\ \cancel{5}3 \\ -29 \\ \hline 2\ 4 \end{array}$$

*See D.E. Smith, *History of Mathematics,* **2** (Lexington, Mass.: Ginn, 1925), p. 95.

Abacus—The abacuses shown next illustrate subtraction with regrouping. The markers on abacus (a) represent 721. In order to compute $721 - 384$, the markers on abacus (a) have been regrouped on abacus (b). One marker on the tens column has been replaced by 10 markers on the units column so that 4 markers can be removed from the units column. One marker on the hundreds column has been replaced by 10 markers on the tens column so that 8 markers can be removed from the tens column. The difference, $721 - 384$, is computed by removing 4 markers, 8 markers, and 3 markers, from the units, tens, and hundreds columns, respectively.

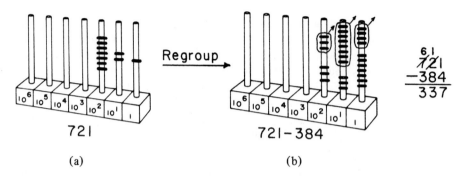

Regroup

721

721 − 384

$$\begin{array}{r} 6\ 1 \\ \not{7}\not{2}1 \\ -384 \\ \hline 337 \end{array}$$

(a)   (b)

Algorithms for Subtraction

There are several algorithms for subtraction and each one has its advantages. Of the three methods described in the following paragraphs, the decomposition method is the most common.

Decomposition Method—This is the method of borrowing or regrouping, which was illustrated by the bundles of sticks model and the abacus in the previous examples. In the accompanying example one of the tens is regrouped from the tens column so that 8 can be subtracted from 14. In the next step, one of the hundreds is regrouped from the hundreds column so that 6 can be subtracted from 11.

$$\begin{array}{r} 4\ 1 \\ \not{5}\not{2}4 \\ -3\ 6\ 8 \\ \hline 1\ 5\ 6 \end{array}$$

Equal Additions Method—The equal additions method is one of the most rapid ways of subtracting. It appeared in Europe in the thirteenth century, and by the fifteenth and sixteenth centuries it had become popular.

As in the decomposition method in the preceding example, 10 is added to the 4 in the units column so that 8 can be subtracted from 14. However, to compensate for this change, the 6 in the tens column is changed to a 7, as shown in the first of the accompanying 3 steps. In the second step, 10 is added to the 2 in the tens column and this is paid for by changing the 3 in the hundreds column to a 4. Then 7 is subtracted from 12. Finally, in the third step, 4 is subtracted from 5.

$$\begin{array}{r} 5\ 2\ 4 \\ -\not{3}\not{6}8 \\ \hline 1\ 5\ 6 \end{array}$$

First Step	Second Step	Third Step
$\begin{array}{r} 5\ 2\ 4 \\ {\scriptstyle 7} \\ -3\not{6}8 \\ \hline 6 \end{array}$	$\begin{array}{r} 5\ 2\ 4 \\ {\scriptstyle 4\ 7} \\ -\not{3}\not{6}8 \\ \hline 5\ 6 \end{array}$	$\begin{array}{r} 5\ 2\ 4 \\ {\scriptstyle 4\ 7} \\ -\not{3}\not{6}8 \\ \hline 1\ 5\ 6 \end{array}$

Complementary Method—The complementary method of subtraction was taught in many schools in the United States in the nineteenth century. As with the equal additions method, this approach has been used since medieval times.

This clever approach to subtraction requires only adding and is perhaps the fastest of the three methods. Beginning with the units column in Example 1, the complement of 7 (for the number 10) is 3. So 3 is added to 5. For each of the remaining lower digits we determine the complement of 9 and add this complement to the above digit. For example, the complement of 6 (for the number 9) is 3, so 3 is added to 1. Next the complement of 9 (for the number 9) is 0, so 0 is added to 4. Finally, the complement of 2 (for the number 9) is 7, so 7 is added to 8. To compensate for the complements that have been added, subtract 1 from the first digit on the left. This last step in Example 1 was accomplished by crossing out the 1 in the ten thousands column. The complementary method is also used in Example 2. In the units column the complement of 1 is 9 and $9 + 2 = 11$. Therefore, a "1" is carried from the units to the tens column. In the final step of the algorithm, 1 is subtracted from 7.

EXAMPLE 1

$$
\begin{array}{r}
8\ 4\ 1\ 5 \\
-\,2\ 9\ 6\ 7 \\
\hline
\cancel{1}\,5\ 4\ 4\ 8
\end{array}
$$

EXAMPLE 2

$$
\begin{array}{r}
1 \\
7\ 4\ 3\ 2 \\
-\,8\ 5\ 1 \\
\hline
\cancel{7}\,5\ 8\ 1 \\
6
\end{array}
$$

Subtraction on Calculators   To find the difference of two numbers on a calculator the numbers and operation are entered in the order in which they appear (from left to right) when written horizontally. The steps for computing $247 - 189$ are shown on the right.

Steps	Displays
1. Enter 247	247.
2. $\boxed{-}$	247.
3. Enter 189	189.
4. $\boxed{=}$	58.

Some calculators will subtract a constant number from any number that is entered into the calculator. Suppose, for example, that you wanted to subtract $26 from each of the weekly paychecks listed. The first step in the next sequence subtracts 26 from 561. To subtract 26 from each of the remaining paychecks it is only necessary to enter the amount of pay and press $\boxed{=}$.

Weekly Paychecks

$561
$452
$608
$513
.
.
.

Steps	Displays
1. 561 $\boxed{-}$ 26 $\boxed{=}$	535.
2. 452 $\boxed{=}$	426.
3. 608 $\boxed{=}$	582.
4. 513 $\boxed{=}$	487.

On calculators that will subtract a constant number, an arithmetic sequence can be generated by entering a number; pressing $-$; entering a second number; and then repeatedly pressing $=$. The first number in the next sequence is 200. The number 6 is being subtracted in steps 4, 5, and 6 on the calculator. A continuation of these steps will produce the arithmetic sequence: 200, 194, 188, 182, 176, ...

Steps	Displays
1. Enter 200	200.
2. $-$	200.
3. Enter 6	6.
4. $=$	194.
5. $=$	188.
6. $=$	182.

SUPPLEMENT (*Activity Book*)

Activity Set 3.2 Subtracting with Multibase Pieces

Additional Sources

Bradford, J.W. "Methods and materials for learning subtraction." *The Arithmetic Teacher*, **25** No. 5 (February 1978), 18–20.

Cleminson, R.A. "Developing the subtraction algorithm." *The Arithmetic Teacher*, **20** No. 8 (December 1973), 634–37.

Locke, F.M. *Math Shortcuts*. New York: John Wiley & Sons, 1972.

Quast, W.G. "Method or justification." *The Arithmetic Teacher*, **19** No. 8 (December 1972), 617–22.

School Mathematics Study Group. *Studies in Mathematics,* **13**. Palo Alto: Stanford University, 1966. pp. 151–64.

Smith, D.E. *History of Mathematics,* **2**. Lexington, Mass: Ginn, 1925. pp. 94–101.

EXERCISE SET 3.2

1. This chart contains the numbers of people living in the world's 25 most populous cities.

City and country	Most recent population	Year
Tokyo, Japan	8,586,900	1976 Estimate
New York City, U.S.	7,567,800	1975 Estimate
Osaka, Japan	2,759,300	1976 Estimate
London, England	7,111,500	1975 Estimate
Mexico City, Mexico	8,628,000	1976 Estimate
Shanghai, China	5,700,000	1970 Estimate
Los Angeles, U.S.	2,745,300	1974 Estimate
West Berlin, W. Germany	2,023,900	1976 Estimate
São Paulo, Brazil	7,198,600	1975 Estimate
Nagoya, Japan	2,071,400	1976 Estimate
Paris, France	2,289,800	1975 Census
Buenos Aires, Argentina	2,977,000	1975 Estimate
Moscow, U.S.S.R.	7,563,000	1976 Estimate
Chicago, U.S.	3,150,000	1975 Estimate
Peking, China	4,800,000	1972 Estimate

(*cont.*)

City and country	Most recent population	Year
Rio de Janeiro, Brazil	4,857,700	1975 Estimate
Cairo, Egypt	6,133,000	1976 Estimate
Calcutta, India	3,148,700	1971 Census
Seoul, South Korea	6,889,400	1972 Census
Jakarta, Indonesia	4,650,000	1975 Estimate
Bombay, India	5,970,500	1975 Estimate
Philadelphia, U.S.	1,824,900	1975 Estimate
Manila, Philippines	1,450,000	1975 Census
Teheran, Iran	4,350,000	1975 Estimate
San Francisco, U.S.	675,600	1974 Estimate

★   a. How many more people live in Tokyo than in Paris?

b. How many more people live in New York City than in Los Angeles and San Francisco combined?

c. Which two cities in this list have populations that are most nearly equal (have the smallest difference)? What is the difference in the numbers of their inhabitants?

2. According to the definition of subtraction, each difference may be found by determining the missing number in a sum. Compute each of the following differences by "counting up" from the smaller number to the larger.

★   a. $247 - 219$ b. $83 - 65$ c. $341 - 250$

3. Addition is often used to check subtraction by adding the two "lower numbers." In this example the sum of the two lower numbers is the top number, 903. In the units column $6 + 7 = 13$, so "1" is carried to the tens column. In the tens column, $1 + 2 + 7 = 10$, so "1" is carried to the hundreds column. Find out which of the following differences is wrong by adding the two lower numbers in each example.

$$\begin{array}{r} 903 \\ -\ 576 \\ \hline 327 \\ 11 \end{array}$$

a. $\quad\begin{array}{r} 436 \\ -\ 197 \\ \hline 239 \end{array}$ b. $\quad\begin{array}{r} 1702 \\ -\ \ 486 \\ \hline 1316 \end{array}$ c. $\quad\begin{array}{r} 500 \\ -\ 164 \\ \hline 336 \end{array}$

4. Try some numbers in parts **a** and **b** to determine if these properties hold.

a. Is subtraction commutative?

$$\square - \triangle \overset{?}{=} \triangle - \square$$

b. Is subtraction associative?

$$(\square - \triangle) - \diagup \overset{?}{=} \square - (\triangle - \diagup)$$

★ 5. One common source of errors in subtraction occurs when elementary school students *add* rather than subtract. For example, when addition is taught *first*, the student learns to respond to the pair of numbers 5 and 3 by writing 8. This response becomes so automatic that later on we find the student writing 8 for the difference $5 - 3$.

$$\begin{array}{r} 5 \\ +3 \\ \hline 8 \end{array} \qquad \begin{array}{r} 5 \\ -3 \\ \hline 8 \end{array}$$

   Try to detect the reason for the error in each of the following computations.

★   a. $\begin{array}{r} 84 \\ -36 \\ \hline 52 \end{array}$ ★   b. $\begin{array}{r} 52 \\ -38 \\ \hline 24 \end{array}$ ★   c. $\begin{array}{r} 46 \\ +27 \\ \hline 73 \end{array}$ ★   d. $\begin{array}{r} 94 \\ -37 \\ \hline 12 \end{array}$

6. Subtraction is illustrated on the number line by arrows that represent numbers. The number being subtracted is represented by an arrow from right to left.

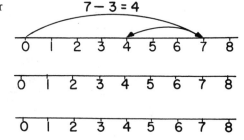

★   a. Illustrate $(6 - 3) - 2$ on the given number line.

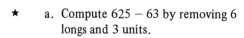

   b. Use this number line to illustrate $6 - 6$.

7. In parts **a** and **b** do the necessary regrouping so that you can remove (or cross out) the required number of objects. Use your results to fill in the missing numbers in the given boxes.

★   a. Compute $625 - 63$ by removing 6 longs and 3 units.

   b. Compute $68 - 29$ by removing 2 bean sticks and 9 beans.

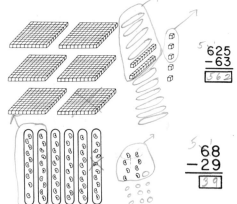

8. In parts **a** and **b** do the necessary regrouping by drawing new sets of markers and chips so that you can remove the required numbers of objects. Use your results to fill in the missing numbers in the given boxes.

a. Compute $1358 - 472$ by regrouping and then removing 2 markers from the units column, 7 markers from the tens column, and 4 markers from the hundreds column.

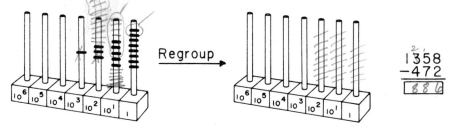

$$\begin{array}{r} 1358 \\ -472 \\ \hline 886 \end{array}$$

★ b. Compute $6245 - 2873$ by removing 3 yellow chips, 7 blue chips, 8 green chips, and 1 red chip.

Red	Green	Blue	Yellow

Regroup

Red	Green	Blue	Yellow

$$\begin{array}{r} 6245 \\ -2873 \\ \hline 3372 \end{array}$$

9. *Equal Additions:* Use the equal additions method to compute the following differences. Explain why this method works.

a.   732
  − 348

b.   1746
  − 382

c.   910
  − 462

10. *Complementary Method:* Use the complementary method to compute the following differences.

a.   347
  − 169

★ b.   8023
  − 476

c.   6023
  − 2184

11. Write the steps to be used on a calculator for obtaining the first five numbers in this arithmetic sequence. Show the numbers that would appear in the calculator display for each of your steps.

$$718, 671, 624, \ldots$$

12. *Calculator Exercise:* Slate and brick are sold by weight. At one company the slate or brick is placed on a loading platform and a forklift moves the total load, platform and all, onto the scales. The weight of the platform is then subtracted from the total weight. Assuming the weight of the platform is 83 kilograms (83 kg), find the weight of the slate for each of these total weights. (If you use a calculator as described in the previous section, the 83 needs only to be entered onto the calculator once.)

Total Load Weight (kg)	Weight of Slate (kg)
748	665
807	
1226	
914	
1372	
655	

★13. *Number Curiosity:* Select any three-digit number whose first and third digits are different, and reverse the digits. Find the difference between these two numbers. By knowing only the units digit in this difference, it is possible to determine the other two digits.

$$\begin{array}{r} 834 \\ -438 \\ \hline \end{array}$$

Try this trick with some three-digit numbers. Look for a pattern in the differences and explain how the trick works.

14. *Faded Documents:* Supply the missing digits.

★ a.
$$\begin{array}{r} \square\,\square\,2 \\ -\ 3\,\square \\ \hline 3\ 7\ 5 \end{array}$$

b.
$$\begin{array}{r} \square\,6\,\square\,5 \\ -\ \square\,4\,\square \\ \hline 2\ 8\ 8\ 1 \end{array}$$

15. *Game of "Diffy"**: Select any four whole numbers and place them in the A circles. Then between each pair of numbers write the difference of the larger minus the smaller in the B circles. Continue this process of taking differences and building inner squares. If you can form six squares before getting all zeros, you win the game. This is not a winning game. Finish forming these squares. How many squares can be formed before they contain all zeros? Find four numbers that can be used for a winning game of "Diffy."

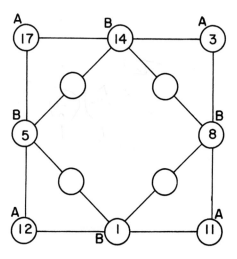

*This game is described by H. Wills, "Diffy," *The Arithmetic Teacher,* **18** No. 6 (October 1971), 402–5.

State office buildings at the Empire State Plaza, Albany, New York

3.3 MULTIPLICATION

The skyscraper in the center of this picture is called the Tower Building. There is an innovative window washing machine mounted on top of this building. The machine lowers a cage on a vertical track so that each column of 40 windows can be washed. After one column of windows is washed, the machine moves to the next column. The rectangular face of the building which can be seen has 36 columns of windows. The total number of windows is $40 + 40 + 40 + \cdots + 40$, a sum in which 40 occurs 36 times. Using multiplication this sum equals the product 36×40, or 1440. We are led to different expressions for the sum and product by considering the rows of windows across the floors. There are 36 windows in each floor on this face of the building and 40 floors. Therefore, the number of windows is 36 + 36 + 36 + ··· + 36, a sum in which 36 occurs 40 times. This sum is equal to 40 × 36, which is also 1440. In sums such as these, where the addition of one number is repeated, multiplication is a shortcut for doing addition rapidly.

Definition of Multiplication   Historically, multiplication was developed to replace certain special cases of addition, namely, the cases of *several equal addends.* This meaning is expressed by the word for "multiply" in Latin, and in several other languages.

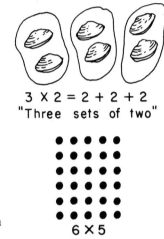

For this reason we usually see multiplication of whole numbers explained and defined as "repeated addition." For whole numbers r and s, $r \times s$ is the sum with s occurring r times.

$$r \times s = s + s + s + \cdots + s$$

$$3 \times 2 = 2 + 2 + 2$$
"Three sets of two"

Another method of viewing multiplication of whole numbers is through rectangular arrays of objects, as suggested by the rows and columns of windows at the beginning of this section. The rectangular array of dots shown here has six rows of five dots each. The number of dots is 6 times 5. In general, $r \times s$ is the number of dots in a rectangular array having r rows with s dots in each row.

6×5

Algorithms for Multiplication   One of the earliest methods of multiplication is found in the Rhind Papyrus. This ancient scroll (ca. 1650 B.C.) is more than 5 meters in length and was written to instruct the Egyptian scribes in computing with whole numbers and fractions. It begins with the words "Complete and thorough study of all things, insights into all that exists, knowledge of all secrets ...," and indicates the Egyptians' respect and awe of mathematics. Although most of its 85 problems have a practical origin, there are some of a theoretical nature. Their algorithm for multiplication was a succession of doubling operations, followed by addition. We call this algorithm the *duplation method.*

Duplation Method—This method depends on the fact that any number can be written as the sum of binary numbers (1, 2, 4, 8, 16 ...). To compute 37 × 52, the 52 is repeatedly doubled as shown here. This process stops when the next binary number in the list is greater than the number you are multiplying by. In this example, we want 37 of

→1 × 52 =	52
2 × 52 =	104
→4 × 52 =	208
8 × 52 =	416
16 × 52 =	832
→32 × 52 =	1664

the 52s, and since 37 is equal to the sum of the binary numbers 32, 4, and 1, the product of 37 × 52 is equal to (32 + 4 + 1) × 52. This is 1664 + 208 + 52, which equals 1924.

Russian Peasant Multiplication—There is a variation of the Egyptian method, called *Russian peasant multiplication,* which was used in medieval Europe. It involves repeatedly doubling and halving the two numbers to be multiplied. In the example shown here, 54 × 17 is computed by repeatedly doubling the 17 and halving the 54. In the halving process, fractions are disregarded. When 1 is reached in the halving column, this process stops. Next, each row with even numbers in the halving column is crossed out. Now, the sum of the remaining numbers in the doubling column is equal to the product of the original two numbers. In this example, 54 × 17 is equal to 34 + 68 + 272 + 544, which is 918.

Halving		Doubling
~~54~~	~~X~~	~~17~~
27		34
13		68
~~6~~		~~136~~
3		272
1		+ 544
		918

Gelosia Method—One of the popular schemes used for multiplying in the fifteenth century was called the *gelosia* or *grating method.* This algorithm was performed in a framework resembling a window grating or jalousie, as shown here. The two numbers to be multiplied, 4826 and 57, are written at the top and right side of the grating. The partial products of these numbers are written in the squares of the grating. The sums of the numbers in the diagonal cells, from upper right to lower left, is the product, 275082, of the original two numbers.

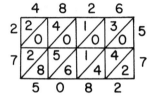

Many of the computing patterns used today in elementary arithmetic, including the algorithm for multiplication, were developed as late as the fifteenth century. Tobias Dantzig, in *Number, The Language of Science,* tells the following story of a fifteenth century German merchant.*

It appears that the merchant had a son whom he desired to give an advanced commercial education. He appealed to a prominent professor of a university for advice as to where to send his son. The reply was that if the mathematical curriculum of the young man was to be confined to adding and subtracting, he perhaps could obtain the instruction in a German university; but the art of multiplying and dividing, he continued, had been greatly developed in Italy, which in his opinion was the only country where such advanced instruction could be obtained.

In the algorithm we now use for computing products, the two numbers being multiplied are placed under each other with units under units, tens under tens, etc. This method first appeared in an Italian text dated 1470. Because of the use of squares resembling a chessboard, the Venetians called this method "per scachiere." By the sixteenth century the squares had disappeared but the "chessboard method" was generally adopted by most writers. This algorithm is illustrated by models on pages 109-11.

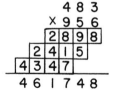

*T. Dantzig, *Number, The Language of Science* (New York: Macmillan, 1954), p. 26.

Number Properties

Distributive Property—When multiplying a sum of two numbers by a third number, we can add the two numbers and then multiply by the third number; or, we can multiply each number of the sum by the third number and then add the two products. For example, to compute $35 \times (10 + 2)$, we can compute 35×12, or add 35 times 10 to 35 times 2. This fact is described by saying that *multiplication is distributive (or distributes) over addition.*

$$35 \times 12 = \underbrace{35 \times (10 + 2)} = \underbrace{35 \times 10} + \underbrace{35 \times 2}$$

Distributive Property

In general, for any three numbers that are used for the placeholders $\square$, $\triangle$, and $\diagup\!\!\!\!\diagdown$

$$\square\,(\triangle + \diagup\!\!\!\!\diagdown\,) = (\square \times \triangle) + (\square \times \diagup\!\!\!\!\diagdown\,)$$

The distributive property can be extended for an arbitrary number of numbers in the sum. For example, $3 \times (17 + 26 + 32) = 3 \times 17 + 3 \times 26 + 3 \times 32$. For sums of more than two numbers this is called the *generalized distributive property.*

Associative Property for Multiplication—Here is an easy way to multiply any number by 5. First divide the number by 2, or multiply by $\frac{1}{2}$, and then multiply by 10. Try this method to compute 5×124. The fact that dividing by 2 and multiplying by 10 produces the same result as multiplying by 5 is a consequence of the *associative property for multiplication.*

$$5 \times 124 = \underbrace{(10 \times \tfrac{1}{2})} \times 124 = 10 \times \underbrace{(\tfrac{1}{2} \times 124)}$$

Associative Property
for Multiplication

In general, for any three numbers that are used for the placeholders $\square$, $\triangle$, and $\diagup\!\!\!\!\diagdown$ the middle number may be grouped or "associated with" either the first number or the third number.

$$(\square \times \triangle) \times \diagup\!\!\!\!\diagdown = \square \times (\triangle \times \diagup\!\!\!\!\diagdown\,)$$

Commutative Property for Multiplication—Before reading beyond this sentence, try computing the product $25 \times 46 \times 4$ in your head. The easy way to do this is by rearranging the order of the numbers so that 25×4 is computed first and then 46×100. Changing the order of two numbers in a product is permitted by the *commutative property for multiplication.* In the following two equations both the commutative and associative properties for multiplication are used to change the order of the numbers in the product.

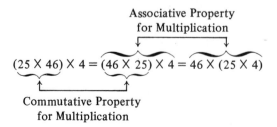

Associative Property
for Multiplication

$$(25 \times 46) \times 4 = \underbrace{(46 \times 25)} \times 4 = 46 \times (25 \times 4)$$

Commutative Property
for Multiplication

In general, for any two number replacements for △ and ☐ the order of multiplication does not change the resulting product.

$$\triangle \times \square = \square \times \triangle$$

Models for Multiplication Algorithm   Physical models for the basic operations can provide an understanding of these operations and suggest or motivate the rules and steps for computing. Bean sticks, bundles of sticks, multibase pieces, Chip Trading, and the abacus are all suitable models for illustrating the multiplication algorithm. Multibase pieces and Chip Trading are used in the following paragraphs.

Multibase Pieces—To multiply 3 times 234, using multibase pieces, the numbers of each type of multibase piece for 234 are tripled. The result is 6 blocks, 9 longs, and 12 units. Regrouping is then needed: 10 units are replaced by 1 long, leaving 2 units; and 10 longs are replaced by 1 flat, leaving 0 longs. The result will be 7 flats, 0 longs, and 2 units.

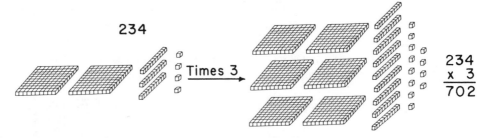

The mathematical justification for tripling the number of each type of multibase piece is based on the distributive property for multiplication over addition. In the first of the following two equations, 234 is written in expanded form. The second equation is obtained from the first by using the distributive property.

$$3 \times 234 = 3 \times (200 + 30 + 4)$$
$$= 3 \times 200 \ + \ 3 \times 30 \ + \ 3 \times 4$$

The next example illustrates how multiplication by 10 can be carried out with multibase pieces. If 10 units are placed together they form 1 long, 10 longs form 1 flat, and 10 flats form 1 block. Therefore, to multiply 234 by 10, each multibase piece for 234 is replaced by the multibase piece for the next higher power of ten. We begin with 2 flats, 3 longs, and 4 units; and we end with 2 blocks, 3 flats, 4 longs, and 0 units.

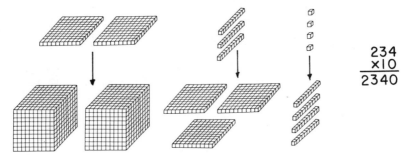

Replacing each multibase piece by the piece for the next higher power of 10 is like computing 10×234 by multiplying 10 times each number in this sum: $100 + 100 + 10 + 10 + 10 + 1 + 1 + 1 + 1$. This product is permitted by the generalized distributive property for multiplication over addition.

$$10 \times 234 = 10 \times (100 + 100 + 10 + 10 + 10 + 1 + 1 + 1 + 1)$$
$$= 10 \times 100 + 10 \times 100 + 10 \times 10 + 10 \times 10 + 10 \times 10 +$$
$$10 \times 1 + 10 \times 1 + 10 \times 1 + 10 \times 1$$

The ideas in the previous two paragraphs can be combined to multiply by any multiple of 10. Multiplication by 30, for example, is accomplished by multiplying by 3 and then by 10. This requires the use of the associative property, as shown in the following equations.

$$30 \times 234 = (10 \times 3) \times 234$$
$$= 10 \times (3 \times 234)$$

Chip Trading—Multiplication by single-digit numbers with the Chip Trading model is done by multiplying the number of chips in each column of the mat and then regrouping when necessary. To compute 4×2063, the chips in each column of mat (a) have been replaced by four times as many chips on the corresponding columns of mat (b). After regrouping there will be 8 red chips, 2 green chips, 5 blue chips, and 2 yellow chips.

(a)　　Times 4　　(b)

Multiplying by 10 is easy when we think in terms of regrouping. Referring to the next diagram, there are 3 yellow chips on mat (c). Multiplying by 10 replaces each yellow chip by 10 yellow chips. Regrouping, each group of 10 yellow chips is replaced by 1 blue chip. Therefore, multiplying by 10 is accomplished by replacing 3 yellow chips by 3 blue chips. Similarly, there are 7 blue chips on mat (c), and multiplying by 10 replaces each blue chip by 10 blue chips. Each group of 10 blue chips can be replaced by 1 green chip. Therefore, the effect of multiplying by 10 is to replace the 7 blue chips by 7 green chips. The final product is represented by the chips on mat (d).

(c)　　Times 10　　(d)

Multiplication by multiples of 10 can be illustrated by combinations of products with single-digit numbers and powers of 10. To multiply 30 times 47, we can first multiply 47 by 3 and then multiply this result by 10.

$$30 \times 47 = (10 \times 3) \times 47 = 10 \times (3 \times 47)$$

Associative Property
for Multiplication

Times 3

Red	Green	Blue	Yellow

Times 10

Red	Green	Blue	Yellow

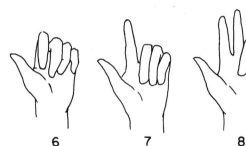

Red	Green	Blue	Yellow

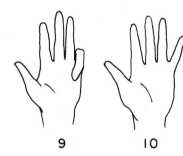

Finger Multiplication   By the Middle Ages, the use of finger numbers and systems of finger computation were widespread. One such system for multiplication, which in recent years was still common in some parts of Europe, uses the finger positions shown here for computing the products of numbers from 6 through 10.

6   7   8   9   10

The two numbers to be multiplied are each represented on a different hand. The hands shown on the right represent 7 and 8. To compute 7×8, the sum of the raised fingers is the number of tens, and the product of the closed fingers is the number of ones. In this case, the 50 from the $2 + 3$ raised fingers and the 6 from the 3×2 closed fingers give a product of 56. Try this method to compute some other products of numbers from 6 through 10.

7   X   8

Napier's Rods   Even as late as the seventeenth century, multiplication of large numbers was a difficult task for all but professional clerks. In order to help people "do away with the difficulty and tediousness of calculations," the Scottish mathematician John Napier (1550–1617) invented a method of using rods for performing multiplication. Napier's rods, or "bones" as they are sometimes called, consist of the 10 rods shown here.

Napier's method is similar to the gelosia or grid method of multiplying. To compute 6 times 479, the rods for the multiples of 4, 7, and 9 are placed side by side, as shown in this set of wooden rods from the Andechs monastery in Bavaria. The sums of the numbers in the diagonal cells of the sixth row (indicated by VI), adding from right to left, give the product 2874.

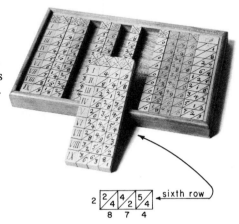

Napier's rods, Andechs monastery in Bavaria

To compute 73 times 61, the rods for the multiples of 6 and 1 are placed side by side. In this case, there are two partial products: one from the third row of the rods for computing 3×61; and one from the seventh row for computing 7×61. These products are 183 and 427. Why must 427 be changed to 4270? The product 73×61 equals $183 + 4270$, or 4453.

To multiply by 3

To multiply by 7

Multiplication on Calculators   The steps for multiplying with a calculator are similar to those for adding and subtracting. The numbers and operations are entered in the order they occur from left to right when the product is written horizontally. The steps for computing $114 \times 238 \times 71$ are listed here on the right.

1. Enter 114
2. $\boxed{\times}$
3. Enter 238
4. $\boxed{\times}$
5. Enter 71
6. $\boxed{=}$

The familiar pencil and paper algorithm for multiplying numbers can be checked with a calculator in several ways. To check the product shown here, first multiply on your calculator to see if the answers agree. If not, the calculator can be used to check the three partial products by the following steps.

$$
\begin{array}{r}
483 \\
\times\,956 \\
\hline
2898 \\
2415 \\
4347 \\
\hline
461748
\end{array}
\left.\phantom{\begin{array}{c}2898\\2415\\4347\end{array}}\right\}
\begin{array}{l}\text{Partial}\\\text{Products}\end{array}
$$

Steps	Displays
1. 6 $\boxed{\times}$ 483 $\boxed{=}$	2898.
2. 50 $\boxed{\times}$ 483 $\boxed{=}$	24150.
3. 900 $\boxed{\times}$ 483 $\boxed{=}$	434700.

$\left.\phantom{\begin{array}{c}1\\2\\3\end{array}}\right\}$ Partial Products

Special care must be taken on some calculators when multiplication is combined with addition or subtraction. The numbers and operations will not always produce the correct answer if they are entered onto the calculator in the order they appear. For example, to compute the expression, $a \times b\ +\ c \times d$, the products $a \times b$ and $c \times d$ must be computed separately and then added. If you compute $a \times b$, then add c, and then multiply by d, what you get will not equal $a \times b\ +\ c \times d$. The distributive property shows that this computation will equal $a \times b \times d\ +\ c \times d$.

$$(a \times b\ +\ c) \times d = a \times b \times d\ +\ c \times d$$

To compute the sum of two products on a calculator, the first product can be computed and stored in memory, by pressing $\boxed{M+}$ or a similar key, while the second product is being computed. The accompanying steps show how to compute $114 \times 238\ +\ 19 \times 605$ by using memory storage. Numbers can be recalled from memory storage by pressing $\boxed{MR}$ or a similar key. Numbers can be cleared from memory storage by pressing $\boxed{MC}$ or a similar key. Note that some calculators *will* produce the correct answer for this example when the numbers and operations

Steps	Displays
1. 114 $\boxed{\times}$ 238 $\boxed{=}$	27132.
2. $\boxed{M+}$	27132.
3. $\boxed{C}$	0.
4. 19 $\boxed{\times}$ 605 $\boxed{=}$	11495.
5. $\boxed{+}$	11495.
6. $\boxed{MR}$	27132.
7. $\boxed{=}$	38627.

are entered from left to right as they appear. When $\boxed{+}$ 19 is entered on this type of calculator, 19 is not added to the product 114×238, but it is multiplied by 605 when $\boxed{\times}$ 605 is entered. Then the answer, 38627, appears by pressing $\boxed{=}$.

On some calculators repeated multiplication by the same number can be carried out just as we have done in the previous sections for addition and subtraction. Suppose we were to multiply the numbers 4107, 631, 916, and 6248 by 214. The 214 $\boxed{\times}$, which is entered in the first step shown here, remains in effect throughout Steps 2, 3, and 4.

Steps	Displays
1. 214 $\boxed{\times}$ 4107 $\boxed{=}$	878898.
2. 631 $\boxed{=}$	135034.
3. 916 $\boxed{=}$	196024.
4. 6248 $\boxed{=}$	1337072.

Geometric sequences are generated by multiplying by a common ratio, similar to the way arithmetic sequences are generated by adding a common difference. Here are the steps for obtaining the first few numbers in the sequence 3, 9, 27, 81, 243, ..., whose common ratio is 3.

Steps	Displays
1. Enter 3	3.
2. $\boxed{\times}$	3.
3. $\boxed{=}$	9.
4. $\boxed{=}$	27.
5. $\boxed{=}$	81.
6. $\boxed{=}$	243.

SUPPLEMENT (*Activity Book*)

Activity Set 3.3 Multiplying with the Abacus and Chip Trading
Just for Fun: Crossnumbers for Calculators

Additional Sources

Boykin, W.E. "The Russian-peasant algorithm: Rediscovery and extension." *The Arithmetic Teacher,* **20** No. 1 (January 1973), 29–32.

Cacha, F.B. "Understanding multiplication and division of multidigit numbers." *The Arithmetic Teacher,* **19** No. 5 (May 1972), 349–54.

Gardner, M. *Mathematical Carnival.* New York: Knopf, 1975. "Tricks of Lightning Calculators," pp. 77–88.

School Mathematics Study Group. *Studies in Mathematics,* **13**. Palo Alto: Stanford University, 1966. pp. 131–50.

Smith, D.E. *History of Mathematics,* **2**. Lexington, Mass: Ginn, 1925. pp. 101–28.

Traub, R.G. "Napier's rods: practice with multiplication." *The Arithmetic Teacher,* **16** No. 5 (May 1969), 363–64.

An exhibit for illustrating multiplication, at the California Museum of
Science and Industry

EXERCISE SET 3.3

1. The children in the picture on the preceding page are computing products of numbers from 1 through 8. Each time three buttons are pressed on the switch box, the product is illustrated by lighted bulbs in the 8 by 8 by 8 cube of bulbs. Buttons 3, 4, and 1 are for the product 3 × 4 × 1. The 12 bulbs in the upper left-hand corner of part **a** will be lighted for this product. Whenever the third number of the product is 1, only those bulbs on the front face of the cube (facing children) will be lighted. Indicate the bulbs in parts **b** and **c** which will be lighted for the given products.

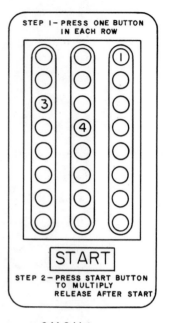

STEP 1 — PRESS ONE BUTTON IN EACH ROW

START

STEP 2 — PRESS START BUTTON TO MULTIPLY RELEASE AFTER START

a. 3 × 4 × 1   ★   b. 7 × 3 × 1   c. 2 × 8 × 1

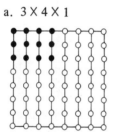

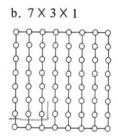

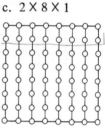

The first two numbers in the product of three numbers determine the rectangular array on the front face of the cube. The third number in the product determines the number of times the array on the front face is repeated in the cube. The 24 bulbs that are boxed-in on the cube in part **d** will be lighted for 3 × 4 × 2. Box-in the bulbs in parts **e** and **f** which will be lighted for the given products.

d. 3 × 4 × 2   e. 6 × 4 × 3   ★   f. 1 × 8 × 8

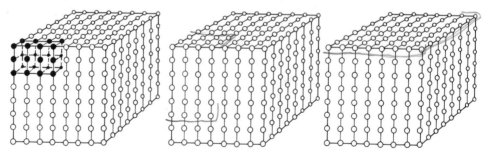

2. One type of computer program for art and amusement prints out figures by typing letters and symbols. The cat in this diagram was produced by such a program. There are 83 horizontal lines of type, and the longest line has 52 symbols.

★ a. It took approximately 3 seconds for the computer to type each row of this cat. How many minutes and seconds did it take to complete this figure?

b. How many letters can be typed in 83 lines if there are 52 letters in each line?

c. Estimate the number of symbols or letter positions in this cat.

3. A preliminary study by the Massachusetts Electric Company on the feasibility of hand delivering bills to cut costs found that if a person could deliver 600 bills a day, it would cost the company only 10 cents apiece, as opposed to 13 cents for a stamp.

★ a. At this rate how much money can one delivery person save the company in a 5-day week? $90.00

b. The program began with 5 people delivering bills. If each person delivers 600 bills a day, how much will the company save in 1 month (4 weeks)? $1,800.00

★ 4. Compute 347 × 605 by using the gelosia method. What advantage is there in using this method?

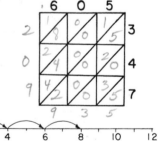

5. Multiplication of whole numbers can be illustrated on the number line by a series of arrows. This top number line shows 4 × 2. Draw arrow diagrams for the products in parts a and b.

a. 3 × 4   b. 2 × 5

c. Use the number line to show that 3 × 4 = 4 × 3.

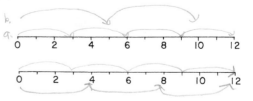

6. Which number property is exemplified in each of the following equalities?

★   a. $3 \times (2 \times 7 + 1) = 3 \times (7 \times 2 + 1)$ *comm. u mult.*

    b. $18 + (43 \times 7) \times 9 = 18 + 43 \times (7 \times 9)$ *ass. u mult.*

·★   c. $(12 + 17) \times (16 + 5) = (12 + 17) \times 16 + (12 + 17) \times 5$ *Distribution*

    d. $4 \times (13 + 22)/(7 + 5) = 4 \times (13 + 22)/(5 + 7)$ *comm. u add.*

    e. $(15 \times 2 + 9) + 3 = 15 \times 2 + (9 + 3)$ *- ass. u add.*

7. Occasionally, when multiplying numbers, it is convenient to replace one number by a difference of two numbers. In the following example, 98 has been replaced by 100 − 2, because multiplying by 100 is so easy. Rather than compute 45 × 98, we can compute 45 × 100 and then subtract 45 × 2. This involves the use of the distributive property.

$$45 \times 98 = 45 \times (100 - 2) = 45 \times 100 \;\; - \;\; 45 \times 2$$

Distributive Property

   Use the fact that *multiplication distributes over subtraction* to compute the following products in your head. Explain what you did in each example.

★   a. $35 \times 19 =$ b. $51 \times 9 =$ c. $30 \times 99 =$

8. This 1-inch-by-1-inch square contains 50 rows and 50 columns of dots.

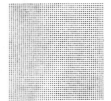

★   a. How many dots are there?

    b. An 8-inch-by-10-inch page would contain a 400 by 500 array of these dots. How many dots would there be on such a page?

★   c. The 1977–78 Manhattan telephone directory of yellow pages has approximately 2000 pages, each with an 8-inch-by-10-inch printed surface. If every page contained the number of dots in part **b**, how many dots would there be in this directory?

    d. The world's population is approximately four billion. If each person is represented by one dot, how many of the telephone directories from part **c** would be required?

9. Use the Egyptian duplation method to compute 27 × 35. Explain why this method works.

    $1 \times 35 = 35$
    $2 \times 35 = 70$
    $4 \times 35 = 140$
    $8 \times 35 = 280$
    $16 \times 35 = 560$

★ 10. Use Russian peasant multiplication to compute
42 X 30. Explain why this method gives the correct
result.

Halving		Doubling
42	X	30
21	X	60
10	X	120
5	X	240
2	X	480
1	X	960

11. In parts **a** (multibase pieces) and **b** (bean sticks) sketch the new sets of objects for the given products. Then do the regrouping and write the numbers represented by the final collection of objects.

a. Multiply by 4:

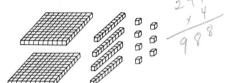

b. Multiply by 3:

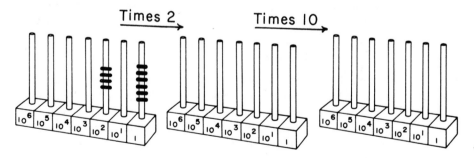

12. The products in parts **a** and **b** are each done in two steps. Sketch the new sets of chips and markers for each mat and abacus. After regrouping, what numbers will be represented by the final sets of chips and markers?

★ a. Multiply by 40:

Times 10 →

Red	Green	Blue	Yellow

Times 4 →

Red	Green	Blue	Yellow

Red	Green	Blue	Yellow

b. Multiply by 20:

Times 2 →

Times 10 →

13. *Calculator Exercise:* Use a calculator to determine which one of the following two products has a mistake. Find the incorrect partial product and correct the mistake.

a.	4367		b.	9736
	X 8954			X 8596
	17468			58416
	21835			87624
	39403			48680
	34936			77888
	39112118			83690656

14. *Calculator Exercise:* Use a calculator to find the next six numbers in this geometric sequence:

$$4, 16, 64, 256, ...$$

15. The sequence of calculator steps, 3 $\boxed{\times}$ 4 $\boxed{+}$ 5 $\boxed{\times}$ 7, will compute $(3 \times 4) + (5 \times 7)$ on some calculators and $(3 \times 4 + 5) \times 7$ on other types of calculators. What are the answers to these two different expressions? If the preceding sequence of steps computes $(3 \times 4) + (5 \times 7)$ on a calculator, which of the following two sequences of steps will compute $(3 \times 4 + 5) \times 7$ on the same calculator?

a. Enter 3
$\boxed{\times}$
Enter 4
$\boxed{+}$
Enter 5
$\boxed{\times}$
Enter 7
$\boxed{=}$

b. Enter 3
$\boxed{\times}$
Enter 4
$\boxed{+}$
Enter 5
$\boxed{=}$
$\boxed{\times}$
Enter 7
$\boxed{=}$

16. *Cyclic Numbers:* The number 142,857 has the property that when it is multiplied by 1, 2, or 3, the products contain the digits 1, 4, 2, 8, 5, and 7 in the same cyclic order.

$1 \times 142857 = 142857$
$2 \times 142857 = 285714$
$3 \times 142857 = 428571$
$4 \times 142857 =$
$5 \times 142857 =$
$6 \times 142857 =$

★ a. Will this continue to be true when multiplying 142,857 by 4, 5, or 6?

The number 142,857 is a cyclic number. A *cyclic number* is a whole number of *n* digits such that when it is multiplied by any whole number from 1 to *n*, the product contains the same *n* digits of the original number in the same cyclic order. This number is also cyclic: 0,588,235,294,117,647.* Compute the following products by making a simple observation in each case. How can each product be found by just multiplying two single-digit numbers?

b. 7 X 0,588,235,294,117,647   ★   c. 8 X 0,588,235,294,117,647

*See M. Gardner, "Mathematical Games," *Scientific American,* **222** No. 3 (March 1970), 121.

17. *Number Tricks:* To multiply a number by 5, you can first divide by 2 and then multiply by 10. Try this method to compute 5 × 142. Devise a similar method for multiplying by 25. Use your method to compute 25 × 36. Explain why these methods work.

★ 18. *Finger Multiplication:* Use your hands to compute 9 × 6, using the finger multiplication system in Section 3.3. Will this method produce the correct results for the special cases of 9 times 10, or 10 times 10?

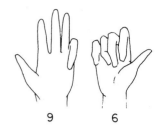

9   6

19. There is a similar method of finger multiplication for computing the products of numbers from 11 through 15. For example, the product of 12 and 14 is 168. Beginning with 100, how can the finger positions for 12 and 14 be used to get 68?*

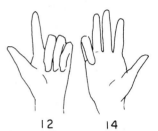

12   14

20. *Napier's Rods:* Use the rods given for multiples of 7, 8, and 4 to compute 6 × 784.

★   a. To compute 8 times 903, the eighth rows of the multiples of 9, 0, and 3 are needed. Write the numbers for these eighth rows and use them to compute 8 × 903.

 eighth row

b. One way to compute 94 times 288 is to use two rods with multiples of 8. Explain how this product can be computed without using any rods with multiples of 8. (*Hint:* Use the commutative property for multiplication.)

*For a description of how this method can be extended to numbers from 10 to 50, see L.P. Alger, "Finger multiplication," *The Arithmetic Teacher,* **15** No. 4 (April 1968), 341-43.

21. *Faded Document:* Supply the missing digits for the
boxes in this faded document problem.

$$
\begin{array}{r}
4\ \square\ \square \\
\times\ \square\ \square\ 7 \\
\hline
\square\ \square\ 8\ 2 \\
1\ 2\ \square\ \square \\
\hline
\square\ \square\ \square\ \square\ \square\ \square
\end{array}
$$

22. *Cryptarithms:* Find the five digits represented by the letters in part **a**. The cryptarithm in part **b** is unusual, in that E and O do not represent only one digit. Each E stands for an even digit (0, 2, 4, 6, or 8) and each O for an odd digit (1, 3, 5, 7, or 9).

★   a.

$$
\begin{array}{r}
S\ T \\
\times\ R\ T \\
\hline
S\ T \\
P\ Q\ R \\
\hline
P\ T\ T\ T
\end{array}
$$

     b.

$$
\begin{array}{r}
E\ E\ O \\
\times\ O\ O \\
\hline
E\ O\ E\ O \\
E\ O\ O \\
\hline
O\ O\ O\ O\ O
\end{array}
$$

23. *Number Pattern:* Continue the pattern of numbers on both sides of these equations. For how many equations will this pattern hold?

$$
\begin{array}{r}
1 \times 9 + 2 = 11 \\
12 \times 9 + 3 = 111 \\
123 \times 9 + 4 = 1111
\end{array}
$$

★ 24. *For the Experts:* In this faded document problem the missing numbers are the prime digits 2, 3, 5, and 7.

$$
\begin{array}{r}
\square\ \square\ \square \\
\times\ \square\ \square \\
\hline
\square\ \square\ \square\ \square \\
\square\ \square\ \square\ \square \\
\hline
\square\ \square\ \square\ \square\ \square
\end{array}
$$

General Motors Terex Titan and Chevrolet Luv pickup

3.4 DIVISION AND EXPONENTS

Division, like multiplication, is a shortcut for computing. One way that division occurs is in comparing two quantities. Consider the relative size of the Terex Titan as compared to the Luv pickup on the Titan's dump body. The Terex Titan can carry 317250 kilograms while the Luv pickup has a limit of 450 kilograms. We can determine how many times greater the Titan's capacity is than the Luv's by dividing 317250 by 450. The answer is 705, which means that the Luv pickup will have to haul 705 loads to fill the Titan just once! If the boy's truck in the photo holds 3 kilograms of sand, how many of its loads will fill the Titan?

Definition of Division   The division example, which compares the size of the Terex Titan to the Luv pickup, can be checked by multiplication. The load weight of the smaller truck times 705 should equal the load weight of the larger truck. The close relationship between division and multiplication can be used to define division in terms of multiplication. This is the most popular method of defining division and perhaps the oldest. J. P. A. Erman, in *Life In Ancient Egypt,* has attributed this definition to the Egyptians. Symbolically, it is stated as follows. For any whole numbers r and s, with $s \neq 0$,

$$r \div s = \square \quad \text{if and only if} \quad r = s \times \square$$

This definition says that division can be done by multiplication. If I asked you to compute $54 \div 9$, you would immediately say 6, because you know that $54 = 9 \times 6$.

There are three common terms in the division process: <u>dividend</u>, <u>divisor,</u> and <u>quotient.</u> In the previous example, 54 is the *dividend,* 9 is the *divisor,* and 6 is the *quotient.*

Over the centuries, division has acquired two meanings or uses. David Eugene Smith, in *History of Mathematics,* speaks of the *twofold nature of division* and gives references to the sixteenth century authors who first clarified the differences.* These two meanings of division, called *measurement* and *partitive,* are illustrated in the following two paragraphs.

*Measurement Concept—*Suppose you had 24 tennis balls and wanted to give 3 tennis balls to as many people as possible. How many people would receive tennis balls? This can be determined by subtracting away, or measuring off, as many sets of 3 as possible. The following arrangement of tennis balls shows the result of this measuring process and illustrates $24 \div 3$. The divisor, 3, is the number of balls in each group, and the quotient, 8, is the number of groups. This is an example of the *measurement* use of division.

*Partitive Concept—*Suppose you had 24 tennis balls, which you wanted to distribute equally among 3 people. How many tennis balls would each person receive? This can be determined by separating (partitioning) the tennis balls into 3 equivalent sets. The following arrangement of tennis balls shows the 24 balls partitioned into 3 groups and illustrates $24 \div 3$. The divisor, 3, indicates the number of groups, and the quotient, 8, is the number of balls in each group. This is an example of the *partitive* use of division.

In both examples the answer to the questions is the same, 8. However, the process of grouping the tennis balls in each case is different. It is important to be aware of these differences, especially when introducing and explaining what is meant by division.

Algorithms for Division   Algorithms for division have historically been the most difficult of the basic operations of arithmetic. The Italian author Pacioli (1494) said, "If a man can divide well, everything else is easy, for all the rest is involved therein."

*D.E. Smith, *History of Mathematics,* **2** (Lexington, Mass: Ginn, 1925), p. 130.

Duplation Method—The ancient Egyptians used an algorithm for division which is similar to their duplation method of multiplying and shows the close relationship between these two operations. For example, to divide 1710 by 90 the algorithm begins by repeatedly doubling the 90s as shown here. This process is stopped when reaching a number greater than 1710. Next, we find the numbers in the right-hand column, whose sum is 1710. Since these numbers correspond to the binary numbers 1, 2, and 16, the answer is 19. This method can also be used when the quotient is not a whole number (see Exercise 7, Exercise Set 3.4).

1	90 ←
2	180 ←
4	360
8	720
16	1440 ←

$$\begin{array}{r} 90 \\ 180 \\ + 1440 \\ \hline 1710 \end{array}$$

Long Division Algorithm—Today's long division algorithm has developed over hundreds of years into an efficient method for computing. It is usually viewed as the process of subtracting off multiples of the divisor, until the remaining number is less than the divisor. One of the forerunners of our long division algorithm is similar to the form shown here, in which each digit of the divisor is multiplied separately by a digit in the quotient. This method has the advantage that the products involve only single-digit numbers.

Forerunner to Long Division Algorithm

$$
\begin{array}{r}
25 \\
38\overline{)952} \\
6 \quad (20 \times 30) \\
\hline
35 \\
16 \quad (20 \times 8) \\
\hline
19 \\
15 \quad (5 \times 30) \\
\hline
42 \\
40 \quad (5 \times 8) \\
\hline
2
\end{array}
$$

Long Division Algorithm

$$
\begin{array}{r}
25 \\
38\overline{)952} \\
76 \quad (20 \times 38) \\
\hline
192 \\
190 \quad (5 \times 38) \\
\hline
2
\end{array}
$$

Models for Division Algorithm

Each of the models used for addition, subtraction, and multiplication are also suitable for illustrating the long division algorithm. The bean sticks model and the abacus are used in the next examples, but similar demonstrations could be given with other models.

Bean Sticks—To compute 78 ÷ 3, using the bean sticks model (see following page), we begin by determining the number of groups of 3 bean sticks in the 7 bean sticks representing 70. Using the measurement concept of division (repeated subtraction) there are 2 groups of 3 bean sticks and 1 left over. The remaining bean stick is regrouped with the 8 beans to give 18 beans. Using the measurement concept again, there are 6 groups of 3 beans each. The quotient of 78 ÷ 3 is determined by the number of each type of group. There are 2 groups of bean sticks and 6 groups of beans, which shows that the quotient is 26.

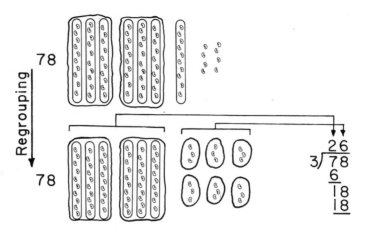

Abacus—The measurement concept is used in the following two examples to illustrate division on the abacus. Consider the example of dividing 639 by 3. Beginning with the hundreds column, measure off and remove 3 markers at a time. To record this activity, 1 marker should be placed above the hundreds column for each group of 3 markers which is removed. In all, 2 markers should be placed above the hundreds column to represent a quotient of 200. We can see the reason for this procedure in the following equations.

$$600 \div 3 = (100 \times 6) \div 3 = 100 \times (6 \div 3) = 100 \times 2$$

We continue dividing by removing 3 markers at a time from the columns and placing 1 marker above a column for each group of 3. The markers that are placed above the columns represent the quotient.

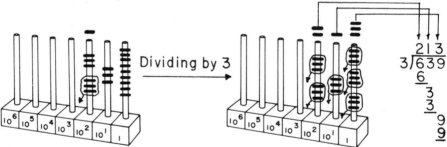

Division on the abacus can be done one column at a time when dividing by a single-digit number. If the number of markers on a column is less than the divisor, the markers can be regrouped. In the illustration shown on the next page for dividing 432 by 6, there are only 4 markers on the hundreds column. These markers can be regrouped to the tens column to give 43 markers, or we can simply think of the 4 markers on the hundreds column and the 3 markers on the tens column as 43 tens. Since there are seven 6s in 42, 7 markers are placed above the tens column, and the remaining marker in the tens column is regrouped with the 2 markers on the units column. For the final step in this example, $12 \div 6 = 2$, so 2 markers are placed above the units column.

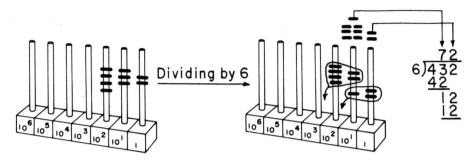

Dividing by 6

$$6\overline{)432}$$
$$\underline{42}$$
$$12$$
$$\underline{12}$$
$$72$$

Division with Zero   The special cases of division involving zero are worth looking at carefully. Perhaps it is the association of zero with nothingness which causes confusion, but whatever the reason, misconceptions involving division and zero are common.

Dividing by a Nonzero Number—The definition of division shows that 0 divided by any nonzero number is 0. In particular, consider dividing 0 by 4. We will use the fact that any number times 0 is equal to 0.

$$0 \div 4 = 0 \quad \text{because} \quad 0 = 4 \times 0$$

Dividing Zero by Zero—The confusion over this case may result from the familiar phrase that "any number divided by itself is 1," which is not true for zero divided by zero. In fact, $0 \div 0$ is never permitted because it leads to an absurd result. For example, we know that $0 \times 6 = 0$ and $0 \times 3 = 0$. Using the fact that division is the inverse of multiplication, these equations could be written as $6 = 0 \div 0$ and $3 = 0 \div 0$, if $0 \div 0$ were permitted. But this would mean that $3 = 6$. We avoid this contradiction by not allowing, or defining, $0 \div 0$ to be a number.

Dividing by Zero—The temptation in this case is to think that a number divided by 0, say, $4 \div 0$, is an infinite number. Actually, no such number exists. Using the definition of division again,

$$4 \div 0 = \square \quad \text{if} \quad 4 = 0 \times \square$$

However, since any number times 0 equals 0, there is no number that can be written into the placeholder, $\square$, to make $4 = 0 \times \square$. Therefore, division by zero is not possible.

Exponents   The large numbers used today were unnecessary a few centuries ago. The word "billion" which is now commonplace wasn't adopted until the seventeenth century. Even now "a billion" means different things to different people. In the United States it represents 1,000,000,000 (a thousand million) and in England it is 1,000,000,000,000 (a million million).

Our numbers are named according to powers of 10. The first, second, and third powers of 10 are the familiar *ten, hundred,* and *thousand.* After this, only the third powers of 10 have new or special names. The first three of these numbers and their names are million, billion, and trillion. Combinations of the words one, ten, hundred, thousand, million, and billion are used to name the remaining powers of 10 which are less than one trillion.

10^0	$= 1$	one
10^1	$= 10$	*ten*
10^2	$= 100$	one *hundred*
10^3	$= 1,000$	one *thousand*
10^4	$= 10,000$	ten thousand
10^5	$= 100,000$	one hundred thousand
10^6	$= 1,000,000$	one *million*
10^7	$= 10,000,000$	ten million
10^8	$= 100,000,000$	one hundred million
10^9	$= 1,000,000,000$	one *billion*
10^{10}	$= 10,000,000,000$	ten billion
10^{11}	$= 100,000,000,000$	one hundred billion
10^{12}	$= 1,000,000,000,000$	one *trillion*

An exponent of 10 indicates the number of 0s which follow the 1. As examples, $10^1 = 10$ (1 followed by one 0), and $10^0 = 1$ (1 followed by zero 0s). A number such as 10^{21}, which is the frequency of gamma rays per second, is a 1 followed by 21 zeros.

Numbers written as b^n are said to be in *exponential form.* We call b^n the *nth power of b* or *b to the nth power,* with the exception that b^2 is called *b squared* and b^3 is *b cubed.* This terminology has been inherited from the early Greeks who pictured the products of numbers as geometric arrays of dots.

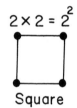

$$2 \times 2 = 2^2$$

Square

In general, for any number b and whole number n, b^n is the product with b occurring n times: $b^n = b \times b \times \cdots \times b$. In case $n = 1$ or 0, $b^1 = b$ and $b^0 = 1$. ($10^0 = 1$, $2^0 = 1$, etc.) The case in which $n = 0$ and $b = 0$, that is, 0^0, is not defined. The number b is called the *base.*

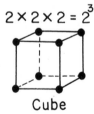

$$2 \times 2 \times 2 = 2^3$$

Cube

Laws of Exponents—Multiplication and division can be performed easily for numbers that are written as powers of the same base. To multiply we add the exponents, and to divide, the exponents are subtracted. Here are two examples that you can verify by multiplying out the powers of 2.

$$2^7 \times 2^3 = 2^{7+3} = 2^{10} \quad \text{and} \quad 2^7 \div 2^3 = 2^{7-3} = 2^4$$

These equations are special cases of the following theorems, which hold for all whole numbers a, n, and m ($a \neq 0$ in the rule for subtracting exponents).

Rule for Adding Exponents $\quad a^n \times a^m = a^{n+m}$
Rule for Subtracting Exponents $\quad a^n \div a^m = a^{n-m}$, for $a \neq 0$

The primary advantage of exponents lies in their convenience for computing, as shown by the following examples.

In our galaxy there are 10^{11} (one hundred billion) stars, and in the observable universe there are 10^9 (one billion) galaxies. If every galaxy had as many stars as ours, there would be $10^9 \times 10^{11}$ stars.

$$10^9 \times 10^{11} = 10^{9+11} = 10^{20}$$

A galaxy similar to the Milky Way, about 100,000 light-years across

If 1 out of every 1000 stars had a planetary system, then there would be $10^{20} \div 10^3$ stars with planetary systems.

$$10^{20} \div 10^3 = 10^{20-3} = 10^{17}$$

To continue this example, if 1 out of every 1000 stars with a planetary system had a planet with conditions suitable for life, there would be $10^{17} \div 10^3$ such stars.

$$10^{17} \div 10^3 = 10^{17-3} = 10^{14}$$

You may wish to continue this line of reasoning to consider the numbers of planets which might have life or perhaps intelligent life. The odds in favor of life in other parts of the universe appear to be overwhelming.

Division and Exponents on Calculators   The quotient of two numbers can be found by entering the numbers and the division operation as they appear from left to right. Here are the steps and displays for computing $27094 \div 7$.

Steps	Displays
1. 27094	27094.
2. $\div$	27094.
3. 7	7.
4. $=$	3870.5714

The sum or product of two whole numbers is always another whole number. This is called the *closure property*, and we say that each of these operations is *closed* in the set of whole numbers because they always produce another whole number. Subtraction and division, on the other hand, are not closed. That is, the difference or quotient of two whole numbers is not always another whole number. In the previous example, $27094 \div 7$ is approximately equal to 3870, but the decimal part of the answer means that there is a remainder. To find this remainder, multiply 7 times .5714 and round off to the nearest whole number, which is 4. (The symbol $\approx$ means "approximately equal to.")

$$7 \times .5714 = 3.9998$$
$$\approx 4$$

To check the quotient, 3870, and the remainder, 4, multiply 3870 by 7 and add 4.

Repeated division by the same number can be carried out on some calculators just as was done for multiplication in the previous section. The sequence of steps on the right shows how to divide the numbers 17385, 228, 685, and 472 by 19. Once we have divided by 19 in the first step, the calculator remembers this and divides each of the remaining numbers by 19.

Steps	Displays
1. $17385 \div 19 =$	915.
2. $228 =$	12.
3. $685 =$	36.052631
4. $472 =$	24.842105

If we begin with Step 1, $17385 \div 19 =$, and repeatedly press $=$, each new number will be divided by 19. The resulting numbers are part of a geometric sequence which will very rapidly approach 0. Theoretically, no number in this sequence will ever equal 0. However, on a calculator without scientific notation (see Section 7.3) 0 will soon appear in the display.

$$915, \quad 48.157894, \quad 2.534626, \quad .1334013, \ldots$$

The concept of division as repeated subtraction can be demonstrated conveniently on a calculator which subtracts constants. To divide 896 by 72, begin by subtracting, $896 - 72$, as shown in Step 1 of the sequence at the right. Then repeatedly press $=$ until there is a positive number in the display which is less than 72, or until the display shows the number 0. The number of times $=$ is pressed will be the whole number part of the quotient. The last number shown in the display will be the remainder.

Steps	Displays
1. $896 - 72 =$	824.
2. $=$	752.
3. $=$	680.
.	.
.	.
.	.

Exponents—Numbers raised to a power can be computed on a calculator, provided they do not exceed the capacity of the calculator's display. On most calculators, the steps on the right will produce the number for 4^{10}, if carried out to Step 10.

Steps	Displays
1. Enter $4 \times$	4.
2. $=$	16.
3. $=$	64.
4. $=$	256.

The number of steps in the previous sequence can be decreased by applying the rule for adding exponents: $a^n \times a^m = a^{n+m}$. To compute 4^{10}, first compute 4^5 on the calculator and then multiply 1024 times 1024 $(4^5 \times 4^5 = 4^{10})$.

Some calculators have exponential buttons for evaluating numbers raised to a power. To compute a number y to some exponential power x, the base y is entered into the calculator first; then the exponential button $\boxed{y^x}$ is pressed; and then the exponent x is entered. The steps for evaluating 4^{10} are shown to the right.

Steps	Displays
1. Enter 4	4.
2. $\boxed{y^x}$	4.
3. Enter 10	10.
4. $\boxed{=}$	1048576.

Numbers that are raised to powers will frequently exceed the calculator display. If you try to compute 4^{15} on a calculator with only eight places in its display, there will not be room for the answer in positional numeration. Some calculators will automatically convert to scientific notation when numbers in positional numeration are too large for the display.

SUPPLEMENT (*Activity Book*)

Activity Set 3.4 Dividing on the Abacus
Just for Fun: Calculator Games and Number Tricks

Additional Sources

Davis, P.J. *The Lore of Large Numbers.* New York: Random House, 1961. pp. 19–28 (exponents and names for powers of 10).

National Council of Teachers of Mathematics, *Twenty-seventh Yearbook, Enrichment Mathematics for the Grades.* Reston, Virginia: NCTM, 1963. pp. 92–100 and 173–79 (shortcuts and tricks and why they work).

School Mathematics Study Group. *Studies in Mathematics,* **13**. Palo Alto: Stanford University, 1966. pp. 165–82.

Smith, D.E. *History of Mathematics,* **2**. Lexington, Mass: Ginn, 1925. pp. 128–44.

Swart, W.L. "Teaching the division-by-subtraction process." *The Arithmetic Teacher,* **19** No. 1 (January 1972), 71–75.

Tucker, B.F. "The division algorithm." *The Arithmetic Teacher,* **20** No. 8 (December 1973), 639–46.

Zweng, M.J. "The fourth operation is not fundamental." *The Arithmetic Teacher,* **19** No. 8 (December 1972), 623–27.

EXERCISE SET 3.4

1. Use the pattern of "a black followed by three whites" to answer parts **a** and **b**.

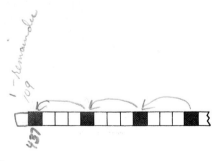

★   a. What is the color of the 29th square?

    b. What is the color of the 437th square? (*Hint:* Divide by 4 and check the remainders.)

★ c. In the pattern "two blacks followed by three whites," what are the colors of the 396th and 404th squares?

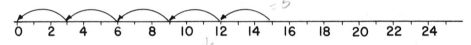

2. Compute each side of the following equation. Does the right side equal the left? Try some other numbers in the square, rhombus, and triangular placeholders. Can you find a case in which division is not distributive over addition?

$$(\boxed{6} + \triangle\!15) \div \cancel{3} = (\boxed{6} \div \cancel{3}) + (\triangle\!15 \div \cancel{3})$$

3. Is division commutative or associative? Try some numbers in the following placeholders. It takes only one counterexample to show that a property does not hold.

★ a. $\square \div \triangle \overset{?}{=} \triangle \div \square$ b. $\square \div (\triangle \div \diagup\!\square) \overset{?}{=} (\square \div \triangle) \div \diagup\!\square$

4. The quotient 15 ÷ 3 is illustrated on the number line in part **a** by beginning at 15 and subtracting 3s. The quotient is the number of arrows.

★ a. Is the partitive or measurement concept of division being used in this illustration?

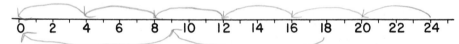

b. Draw arrow diagrams for 24 ÷ 4 and 18 ÷ 9.

5. Encircle groups of bean sticks and beans to illustrate the given quotients. Cross out and sketch new beans if regrouping is needed.

a. 86 ÷ 2 = 43

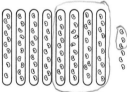

b. 96 ÷ 4 = 24

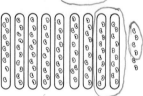

6. Use the abacus in parts **a** and **b** to compute the given quotients. Use the long division algorithm and compare the steps used on the abacus with those used in the algorithm.

★ a. Illustrate $8064 \div 2$ by drawing markers on the second abacus and then encircling groups of 2 markers. How is the quotient determined by the groups of markers?

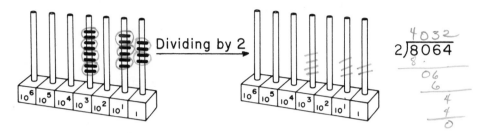

Dividing by 2

b. Illustrate $4272 \div 3$ by drawing markers on the second abacus and then encircling groups of 3 markers. How is the quotient determined by the groups of markers? (Note: Regrouping is needed.)

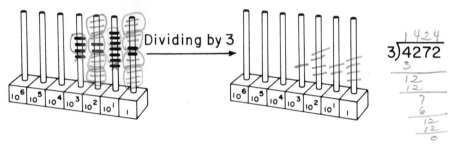

Dividing by 3

7. Use the duplation method to compute the following quotients. This method requires a little trial and error to find the numbers for the correct sum. The exact sum cannot be found for one of these examples. In this case, a fraction will be needed in the answer.

★ a. $1232 \div 112$

1	112
2	224
4	448
8	896

b. $300 \div 24$

1	24
2	48
4	96
8	192

8. Use a calculator to evaluate the following expressions. Which one of these numbers will not fit onto a calculator with an eight-place display?

★ a. 3^{15} b. 15^6 ★ c. 4^{14} d. 2^{25}

9. During the Buddhistic period (sixth century to first century B.C.) the Hindus were particularly interested in large numbers. B.L. Van Der Waerden describes this scene from the book *Lalilavistara*.* Prince Guatama (Buddha) asks the prince Dandapani for the hand of his daughter Gopa. He is required to compete with five other suitors in writing, wrestling, archery, running, swimming, and arithmetic. The great mathematician Arjuna questions him.

*B.L. Van Der Waerden, *Science Awakening* (Groningen, Holland: P. Noordhoff, 1954), pp. 51–52.

Oh, young man, do you know how the numbers beyond the koti continue by hundreds?

I know it.

How then do the numbers beyond the koti continue by hundreds?

One hundred kotis are called *ayuta,* one hundred ayutas *niyuta,* one hundred niyutas *kanikara,* one hundred kanikaras *vivara* ...

★ a. A *koti* is $10^2 \times 10^5$ (one hundred times one hundred thousand). Compute this product and use an exponent to write your answer as a power of 10.

b. Write each of the following Hindu numbers as a power of 10.

ayuta niyuta kanikara

★ c. Prince Guatama continues through 24 stages, where koti is the first stage, ayuta is the second, niyuta is the third, etc. Write the number for the 24th stage as a power of 10.

d. Beyond this there are eight other series. If each series increases the numbers by a multiple of 10^{48}, the total increase for all eight series will be $10^{48} \times 10^{48} \times 10^{48} \times 10^{48} \times 10^{48} \times 10^{48} \times 10^{48} \times 10^{48}$. Compute this product and write your answer as a power of 10.

10. *Calculator Exercise:* A calculator has been used to compute the following quotients. Convert the decimal part of each quotient into a remainder. Check your answer by multiplying the divisor by the whole number part of the quotient and adding the remainder.

★ a. $47208 \div 674 = 70.041543$ b. $107253 \div 86 = 1247.1279$

c. $13738 \div 24 = 572.41666$

11. *Number Curiosity:* Select any four-digit number. Rewrite that number keeping the digits in the same order but putting the left-most digit at the right. Repeat this process two more times and then add the four numbers. Now divide the total by the sum of the digits in the original number, which in this example is 19. You will get a very special result.

```
  2917
  9172
  1729
+ 7291
 21109
```

Sum of Digits

$2 + 9 + 1 + 7 = 19$

a. Try this process for some other four-digit numbers. Make a conjecture about what will happen.

b. Will there be a similar result if this process is used on a three-digit or five-digit number?

yes yes

★ 12. *All Four Basic Operations:* Using exactly four 4s and only addition, subtraction, multiplication, and division, try to write an expression that equals each of the numbers from 1 to 10. For example, $2 = (4 \div 4) + (4 \div 4)$. You do not have to use all operations, and expressions such as 44 are allowable.*

*Similar equations exist for five 5s, six 6s, etc. See R. Crouse and J. Shuttleworth, "Playing with Numerals," *The Arithmetic Teacher,* **21** No. 5 (May 1974), 417–19.

13. Compute these products and quotients. Leave your answers in exponential form.

★   a. $5^{14} \times 5^{20}$ b. $10^{12} \times 10^{10}$ ★ c. $10^{32} \div 10^{15}$ d. $3^{22} \div 3^{8}$

14. Find numbers for m and n to show that
   a. $2^n - 2^m \neq 2^{n-m}$
   b. $2^n + 2^m \neq 2^{n+m}$

15. *"Krypto"*: "Krypto" is a commercially produced card game with numbers from 1 through 25. The object is to combine the numbers on 5 cards that are dealt to obtain the number on a sixth card, the target number. Any of the 4 basic operations may be used, but each of the 5 cards must be used once and only once. Each of the following sets of cards can be combined by using all 4 operations to obtain the target number.

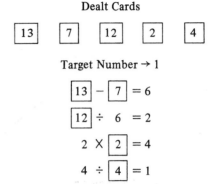

Dealt Cards

13   7   12   2   4

Target Number → 1

13 − 7 = 6
12 ÷ 6 = 2
2 × 2 = 4
4 ÷ 4 = 1

★   a. 22   19   2   14   10 → 7

      b. 21   2   3   12   7 → 20

★ 16. If a piece of paper is cut in half and then these two pieces are placed on top of each other and cut in half again, there will be four pieces. If this process is continued, how many pieces will there be after 20 cuts? Write your answer as a number raised to a power.

17. The chart below shows the approximate frequencies of some common types of waves. Visible light waves, for example, are between 10^{14} and 10^{15} waves or cycles per second.

★   a. The frequency of television is 10^8 cycles per second. If a type of X ray has a frequency that is 10^{11} times greater, what is the X-ray frequency?

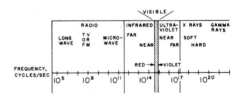

      b. If the frequency for infrared is 10^{13} and it is 1000 times greater than the frequency for microwaves, what is the microwave frequency?

★   c. If a radio frequency is 10^8 and gamma rays have a frequency of 10^{21}, how many times greater is the gamma ray frequency?

★ 18. *Calculator Exercise:* Beneath each of the following equations there is a sequence of calculator steps. Which sequences produce the correct answers? If a sequence does not produce the correct answer, write the equation which corresponds to the sequence of steps.

a. $8 \times (12 \div 3) = 32$
   1. Enter 8
   2. $\boxed{\times}$
   3. Enter 12
   4. $\boxed{\div}$
   5. Enter 3
   6. $\boxed{=}$

b. $3 \times 4 + 7 = 19$
   1. Enter 3
   2. $\boxed{\times}$
   3. Enter 4
   4. $\boxed{+}$
   5. Enter 7
   6. $\boxed{=}$

c. $17 - 3 \times 5 = 2$
   1. Enter 17
   2. $\boxed{-}$
   3. Enter 3
   4. $\boxed{\times}$
   5. Enter 5
   6. $\boxed{=}$

★ 19. *Faded Document:* Supply the missing digits.

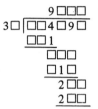

20. *"Division Game"* (3 or 4 players): One player selects a number less than 100 and the other players take turns trying to guess the number by asking questions involving division by numbers which are less than or equal to 10. Each player gets one question and one guess, each time it is his or her turn. The first player to guess the number wins. Use the following questions and answers to determine the number: What is the remainder when dividing by 5? (Answer, 3); What is the remainder when dividing by 2? (Answer, 0); What is the remainder when dividing by 4? (Answer, 2); What is the remainder when dividing by 9? (Answer, 6).

21. *Number Trick:* You can use a variation of the "Division Game" as a number trick. Ask a friend to choose any number less than 105. Have him/her divide it by 3, 5, and 7, giving you only the remainders x, y, and z for each division. For example, if he/she chose 46, then $x = 1, y = 1$, and $z = 4$. When the expression $70x + 21y + 15z$ is evaluated and divided by 105, the whole number remainder is the chosen number. Find the number if the remainders are 1, 0, and 6 when dividing by 3, 5, and 7, respectively.

© 1968 United Feature Syndicate, Inc.

3.5 PRIME AND COMPOSITE NUMBERS

Do you have a favorite number? The number 3 is a popular choice and there may be historical reasons for this preference. For example, in the French "très bien," which means good, "très" is derived from the word for three. One of the oldest superstitions is that odd numbers are lucky. One exception to this is the common fear of 13, called *triskaidekaphobia.* Often, hotels will not have a floor that is numbered 13, and motels will not have a room 13. Some people have a particular fear of Friday, the thirteenth. These people will be unhappy to know that the 13th day of the month falls more frequently on a Friday than on any other day of the week.*

Number Mysticism   The Pythagoreans (ca. 500 B.C.), a brotherhood of mathematicians and philosophers, believed that numbers had special meanings which could account for all aspects of life. The number 1 represented reason, 2 stood for opinion, 4 was symbolic of justice, and 5 represented marriage. Even numbers were weak and earthly, and odd numbers were strong and heavenly. The numbers 1, 2, 3, and 4 also represented fire, water, air, and earth, and that fact that $1 + 2 + 3 + 4 = 10$ had many meanings. When only nine heavenly bodies could be found, including the earth, sun, moon, and the sphere of stars, they imagined a tenth to "balance the earth."**

Through the ages, numbers have played an important role in astrology and in the casting of horoscopes. Since many numeration systems used letters, it was natural to substitute the number value for the letters in a name. This practice, which is called *gematria,* was popular with the ancient Hebrews and Greeks. The following table shows the letter symbols the Greeks used for numbers.

α	β	γ	δ	ϵ	ς	ζ	η	θ
1	2	3	4	5	6	7	8	9
ι	κ	λ	μ	ν	ξ	o	π	a
10	20	30	40	50	60	70	80	90
ρ	σ	τ	υ	φ	χ	ψ	ω	λ
100	200	300	400	500	600	700	800	900

*J.O. Irwin, "Friday 13th," *Mathematical Gazette,* **55** (1971), 412–15.

**M. Kline, *Mathematics in Western Culture* (New York: Oxford University Press, 1953) p. 77.

The Greek word for amen is $\alpha\mu\eta\nu$, and from the table this word corresponds to $1 + 40 + 8 + 50$, or 99. In some old editions of the Bible the number 99 appears for the word amen. During the Middle Ages gematria was revived and used in the art of "beasting." This consisted of showing that the number of the beast, 666, could be associated with certain people by assigning numbers to the letters of their name. In one case, the Catholic theologian Peter Bungus showed that a form of the name Martin Luther was numerically equivalent to 666.

Primes and Their Distribution   If one whole number divides a second whole number evenly, that is, leaving only a zero remainder, then the first number is a *factor* (or *divisor*) of the second number. Any whole number greater than 1 which has only itself and 1 as factors is called a *prime.* All other whole numbers greater than 1 are called *composite.* One of the earliest number distinctions involved primes. The number 7, for example, was associated with purity and the maiden goddess Athena, because it cannot be broken down into a product of smaller factors.

There are 25 primes less than 100. Here are the first few.

$$2 \quad 3 \quad 5 \quad 7 \quad 11 \quad 13 \quad 17 \quad 19 \quad 23 \quad 29$$

Some very large primes have been discovered. From 1876 to 1951, this 39-digit number was the largest known prime:

$$170,141,183,460,469,231,731,687,303,715,884,105,727$$

Now with high-speed computers, someone finds a larger prime every few years. In 1971, for example, at the IBM Research Center in Yorktown, New York, Bryant Tuckerman found a prime with 6002 digits. We know that there are an infinite number of primes. This was proven by Euclid (see Exercise 14 of Exercise Set 3.5). On the other hand, there are arbitrarily large stretches of consecutive whole numbers with no primes! For example, between the two prime numbers 396,733 and 396,833 there are 99 composite numbers. It is possible to find a string of one million, one billion, etc., consecutive numbers that contain no prime numbers (see Exercise Set 3.5, Exercise 13). No wonder mathematicians have been unable to find a formula that will give all primes less than any given number.

Prime Number Test   How can we determine whether 421 is a prime number? A natural but time-consuming approach is to try dividing by smaller whole numbers, 2, 3, 4, 5, 6, 7, 8, etc. This method can be improved by noticing that it is unnecessary to divide by the composite numbers 4, 6, 8, 9, etc. For example, if 2 is not a factor of a number, then 4, 6, and 8 will not be factors. In other words, to determine whether a number is prime or composite, it is necessary only to try dividing by smaller numbers that are prime.

A second improvement for determining whether or not a number is prime comes from the observation that composite numbers have at least one prime factor less than or equal to the positive square root of the number. Compare the prime factors of the numbers in the first column of this table to the numbers in the third column of the table. In some cases, there is a prime factor greater than the square root of the

n	Prime Factorization	$\sqrt{n}$
51	3×17	between 7 and 8
70	$2 \times 5 \times 7$	between 8 and 9
121	11×11	11
115	5×23	between 10 and 11
195	$3 \times 5 \times 13$	between 13 and 14
357	$3 \times 7 \times 17$	between 18 and 19

number, but there is always at least one prime factor that is less than this square root. Thus, to determine whether a number is prime, it is necessary only to try dividing by the primes that are less than or equal to the square root of the number.

Let's return to the question of whether or not 421 is a prime. The positive square root of 421 is between 20 and 21. Since 2, 3, 5, 7, 11, 13, 17, and 19 are not factors of 421, this number is a prime. We do not have to try dividing by primes greater than 20 because there are no prime factors of 421 which are less than 20, and the product of two primes that are both greater than 20 would be greater than 421.

Sieve of Eratosthenes   One way of finding all the primes less than a given number is to eliminate those numbers which are not prime. This method was first used by the Greek mathematician Eratosthenes (ca. 230 B.C.) and is called the *Sieve of Eratosthenes.* It is illustrated in the following array of numbers to find the primes that are less than 120.

The process begins by circling the smallest prime, 2, and crossing out all the remaining multiples of 2 (4, 6, 8, 10, ...). Then 3 is circled and all of the remaining multiples of 3 are crossed out (6, 9, 12, 15, ...). Continue this process by circling 5 and 7 and crossing out their multiples. With the exception of 1, the numbers that are not crossed out will be prime. Since every composite number less than 120 must have at least one prime factor less than 11 (11 $\times$ 11 = 121), it is unnecessary to cross out the multiples of primes greater than 7.

1	2	3	4	5	6	7	8	9	10
11	12	13	14	15	16	17	18	19	20
21	22	23	24	25	26	27	28	29	30
31	32	33	34	35	36	37	38	39	40
41	42	43	44	45	46	47	48	49	50
51	52	53	54	55	56	57	58	59	60
61	62	63	64	65	66	67	68	69	70
71	72	73	74	75	76	77	78	79	80
81	82	83	84	85	86	87	88	89	90
91	92	93	94	95	96	97	98	99	100
101	102	103	104	105	106	107	108	109	110
111	112	113	114	115	116	117	118	119	120
121	122	123	124	125	126	127	128	129	130

Fundamental Theorem of Arithmetic   Composite numbers can always be written as a product of primes. For this reason, prime numbers are often referred to as the building blocks of whole numbers.

One method of finding all the prime factors of a composite number is through the use of a *factor tree*. This is a series of steps in which a number is broken down into smaller and smaller factors until all the final factors are prime numbers.

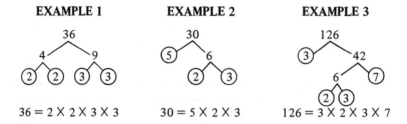

EXAMPLE 1   EXAMPLE 2   EXAMPLE 3

36 = 2 × 2 × 3 × 3   30 = 5 × 2 × 3   126 = 3 × 2 × 3 × 7

A factor tree can be started with any two factors of a number. If the number is even, we can always begin by dividing by 2 to get a second factor. In one of the factor trees for 84 the first two factors are 2 and 42. In another factor tree for 84 the first two factors are 7 and 12.

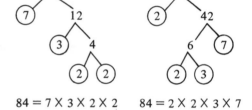

84 = 7 × 3 × 2 × 2   84 = 2 × 2 × 3 × 7

The same number may have several different factor trees, but the prime factors in each will be the same except for the order in which they appear. This fact is due to the following important theorem.

Fundamental Theorem of Arithmetic: Every whole number greater than 1 is either a prime or a product of primes, and the product is unique, except for the order in which the factors occur.

Models for Factors and Multiples   Colored rods are useful for illustrating the concepts of factor, multiple, and prime and composite numbers. The rods pictured here are Cuisenaire rods. While these ten rods can be assigned different values, they are often used to represent the whole numbers from 1 to 10, as indicated.

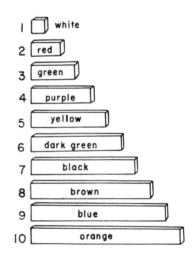

Numbers greater than 10 are represented by placing rods end to end to form a linear figure called a *train.* Here are two examples.

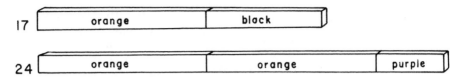

To determine the factors of a number, we can build trains with only one type of rod, called *one-color trains.* For example, the number represented by the top train in the following diagram has the number 2 as a factor because there is an all-red train equal to the length of this train. Furthermore, the fact that there are 13 red rods tells us that 13 is also a factor of the number represented by the top train. The number 3, on the other hand, is not a factor because a train of just green rods does not equal the length of the top train.

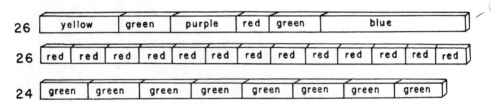

Every number can be represented by a one-color train of white rods because 1 is a factor of every number. When a number can be represented only by a one-color train of white rods, it is a prime number. The first of the following trains represents a prime number because one-color trains of red, green, and yellow will not equal its length. The top train in the preceding illustration represents a composite number because it is equal to the length of an all-red train.

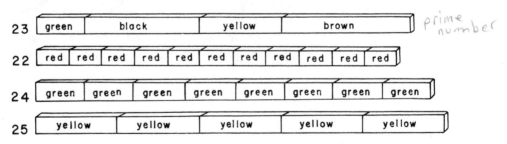

SUPPLEMENT (*Activity Book*)

Activity Set 3.5 Patterns on Grids (Sieves for locating primes, factors, and multiples)

Just for Fun: Spirolaterals

Additional Sources

Barnett, I.A. "The Fascination of Whole Numbers." *The Mathematics Teacher,* **64** No. 2 (February 1971), 103–8.

Beck, A., M.N. Bleicher, and D.W. Crowe. *Excursions into Mathematics.* New York: Worth, 1969. pp. 81–143 (prime numbers, factorization techniques, and perfect numbers).

Dantzig, T. *Number, the Language of Science,* 4th ed. New York: Macmillan, 1954. "Number Lore," pp. 36–56.

Eves, H.W. *Mathematical Circles Revisited.* Boston: Prindle, Weber and Schmidt, 1971. p. 34 (examples of beasting).

Gullen, III., G. "The Smallest Prime Factor of a Natural Number." *The Mathematics Teacher,* **67** No. 4 (April 1974), 329–32.

Schatz, M.C. "Of the Infinitude of Primes." *The Mathematics Teacher,* **68** No. 8 (December 1975), 676–77.

"I TEND TO AGREE WITH YOU — ESPECIALLY SINCE $6 \cdot 10^{-9}\sqrt{t_c}$ IS MY LUCKY NUMBER."

EXERCISE SET 3.5

1. Numerology is the study of the occult significance of numbers. It is taken lightly by most people, who consider it to be either a recreational diversion or a fraud. Numerology buffs sometimes choose names depending on the number of their name, with $a = 1$, $b = 2$, etc., according to this chart. Use the name by which you are best known and add the numbers corresponding to the letters.

1	2	3	4	5	6	7	8	9
A	B	C	D	E	F	G	H	I
J	K	L	M	N	O	P	Q	R
S	T	U	V	W	X	Y	Z	

a. Is your number prime or composite? Compare your number to that of a friend. Is the sum of these numbers a prime?

b. Add the digits of your number until there is only one digit left. This final digit is called the *digital root* of the number. The digital root of 84 is 3, because $8 + 4 = 12$ and $1 + 2 = 3$. The following table lists the numbers you are most compatible with.

Compatibility with other numbers

Your number	1	2	3	4	5	6	7	8	9
Best chance of happiness	2,6	4,6	1,8	7,9	5,3	6,2	9,4	6,2	4,7

★ 2. The following sequence of numbers increases by 2, then by 4, then 6, etc. Continue this sequence until you reach the first number that is not a prime.

17 19 23 29 37

3. The numbers 2, 3, 5, 7, 11, and 13 are not factors of 173. Explain why it is possible to conclude that 173 is prime without checking for more factors.

★ 4. The first ten prime numbers are 2, 3, 5, 7, 11, 13, 17, 19, 23, and 29. Which of these primes would you have to consider as possible factors of 367 to determine whether 367 is prime or composite?

★ 5. Which of the following numbers are prime?
   a. 231   b. 227   c. 187   d. 431

6. Sketch a factor tree that contains the prime factors for each of these numbers. Express each number as a product of primes.
★   a. 924   b. 364   c. 864

★ 7. It has been estimated that life began on earth 1,000,000,000 (one billion) years ago. How can the Fundamental Theorem of Arithmetic be used to show that 7 is not a factor of this number?

8. The top train is equal in length to the one-color train of green rods.

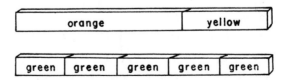

★   a. Name two other one-color trains that equal the length of this train.
   b. Name two factors of a number that is represented by seven blue rods (9 units).

★ c. If a number can be represented by an all-red train (2 units), an all-green train (3 units), and an all-black train (7 units), it has at least eight factors. Name these factors.

9. It is likely that no one will ever find a formula that will give all the primes less than an arbitrary number. The following formulas produce primes for awhile, but eventually composite numbers are obtained.

★ a. For which of the whole numbers for $n = 2$ to 7 is $2^n - 1$ not a prime?

b. The formula $n^2 - n + 41$ will give a prime for n equal to the whole numbers 1, 2, 3, ... up to 40 but not for 41. Which of the primes less than 100 are given by this formula?

10. *Conjectures:* There are many conjectures involving primes.

★ a. Goldbach conjectured that every odd number greater than 5 is the sum of three primes. Verify this conjecture for the following numbers.
   21 =   27 =   31 =

b. Arthur Hamann, a seventh grade student, conjectured that every even number is the difference between two primes.* Express the following numbers as the difference of two primes.
   10 = $13-3$ 12 = $17-5$ 14 = $19-5$

★ c. In 1845 the French mathematician Bertrand conjectured that between any whole number greater than 1 and its double there exists at least one prime. After 50 years this conjecture was proven true by the Russian mathematician Tchebyshev. For the numbers greater than 5 and less than 50, is it true or false that there are at least two primes between every number and its double?

11. The Italian mathematician Tartaglia (ca. 1499–1557) claimed that the sums $1 + 2 + 4 = 7, 1 + 2 + 4 + 8 = 15, 1 + 2 + 4 + 8 + 16 = 31, 1 + 2 + 4 + 8 + 16 + 32 = 63$, etc., are alternately prime and composite. Compute a few more of these sums. Prove that Tartaglia was wrong by finding two consecutive sums that are both composite.

12. The French mathematician Pierre Fermat (1601–1665) is referred to as the founder of modern number theory because of his many contributions to this subject. One of his theorems states that every prime of the form $4n + 1$ is the sum of two square numbers in one and only one way. For example, $29 = 4(7) + 1$, and it is the sum of the square numbers 4 and 25.

★ a. List the primes less than 50 which are of the form $4n + 1$, where n is a whole number.

b. Express each prime in part **a** as the sum of two square numbers.

*See M.R. Frame, "Hamann's conjecture," *The Arithmetic Teacher*, **23** No. 1 (January 1976), 34–35. A more general conjecture by A. de Polignac states that every even number is the difference of two consecutive primes in an infinite number of ways.

13. There are arbitrarily long sequences of consecutive whole numbers with no primes. The following questions will help you to see how such a sequence can be constructed.

★   a. The product $1 \times 2 \times 3 \times 4 \times 5 \times 6 \times 7 \times 8 \times 9 \times 10 = 3628800$ has the numbers from 2 through 10 as factors. Explain why the nine consecutive numbers 3628802, 3628803, etc., to 3628810, are all composite. (*Hint:* If a number is a factor of r and s, then it is a factor of $r + s$.)

   b. Denote the product of the first 100 consecutive whole numbers greater than zero by p. Use the hint in part **a** to explain why the following numbers are composite: $p + 19, p + 37$, and $p + 21$.

★   c. Explain how the product of the first 100 consecutive whole numbers greater than zero can be used to form a sequence of 99 consecutive composite whole numbers.

   d. Explain how to form a sequence of 1000 consecutive composite whole numbers.

14. *Infinitude of Primes:* The fact that there are infinitely many primes was proven by Euclid in the third century B.C. This classic proof was accomplished by showing that if you assume there is only a finite number of primes, this leads to a contradiction. This type of proof is called an *indirect proof*. Follow Euclid's reasoning by answering the following questions. We will assume that there are a finite number of primes, and denote them by $p_1, p_2, p_3, ..., p_n$.

★   a. A new number K is formed by multiplying all the primes together and adding 1 $(p_1 \times p_2 \times p_3 \times \cdots \times p_n + 1)$. Explain why K must be a composite number.

★   b. Use the Fundamental Theorem of Arithmetic to explain why at least one of the primes must divide K.

   c. If some prime divides K, explain why this prime must also divide 1.

   d. Since we know that no prime can divide 1, there is a contradiction in part **c**. Why can we conclude that there are an infinite number of primes?

15. *Hit and Run:*

Another American every 21 seconds

2 15,000,000

OUR CHANGING POPULATION

The Population Clock above registers the change of the American population— a summary of the components detailed below.

A BIRTH
every 10 seconds

A DEATH
every 16 seconds

AN IMMIGRANT
every 81 seconds

AN EMIGRANT
every 15 minutes

Census Clock, United States Department of Commerce Building, Washington, D.C.

3.6 FACTORS AND MULTIPLES

The census clock pictured above is located in the lobby of the U.S. Department of Commerce Building and is regulated by the Bureau of Census. This clock shows the estimated population of the United States at any given moment. Four illustrated clock faces display the components of population change: births, deaths, immigration, and emigration. Plus and minus signs above the clock faces light up to show the inputs of these components: a birth every 10 seconds, a death every 16 seconds, the arrival of an immigrant every 81 seconds, and the departure of an emigrant every 15 minutes.

The times when these lights flash together can be determined by using multiples of the different time periods of each clock. The plus sign above the birth clock lights at 10-second intervals, and the minus sign for the death clock lights every 16 seconds. If these two indicators both flashed at the same time, then they would flash together again after 80 seconds.

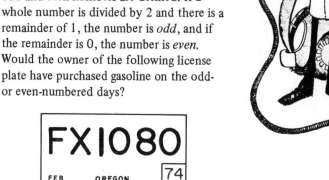

Birth clock intervals

0 10 20 30 40 50 60 70 80

Death clock intervals

0   16   32   48   64   80

Divisibility Tests   During the gasoline shortage of 1974, the state of Oregon adopted the "odd and even system." A car owner with an odd-numbered license plate could get gasoline on the odd-numbered days of the calendar, and those with even-numbered plates got gasoline on the even-numbered days. Some people whose license plate numbers ended in zero were confused as to whether their numbers were odd or even. The solution to this problem is in the way odd and even numbers are defined. If a whole number is divided by 2 and there is a remainder of 1, the number is *odd*, and if the remainder is 0, the number is *even*. Would the owner of the following license plate have purchased gasoline on the odd- or even-numbered days?

FX1080

FEB   OREGON   74

There are a few simple divisibility tests for determining whether or not a number is divisible by 2, 3, 4, 5, 6, or 9, without carrying out the division. These tests and some indications as to why they work are given in the following paragraphs.

Divisibility by 2 or 5—A number is divisible by 2 or 5 if the number represented by the units digit is divisible by 2 or 5. This means that a number is divisible by 2 if its units digit is 0, 2, 4, 6, or 8, and it is divisible by 5 if its units digit is 0 or 5.

To understand why these tests work, let's look at the expanded form for 5273.

$$5273 = 5 \times 10^3 + 2 \times 10^2 + 7 \times 10 + 3$$

Since 2 divides 10^3, 10^2, and 10, it will divide the bracketed portion of this equation. Therefore, 2 will divide the right side of the equation if and only if it divides the units digit, 3. Since 2 does not divide 3, it does not divide 5273.

Similarly, 5 divides the bracketed portion of the previous equation, but it does not divide 3. Therefore, 5 does not divide 5273.

Divisibility by 3 or 9–A number is divisible by 3 if the sum of its digits is divisible by 3. For example, 2847 is divisible by 3 since 3 divides $2 + 8 + 4 + 7$. Let's examine this test by looking at the expanded form for 2847.

$$2847 = 2 \times 10^3 \ + \ 8 \times 10^2 \ + \ 4 \times 10 \ + \ 7$$
$$= 2 \times (999 + 1) \ + \ 8 \times (99 + 1) \ + \ 4 \times (9 + 1) \ + \ 7$$
$$= \underbrace{2 \times 999 \ + \ 8 \times 99 \ + \ 4 \times 9} \ + \ (2 + 8 + 4 + 7)$$

The bracketed portion of this last equation is divisible by 3. Why? Therefore, to determine if 2847 is divisible by 3, it is necessary only to determine if the remaining portion of the equation, $2 + 8 + 4 + 7$, is divisible by 3.

The expanded form for 2847 also shows why the same test works for divisibility by 9. Since 9 divides the bracketed portion of the previous equation, it will divide 2847, if it divides $2 + 8 + 4 + 7$. In this case, 9 does not divide this sum and so it does not divide 2847.

Divisibility by 6–If a number is divisible by both 2 and 3, then it is divisible by 6, and if it is not divisible by both 2 and 3, then it is not divisible by 6. Apply this test to the following numbers. The first of these numbers is odd, and so it is not divisible by 2. Therefore, it is not divisible by 6. The test shows that the second number is not divisible by 6 because it is not divisible by 3. Test the remaining two numbers for divisibility by 6.

$$4,328,561,785 \qquad 1,000,000,000 \qquad 2,100,000,000 \qquad 123,090,534$$

Divisibility by 4–If the number represented by the last two digits of a number is divisible by 4, the number will be divisible by 4. The number 65,932 is divisible by 4 because 4 is a factor of 32. The expanded form shows why this test works for 65,932.

$$65,932 = \underbrace{6 \times 10^4 \ + \ 5 \times 10^3 \ + \ 9 \times 10^2} \ + \ 3 \times 10 \ + \ 2$$

Since 4 is a factor of 10^4, 10^3, and 10^2, it will divide the bracketed portion of the expanded form. Thus, whether or not 4 divides 65,932 depends only upon whether or not it divides 32.

Greatest Common Factor   For any two numbers there is always a number that is a factor of both. The numbers 24 and 36 both have 6 as a factor. When a number is a factor of two numbers, it is called a *common factor*. There are several common factors for both 24 and 36: 1, 2, 3, 4, 6, and 12. Among the common factors of two numbers there will always be a largest number, which is called the *greatest common factor.* The greatest common factor of 24 and 36 is 12. This is sometimes written, g.c.f. $(24,36) = 12$.

The greatest common factor of two numbers can be found from the prime factors of the two numbers. In this example, 2 occurs as a factor twice and 3 occurs as a factor once in *both* 60 and 72. Therefore, their greatest common factor is 2 × 2 × 3, or 12. For any two numbers the greatest common factor can be found by using the prime factors the greatest number of times they occur in both numbers.

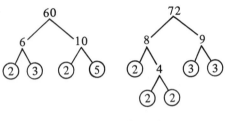

g.c.f. (60,72) = 12

Two numbers such as 24 and 35, which have no prime factors in common, are called *relatively prime.* In other words, two numbers are relatively prime if their greatest common factor is 1.

Least Common Multiple   When one number is a factor of a second, the second number is called a *multiple* of the first. Every number has an infinite number of multiples. Here are a few multiples of 5.

5   10   15   20   25   35   70   300   9000

A number is called a *common multiple* of two numbers if it is a multiple of both. The following numbers are multiples of 7, and the ones that are circled are multiples of both 7 and 5.

7   14   21   28   (35)   42   (70)   (280)   (7000)

Every pair of numbers has an infinite number of common multiples. For instance, 35, 70, 280, and 7000 are only a few of the multiples of both 5 and 7. One of these multiples is just the product of the two numbers. The smallest multiple of both 5 and 7 is 35. This can be illustrated on the number line by intervals or jumps of 5 and of 7. Beginning at 0, jumps of 5 and jumps of 7 do not coincide at the same point until after 35 intervals.

The smallest of the common multiples of two numbers is called the *least common multiple.* This can be found by listing the consecutive multiples of both numbers and selecting the first number that is a multiple of both. In the example shown here, the smallest multiple of both 14 and 18 is 126. This is written, l.c.m. (14,18) = 126.

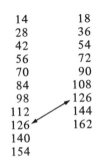

Sometimes it is easier to find the least common multiple of two numbers by building it up from the prime factors of the numbers. For example, 2, 3, 5, and 7 are factors of 210, and so any multiple of 210 must have these prime factors occurring at least once. Similarly, any multiple of 198 must have at least one factor of 2, two factors of 3, and one factor of 11. The smallest number satisfying these conditions for both 210 and 198 is $2 \times 3 \times 3 \times 5 \times 7 \times 11$ or 6930. That is, l.c.m. (210,198) = 6930. In general, each prime factor of the least common multiple must occur the maximum number of times it occurs in the two given numbers.

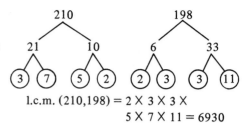

l.c.m. (210,198) = $2 \times 3 \times 3 \times$
$5 \times 7 \times 11 = 6930$

Once you have found the prime factors of two numbers, you may find the following schemes helpful for determining the greatest common factor and least common multiple.

$$198 = 2 \times 3 \times 3 \qquad\quad \times 11$$
$$210 = 2 \times 3 \quad\ \times 5 \times 7$$
$$\text{g.c.f.} = 2 \times 3$$

$$198 = 2 \times 3 \times 3 \qquad\quad \times 11$$
$$210 = 2 \times 3 \quad\ \times 5 \times 7$$
$$\text{l.c.m.} = 2 \times 3 \times 3 \times 5 \times 7 \times 11$$

Models for G.C.F. and L.C.M.   The concepts of greatest common factor and least common multiple can be illustrated by forming trains of rods. The following multicolored trains, representing 15 and 24, have the same lengths as two single-colored green trains. This shows that 3 is a factor of both 15 and 24. Since the green rods are the longest rods that can be used for making one-color trains whose lengths equal the lengths of the trains for 15 and for 24, the greatest common factor of these two numbers is 3.

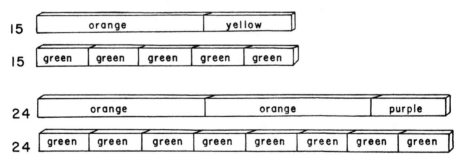

Least common multiple is illustrated by forming two different one-color trains of equal length. The train of yellow rods (5 units) and the train of dark green rods (6 units) shown next have the same length. Therefore, the number represented by these trains is a multiple of both 5 and 6. Since there are no shorter one-color trains of yellow and dark green rods which equal the same length, 30 is the least common multiple of both 5 and 6.

Casting Out Nines   Before the advent of calculators, it was a customary business practice to check the sums of columns of numbers by "casting out nines." This procedure was also used by schoolchildren to check computations involving addition, subtraction, and multiplication.

Casting out nines is the process of determining the remainder when a number is divided by 9. This remainder will be called the *nines excess* of the number. The nines excess can be easily found by using the test for divisibility by 9. For example, the nines excess for 47,281 is 4, because $4 + 7 + 2 + 8 + 1$ leaves a remainder of 4 when divided by 9.

The nines excess can be used to check the accuracy of any computation involving only addition, subtraction, and multiplication. For example, if addition is carried out with the nines excesses in place of the original numbers, the nines excess of this sum will equal the nines excess of the sum of the original numbers. A similar check works for subtraction and multiplication. Here are three examples.

	Excess
4728	3
3147	6
2096	8
+ 4731	+ 6
14702	23

The nines excess for both 14,702 and 23 is 5.

	Excess
2865	3
× 778	× 4
2228970	12

The nines excess for both 2,228,970 and 12 is 3.

	Excess
43891	7
− 6237	− 0
37654	7

The nines excess for both 37,654 and 7 is 7.

If there is a mistake in a computation, the chances are fairly good that it can be found by casting out nines. It is possible, however, for a mistake to occur and not be detected by this method. This will happen when the computed result differs by a multiple of 9 from the correct answer. In the example shown here, casting out nines does not indicate an error because the correct product, 41667, and the answer shown, 41217, differ by 450, which is a multiple of 9.

	Excess
731	2
× 57	× 3
41217	6

The nines excess for both 41217 and 6 is 6.

SUPPLEMENT (*Activity Book*)

Activity Set 3.6 Factors and Multiples with Cuisenaire Rods

Just for Fun: Star Polygons

Additional Sources

Beiler, A.H. *Recreations in the Theory of Numbers.* New York: Dover, 1966.

Dubisch, R. "Generalizing a property of prime numbers." *The Arithmetic Teacher,* **21** No. 2 (February 1974), 93–94 (determining the total number of factors of a number).

Madachy, J.S. *Mathematics on Vacation.* New York: Charles Scribner's Sons, 1966. "Number Recreations," pp. 178–200.

Morton, R.L. "Divisibility by 7, 11, 13, and Greater Primes." *The Mathematics Teacher,* **61** No. 4 (April 1968), 370–73.

Smith, F. "Divisibility rules for the first fifteen primes." *The Arithmetic Teacher,* **18** No. 2 (February 1971), 85–87.

EXERCISE SET 3.6

1. Two new skyscrapers in Boston and Chicago will have double-deck elevators to minimize the number of necessary elevator shafts. People entering the buildings will use the street level if they are going to odd-numbered floors, or take a moving stairway to a mezzanine for even-numbered floors.

   a. Suppose you had to deliver packages to floors 11, 26, 35, and 48. How can this be done by doing the least amount of elevator riding and walking only one flight of stairs?

   ★ b. Describe an efficient scheme for delivering to any number of odd- and even-numbered floors.

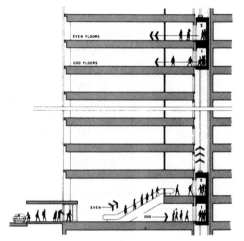

2. The United States Census Clock has flashing light signs to indicate gains and losses in the population. Here are the time periods of these flashes in seconds: Birth, 10; Death, 16; Immigrant, 81; and Emigrant, 900. If these lights all started flashing at the same moment, the times when they flash together can be computed by using the concept of least common multiple.

   ★ a. If you saw the birth and emigrant signs light up at the same time, how many seconds would pass before they would both light together again?

   b. Suppose you saw the immigrant and emigrant signs both light at the same time. What is the shortest time before they will both be lighted together again?

   ★ c. It is possible for the birth and death signs to light together every 80 seconds. What is the increase in population during each 80-second period due to births and deaths alone?

3. The trains for both 24 and 18 are each equal to the lengths of all-red trains (a red rod represents 2).

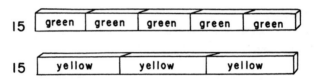

★   a. Name another type of rod that can be used to represent both 24 and 18 by one-color trains.

b. What kind of rods can be used to form one-color trains that represent both 24 and 18?

★   c. What can be said about the relationship between two numbers if the only one-color trains representing these numbers contain white rods (a white rod represents 1)?

4. The following all-green train equals the length of the all-yellow train. Does this give information about common factors or common multiples?

| 15 | green | green | green | green | green |

| 15 | yellow | yellow | yellow |

★   a. If the purple rods (4 units) and the black rods (7 units) are used to form the shortest possible one-color trains that both have the same length, how many purple rods and how many black rods will be required?

b. If an all-brown train (brown represents 8 units) equals the length of an all-orange train (orange represents 10 units), what can be said about the number of brown rods?

c. If the two trains in part b are the shortest possible, how many brown rods will be required?

5. Each rectangular array of squares gives information about the number of factors of a number. Two rectangles can be formed for the number 6, which shows that 6 has factors of 2, 3, 1, and 6. Sketch as many different rectangular or square arrays as possible for each of the following numbers.

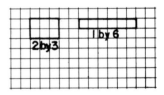

15   16   30   25   11

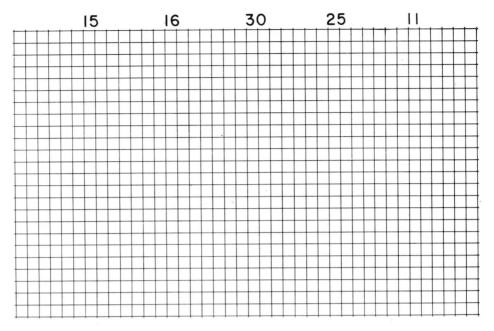

★   a. What kinds of numbers will have only one rectangular array?

       b. Which three of the given numbers have an even number of factors? *15, 30, 11*

★   c. Two of the given numbers have square arrays. Make a conjecture about the number of factors for square numbers.

       d. *Challenge:* Find a number with 9 factors. Explain how numbers with 11, 13, 15, or any other odd number of factors can be found.

6. List the factors for the numbers in this table. Include 1 and the number itself as factors.

★   a. Find the numbers having only two factors. What kind of numbers are these?

       b. Which numbers have an odd number of factors? What kind of numbers are these?

7. The factors of a number which are less than the number are called *proper factors*. The Pythagoreans classified numbers as <u>deficient</u>, <u>abundant</u>, or <u>perfect</u>, depending upon the sum of their proper factors.

Number	Factors	Number of factors
1	1	1
2	1,2	2
3	1,3	2
4	1,2,4	3
5	1,5	2
6	1,2,3,6	4
7	1,7	2
8	1,2,4,8	4
9	1,3,9	3
10	1,2,5,10	4
11	1,11	2
12	1,2,3,4,6,12	6
13	1,13	2
14	1,2,7,14	4
15	1,3,5,15	4
16	1,2,4,8,16	5
17	1,17	2
18	1,2,3,6,9,18	6
19	1,19	2
20	1,2,4,5,10,20	6

★ a. If the sum of the proper factors of a number is less than the number, it is called *deficient*. Circle all of the deficient numbers in the table.

b. If the sum of the proper factors of a number is greater than the number, it is called *abundant*. Which numbers in the table are abundant?

★ c. If the sum of the proper factors of a number is equal to the number, it is called *perfect*. Find the one perfect number in this table.

8. Circle the numbers that are divisible by 3. Can the test for divisibility by 3 be used to determine the remainder when dividing by 3?

   a. 465,076,800   b. 100,101,000   c. 907,116,341

9. Which number in Exercise 8 is divisible by 9?

★ a. If a number is divisible by 3, is it divisible by 9?

   b. If a number is divisible by 9, is it divisible by 3?

★ 10. Which of the following numbers are divisible by 4? How can the last two digits be used to determine the remainder when dividing by 4?

   47,382   512,112   14,710   4,328,104,292

11. To test for divisibility by 11, alternately add and subtract the digits from right to left beginning with the units digit, that is, units digit minus tens digit plus hundreds digit, etc. If the result is divisible by 11, then the original number will be divisible by 11. Use this test on the following numbers.

   a. 63,011,454   b. 19,321,488   c. 4,209,909

★ d. Will the test for divisibility by 11 work if the digits are alternately added and subtracted from left to right?

12. Here is a test for divisibility by 7, which works quite nicely with a little practice.

   Remove the last digit, double it, and subtract it from the remaining number. Continue this process until one or two digits remain. If the remaining digits are divisible by 7, then the original number will be also.

$$
\begin{array}{r}
173\!\!\!\diagup \\
12 \\
\hline
16\!\!\!\diagup \\
2 \\
\hline
14
\end{array}
$$

Since 14 is divisible by 7, 1736 is also.

   Try this test on the following numbers. Check your answers by dividing each number by 7.

   a. 3941   b. 14021   c. 17976

13. Draw factor trees for the numbers in parts **a** and **b**, and compute the greatest common factor for each pair of numbers.

★   a.   198   165   b.   280   168

     g.c.f. (198,165) =   g.c.f. (280,168) =

14. Draw factor trees for these numbers and compute the least common multiple for each pair of numbers.

★   a.   30   42   b.   22   56

     l.c.m. (30,42) =   l.c.m. (22,56) =

15. Find two numbers for part **a** and two numbers for part **b** which can be placed in the boxes to make true statements.

★   a.   g.c.f. (18,□) = 3

         b.   l.c.m. (□,9) = 45
                 5 + 15

★ 16. What is the smallest number divisible by 1, 2, 3, 4, 5, 6, 7, 8, 9, and 10?

17. If 8 and 12 both divide a number k, name four other numbers, different from 1, which are factors of k.

18. *Casting Out Nines:* In each of the two examples in parts **a**, **b**, and **c**, one answer is correct and one is incorrect. Use casting out nines to determine the incorrect example.

★ a.
      64796   324862
    × 1560   × 316
  101181760   102656392

  b.
     496742317   463798260
     341264682   281417863
   + 284651906   + 945516792
   1122658905   1680732915

  c.
     681728906   361972482
   − 419847268   − 129865396
    261481638   232107086

★ d. In positional numeration, 2^{64} has 20 digits. Use casting out nines to determine the remainder when this number is divided by 9.

19. *Number Curiosity:* Take any five-digit number and rearrange its digits in any order to form another number. The difference between the two numbers is always divisible by 9. Will this be true for numbers with three or four digits?

      73108
   −30871
    42237

20. *Birthdate Trick:* Add the numbers of the day, the month, and the year you were born: Subtract 23 times the day; add 21 times the month; add your age on December 31, 1980. If this result is divisible by 11, then you were born on a lucky date.

21. *Card Puzzle:* Place ten cards face up as shown.

    | 1 | | 2 | | 3 | | 4 | | 5 | | 6 | | 7 | | 8 | | 9 | | 10 |

Turn over every card, then every second card (Cards 2, 4, 6, etc.), then every third card, and so on, until every tenth card is turned over. Which cards will be face down?

★ a. Repeat this process for 16 cards. Which cards will be face down?

  b. If this process is repeated for any number of cards, which cards will be face down?

22. *Multiples Game* (2 Players): Place 25 objects on a table. The players take turns removing either 1, 2, or 3 objects until there are no objects left. The player who takes the last object loses the game. Describe a winning strategy for this game.

★ 23. *Locker Room Puzzle:* In a new school built for 1000 students there were 1000 lockers that were all left closed. As the students entered the school, they decided on the following plan. The first student who entered the building opened all of the 1000 lockers. The second student closed all lockers with even numbers. The third student "changed" all lockers that were numbered with multiples of 3 ("changed" means opening those that were closed or closing those that were open). The fourth student changed all lockers that were numbered with multiples of 4, the fifth changed multiples of 5, etc. After 1000 students had entered the building and "changed" the lockers according to this pattern, which lockers were left open?

CHAPTER FOUR

GEOMETRIC FIGURES

*It is nature herself, and not the mathematician,
who brings mathematics into natural philosophy.*

Immanuel Kant

Crystals of calcite

4.1 PLANE FIGURES

The assorted hexagonal prisms pictured above are crystals of calcite. The sharp edges and corners of these six-sided figures occur naturally in this mineral. The flat, thin crystals in the photo at the top of the next page are of the mineral wulfenite. This mineral usually forms these thin plates with bevelled edges. The angles between the bevelled edges and the flat faces of this crystal are always equal.

The study of the relationships between lines, angles, surfaces, and solids is part of geometry. This is one of the earliest branches of mathematics. More than 4000 years ago the Egyptians and Babylonians were using geometry in surveying and in architecture.

These ancient mathematicians discovered geometric facts and relationships by experimentation and inductive reasoning. Because of this approach they could never be sure of their conclusions, and in some cases their formulas for area and volume were inaccurate. The Greeks, on the other hand, viewed points, lines, and figures as abstract concepts about which they could reason deductively. They were willing to experiment in order to formulate ideas, but final acceptance of a mathematical statement depended on proof by deductive reasoning.

Crystals of wulfenite

Points, Lines, and Planes The fundamental notion in geometry is that of a *point.* All geometric figures are sets of points. Points are abstract ideas which we illustrate by dots, corners of boxes, and tips of pointed objects. These concrete illustrations have width and thickness, but points have no dimensions. The following description of a point, from *Mr. Fortune's Maggott,* by Silvia Townsend Warner, indicates some of the pitfalls associated with this concept.*

> Calm, methodical, with a mind prepared for the onset, he guided Lueli down to the beach and with a stick prodded a small hole in it.
> "What is this?"
> "A hole."
> "No, Lueli, it may seem like a hole, but it is a point."
> Perhaps he had prodded a little too emphatically. Lueli's mistake was quite natural. Anyhow, there were bound to be a few misunderstandings at the start.
> He took out his pocket knife and whittled the end of the stick. Then he tried again.

*J. R. Newman, *The World of Mathematics,* **4** (New York: Simon and Schuster, 1956), p. 2254.

"What is this?"

"A smaller hole."

"Point," said Mr. Fortune suggestively.

"Yes, I mean a smaller point."

"No, not quite. It is a point, but it is not smaller. Holes may be of different sizes, but no point is larger or smaller than another point."

A *line* is a very specific set of points which we some- times intuitively describe as "straight" and by saying it

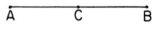

extends indefinitely in both directions. The two parallel rows which Peter is drawing in the previous cartoon, the edges of boxes, and taut pieces of string or wire are all models of lines. The above line passes through points A and B and is denoted by $\overrightarrow{AB}$. The arrows indicate that the line continues indefinitely in both directions. If two or more points are on the same line, they are called *collinear*.

A *plane* is another basic set of points which we will not define precisely. Intuitively, we describe a plane as being "flat," like a sheet of paper and extending indefinitely. The tops of tables and the surfaces of floors and walls are common models for planes.

Points, lines, and planes are undefined terms in geometry which are used to define other terms and geometric figures. The following paragraphs contain some of the more common definitions.

Line Segments, Rays, and Angles

Line Segments—A *line segment* consists of two points on a line and all the points between them. The line segment with *end points* A and B is denoted by $\overline{AB}$. To *bisect* a line segment means to divide it into two parts of equal length. The *midpoint C* bisects $\overline{AB}$.

Rays—A *ray* consists of a point on a line and all the points on one side of this point. The ray which begins at D and contains point E is denoted by $\overrightarrow{DE}$. The point D is the *end point* of $\overrightarrow{DE}$.

Angles—An *angle* is the union of two rays having a com- mon end point. This end point is called the *vertex,* and the rays are called the *sides of the angle.* The angle with vertex G, whose sides contain points F and H, is denoted by $\angle FGH$. Point K is in the *interior* of $\angle FGH$.

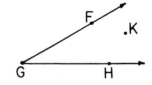

Referring to the accompanying photo, which of these geometric terms are illustrated in this art form by Kenneth Snelson?

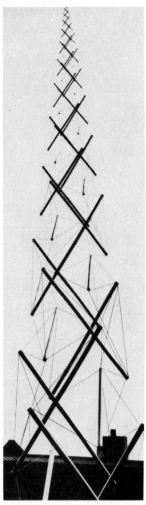

Needle Tower—Aluminum and stainless steel rods and wires, by Kenneth Snelson

Measure of Angles—The ancient Babylonians devised a method for measuring angles by dividing a circle into 360 equal parts. Each angle has a measure, called *degrees,* which is some number between 0 and 360. The circular protractor in the diagram on the right shows that there are 20 degrees (20°) in ∠*KET.* To measure any angle, the vertex is placed at the center of the protractor with one side of the angle lying on the center line. If an angle has a measure of 90°, its sides are *perpendicular* and it is

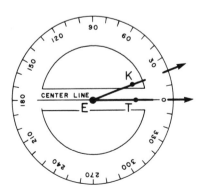

called a *right angle.* If it is less than 90° it is called an *acute angle,* and if it is greater than 90° and less than 180° it is called an *obtuse angle.* Occasionally, we will need measures of angles which are greater than 180°. For example, angles in polygons are sometimes greater than 180°, as shown by ∠ *ABC* in this figure. Such an angle is called a *reentrant angle.* To indicate an angle greater than 180°, a circular arc will be drawn to connect the two sides of the angle. If the sum of two angles is 90°, the angles are called *complementary;* if their sum is 180°, they are called *supplementary.*

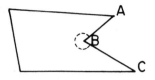

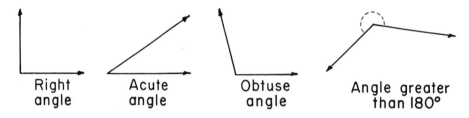

Right angle   Acute angle   Obtuse angle   Angle greater than 180°

Curves and Convex Sets   A *curve* is a set of points which can be drawn by a single continuous motion.

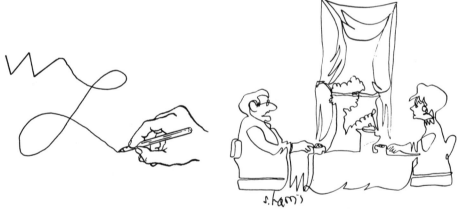

"You don't see many drawings made with one continuous line anymore."

Several types of curves are shown following this paragraph. Curve A is called a *simple curve* because it starts and stops without intersecting itself. Curve B is a *simple closed curve* because it is a simple curve which starts and stops at the same point. Curve C is a *closed curve,* but since it intersects itself it is not a simple closed curve.

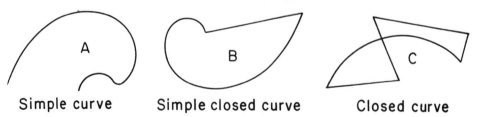

Simple curve   Simple closed curve   Closed curve

There is a well-known theorem in mathematics, called the *Jordan curve theorem,* which states that every simple closed curve partitions the plane into three disjoint sets: the points on the curve, the points in the interior, and the points in the exterior. This means that if K is in the interior and M is in the exterior, then $\overline{KM}$ will intersect the curve.

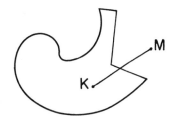

Convex Sets—The union of a simple closed curve and its interior is called a *plane region.* These regions can be classified as nonconvex and convex. You may have heard the word "concave," rather than "nonconvex." An object is concave if it is "caved in," such as set F, and convex if it is not, such as set G.

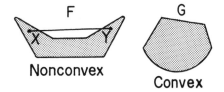

Nonconvex

Convex

To be more mathematically precise, we say that a set is *convex* if the line segment joining any two points of the set lies completely in the set. Set F is *nonconvex* because $\overline{XY}$ is not completely in the set. An intuitive way of testing for convexity of plane regions is to think about stretching a rubber band around the boundary of the set. If it touches all points on the boundary (as it will for set G), the set is convex, and if not, as for set F, the set is nonconvex.

A *circle* is a special case of a simple closed curve whose interior is a convex set. Each point on a circle is the same distance from a fixed point called the *center.* This distance is the *radius* of the circle, and twice this distance is the *diameter* of the circle. A line segment from a point on the circle to its center is also called a *radius,* and a line segment from two points on the circle which passes through the center is a *diameter.* The distance around the circle is the *circumference.* The union of a circle and its interior is called a *disc.*

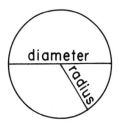

Polygons   A *polygon* is a simple closed curve which is the union of line segments. Polygons are classified according to the number of line segments. Here are a few examples.

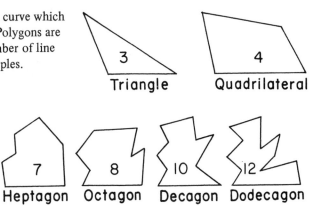

The line segments of a polygon are called *sides,* and the end points of these segments are *vertices.* Any line segment connecting two nonadjacent vertices is a *diagonal.*

Triangles have been the most thoroughly studied of all the polygons. One reason for this is the importance of triangular supports in buildings, bridges, and other structures. A triangle is more "rigid" than the other polygons, in that its shape is determined by its three sides. This can be illustrated by linkages, such as those shown here. The shapes of all these polygons can be changed except for that of the triangle. For example, the pentagon can be made nonconvex by pushing one of its vertex points into the interior of the polygon. Similarly, the hexagon in the illustration can be reshaped into a convex hexagon. The triangle is the only one of these linkages whose shape cannot be changed.

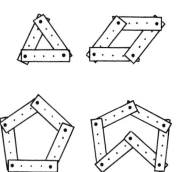

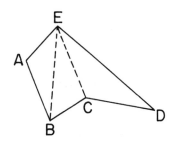

Another reason for the extensive study of triangles is that every polygon of four or more sides can be subdivided into triangles by drawing diagonals. Polygon *ABCDE* is subdivided into 3 triangles by diagonals $\overline{BE}$ and $\overline{CE}$. Information about these triangles, such as angle sizes or area, can be used to obtain similar information about the polygons.

Several types of triangles and quadrilaterals occur often enough to be given special names.

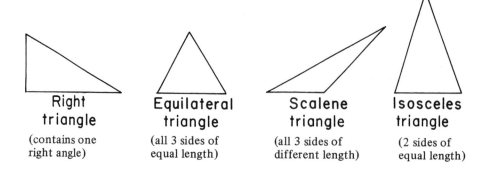

Right triangle	Equilateral triangle	Scalene triangle	Isosceles triangle
(contains one right angle)	(all 3 sides of equal length)	(all 3 sides of different length)	(2 sides of equal length)

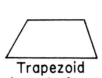

Trapezoid
(one pair of
opposite parallel
sides)

Rhombus
(opposite sides
parallel and
all sides of
equal length)

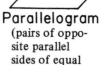

Parallelogram
(pairs of oppo-
site parallel
sides of equal
length)

Rectangle
(pairs of oppo-
site sides paral-
lel and of equal
length, and all
right angles)

Square
(all sides of
equal length
and all
right angles)

Congruence   The intuitive idea of congruence is quite simple: Two plane figures are congruent if one can be placed on the other so that they coincide. Another way to describe congruent plane figures is to say that they have the same size and shape. (See Section 9.1 for more details on congruence.)

For line segments and angles we can be more precise about congruence. Two *line segments are congruent* if they have the same length, and two *angles are congruent* if they have the same number of degrees.

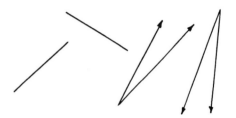

SUPPLEMENT *(Activity Book)*

Activity Set 4.1 Rectangular and Circular Geoboards
Just for Fun: Tangram Puzzles

Additional Sources

Bruni, J.V. *Experiencing Geometry.* Belmont, California: Wadsworth, 1977. "From Points to Polygons," pp. 1–25.

Madachy, J.S. *Mathematics on vacation.* New York: Charles Scribner's Sons, 1966, pp. 15–33 (geometric dissections).

Meggison, G.W. "Rays and angles." *The Arithmetic Teacher,* **21** No. 5 (May 1974), 433–35.

Murrow, G. "A Geometric Application of the Sheperd's Principle." *The Mathematics Teacher,* **64** No. 8 (December 1971), 756–58.

Olson, A.T. "Some Suggestions for an Informal Discovery Unit on Plane Convex Sets." *The Mathematics Teacher,* **66** No. 3 (March 1973), 267–69.

"I am not sure it is of great value in life to know how many diagonals an *n*-sided figure has. It is the method rather than the result that is valuable."

W. W. Sawyer

EXERCISE SET 4.1

1. This picture of a cross section of natural sapphire shows angles that each have the same number of degrees.

   a. Are these angles acute or obtuse?

   ★ b. Approximately how many degrees are there in these angles?

Natural sapphire

2. The next picture shows three crystals of the mineral staurolite. The crystal on the right is known as the "Fairy Stone" of the Appalachian Mountains. These stones are found in all parts of the world. They are especially common in the Shenandoah Valley. This form and the one on the left are often imitated by jewelers.

   a. The ridges on the top of the Fairy Stone form lines that intersect in equal angles. How many degrees are in each of these angles?

   ★ b. The ridges on the top of the crystal on the left form three lines that intersect in six equal angles. How many degrees are in each of these angles?

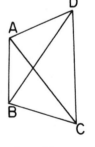

Crystals of staurolite

3. For the points *A*, *B*, *C*, and *D* there are 6 line segments which have these points as end points.

   ★ a. Draw line segments from *E* to *A*, *B*, *C*, and *D*. How many new line segments are there? What is the total number of line segments for these 5 points?

   ★ b. How many more line segments will there be if point *F* is used? What is the total number of line segments having 6 points as end points?

   c. Find a pattern and complete the table to show the number of line segments which have the given numbers of points as end points.

Number of points	2	3	4	5	6	7	8	9	10	15
Number of segments	1	3	6							

d. You probably recognize the numbers 1, 3, 6, etc., as being the triangular numbers (see page 8). The nth triangular number is $n(n + 1)/2$. Use this formula to find the 98th triangular number.

★ 4. Draw the diagonals for each polygon. Find a pattern and complete the table.

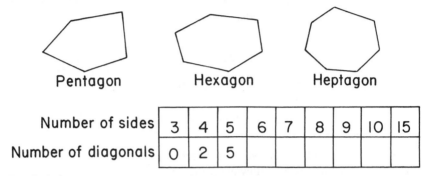

Pentagon   Hexagon   Heptagon

Number of sides	3	4	5	6	7	8	9	10	15
Number of diagonals	0	2	5						

Compare this table with the one in Exercise 3. How can the number of diagonals for a polygon be found by knowing the number of line segments between its vertices?

5. The number of line segments between a given number of points has many practical applications. One of these involves the remarkable accomplishments of the Bell System. The fundamental problem was that of connecting two subscribers who wanted to talk. This was done by cords and plugs.

★ a. In 1884 Ezra T. Gilliland devised a mechanical system that would allow 15 subscribers to reach each other without the aid of an operator. How many line segments are needed to connect 15 points?

b. In 1891 Almon B. Strowger patented a dial machine which connected up to 99 subscribers. How many different two-party calls does this permit?

6. Three lines in a plane may intersect in 0, 1, 2, or 3 points.

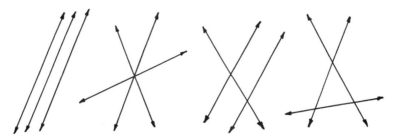

★ a. What are the possible numbers of points of intersection for 4 lines in a plane?

b. What are the possible numbers of points of intersection for 5 lines in a plane?

7. Here are 3 lines which intersect in 3 points.

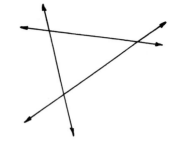

★ a. Draw a fourth line which intersects each of these lines in distinct points. How many new points of intersection are there? What is the maximum number of points of intersection for 4 lines?

★ b. Draw a fifth line which intersects each of these 4 lines in new points of intersection. What is the maximum number of points of intersection for 5 lines?

c. Complete the table to show the maximum number of points in which the given numbers of lines will intersect.

Number of lines	2	3	4	5	6	7	8	15
Number of intersections	1	3						

8. The 2 intersecting lines shown here partition this page into 4 disjoint regions.

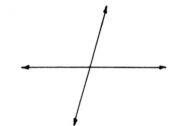

★ a. Draw a third line which intersects these 2 lines in 2 new points. How many more regions are created by the third line? What is the total number of regions?

★ b. Draw a fourth line which intersects these 3 lines in 3 new points. How many additional regions are created? What is the total number?

c. Find a pattern and use it to determine the maximum number of regions for each number of lines in this table.

Number of lines	1	2	3	4	5	6	7	8	15
Number of regions	2	4							

9. The white path in this ornament from the Middle Ages is a curve. Is it a simple curve? Is it closed?

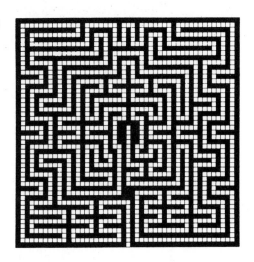

10. Use the angles in these polygons to answer the questions that follow.

Which angles are:

★   a. Acute? b. Obtuse?

★   c. Right angles?

    d. Which angle is a reentrant angle (greater than 180 degrees)?

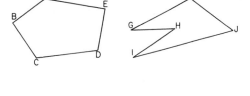

★11. Figure (a) has 3 disjoint regions which are bounded by polygons. This figure has 7 sides and 5 vertices. Fill in the table for Figures (b) through (e), and find a formula relating the numbers of regions, sides, and vertices. Draw a few more figures and check your conclusion.

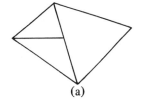

(a)

	(a)	(b)	(c)	(d)	(e)
Regions	3				
Sides	7				
Vertices	5				

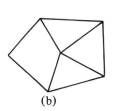
(b)

(c)

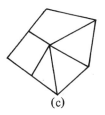
(d)

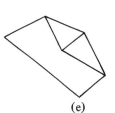
(e)

★ 12. *Matchstick Puzzle:* Using only four more matchsticks, divide this region into four congruent regions. (*Hint:* Some of the matchsticks may be broken.)

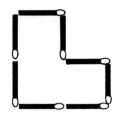

13. *Tracing Puzzle:* This figure is a curve, because it can be traced without lifting your pencil from the paper. Find a way to do this without retracing any lines. Indicate your starting point and path by arrows.

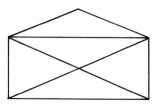

14. *Utilities Puzzle:* A, B, and C are houses and E, G, and W represent sources of electricity, gas, and water. Try connecting these houses with each utility by drawing lines or curves so that they do not cross each other. It is possible to make only 8 of the 9 connections. Draw these 8 connections.

★   a. Some of your connections will form a simple closed curve for which the remaining unconnected house and utility are on opposite sides of this curve. Find this curve and mark it with dark lines.

b. How does the Jordan curve theorem show that all 9 connections cannot be completed?

15. *Dissection Puzzle:* Trace and cut out the five pieces of this polygon. Reassemble these pieces to form a square.

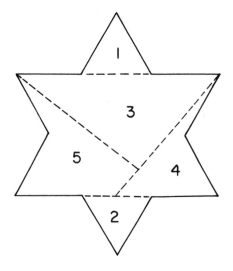

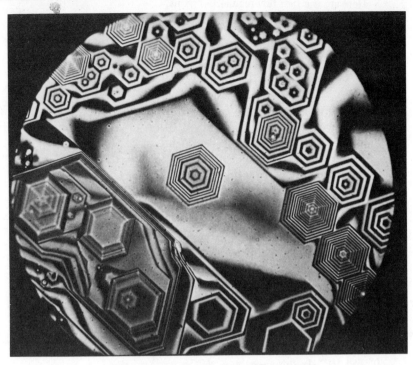

Cross section of cadmium sulfide crystals

"The Great Architect of the Universe now appears as a pure mathematician."
James H. Jeans

4.2 PROPERTIES OF POLYGONS

Regular Polygons   We have become so accustomed to hearing about the regularity of patterns in Nature that it is often taken for granted. Still, it is a source of wonder to see polygons with straight edges and uniform angles such as those which occur in these photos of cadmium sulfide and tourmaline. The details and symmetry of the hexagons in the cadmium sulfide remind us of snow crystals. Equally impressive are the triangles which occur in the cross section of tourmaline. In each of these hexagons and triangles, the sides have equal length and the angles have the same number of degrees. Polygons satisfying these two requirements are called *regular polygons*.

Cross section of the gem tourmaline

Here are the first few regular polygons.

Equilateral   Square   Regular   Regular   Regular
triangle   pentagon hexagon heptagon

It is necessary to require both of the following conditions for regular polygons, namely: 1) sides of equal length; and 2) angles with the same number of degrees. The rhombus is not a regular polygon, because it satisfies condition 1 but not condition 2. On the other hand, although the hexagon satisfies condition 2, it is not regular because it does not satisfy condition 1.

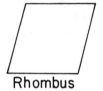 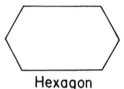

Rhombus   Hexagon

Angles in Polygons   The angles in a polygon with four or more sides can be any size between $0°$ and $360°$. In this hexagon, $\angle ABC$ is less than $20°$ and $\angle EDC$ and $\angle FAB$ are both greater than $180°$. In spite of this degree of freedom, there is a relationship between the sum of the angles in a polygon and its number of sides. Let's begin by considering the triangle.

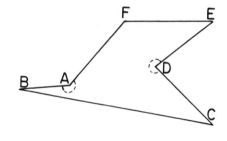

The sum of the angles in a triangle is $180°$. This fact was proven by the Greek mathematicians of the fourth century B.C. One method of illustrating this theorem is to draw an arbitrary triangle and cut off its

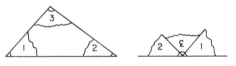

angles. When these angles are placed side by side, with their vertices at a point, they form one-half of a revolution ($180°$) about the point.

The sum of the angles in a polygon of four or more sides can be found by subdividing the polygon into triangles. This quadrilateral is partitioned into two triangles whose angles are numbered from 1 through 6. The sum of all six angles is 2 × 180 or $360°$. The sum of the four angles of the quadrilateral is also $360°$ since its angles are made up of these six angles.

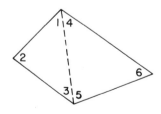

There is an infinite variety of quadrilaterals, some convex and others nonconvex. However, since each quadrilateral can be partitioned into two triangles, the sum of the angles will always be 360°. A similar approach can be used to find the sum of the angles in any polygon.

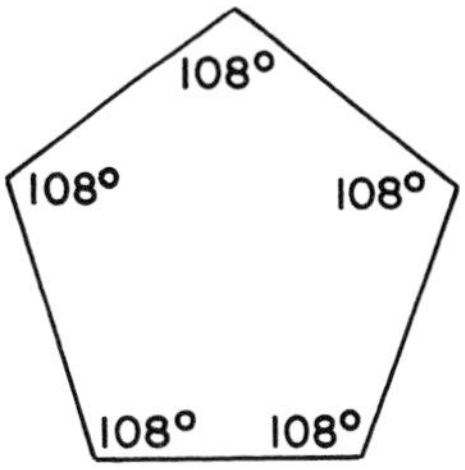

Constructing Regular Polygons   The sum of the angles in a polygon can be used to compute the number of degrees in each angle of a regular polygon. Simply divide the sum of the angles by the number of angles. For example, the sum of the angles in a pentagon is 3 × 180, or 540, because it can be subdivided into three triangles. Therefore, each angle in a regular pentagon is 540 ÷ 5, or 108°. The regular pentagon which is shown here was constructed by using a protractor to draw angles of 108°. Here are the first three steps in constructing a regular pentagon.

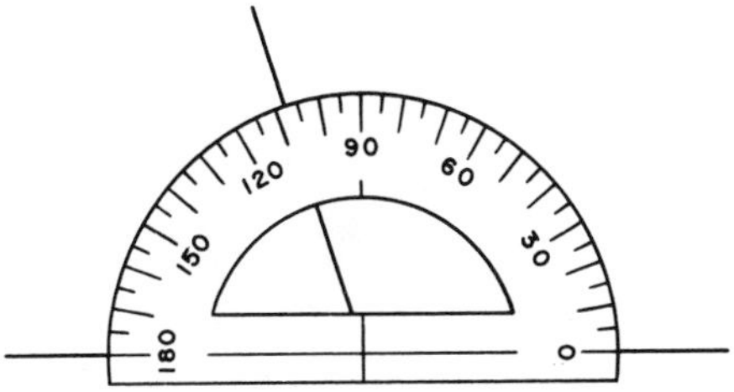

Step I— Measure off 108° angle

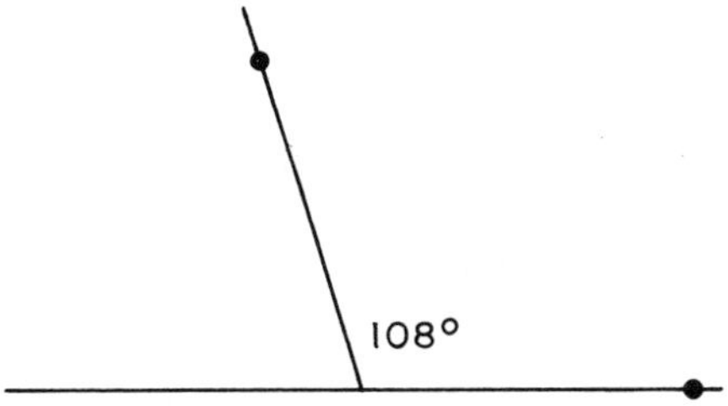

Step 2—Mark off two sides of equal length

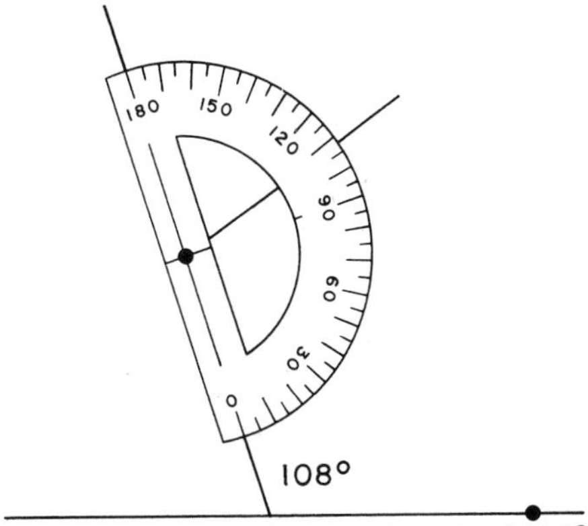

108°

Step 3—Measure off second angle of 108°

For another approach to constructing regular polygons, we can begin with a circle and use central angles. The number of degrees in a *central angle* of a regular polygon is the number of sides of the polygon divided into 360. For a decagon, each central angle is 360 ÷ 10, or 36°.

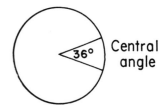
Central angle

Here is a four-step sequence for constructing a regular decagon.

 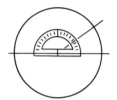 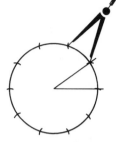 

Draw circle
with compass

Measure off
36° angle

"Pace off" 10
equal arcs

Connect the
10 points

Tessellations with Polygons   The hexagonal grid of the honeycomb provides another remarkable example of regular polygons in Nature. Bee eggs can be seen in the cells of the comb in the accompanying photo. These cells show that regular hexagons can be placed side by side with no uncovered gaps between them. Any arrangement of nonoverlapping figures that can be placed together to entirely cover a region is called a *tessellation.* Floors and ceilings are often *tessellated* or *tiled* with material that is square, because squares can be joined together without gaps or overlapping. Equilateral triangles are also commonly used for tessellations.

Honeycomb with bee eggs.
Courtesy of the American Museum of
Natural History.

From ancient times, tessellations have been used as patterns for rugs, fabrics, pottery, and architecture. The Moors of Spain were masters of tessellating walls and floors with colored geometric tiles. Some of their work is shown in the following picture of a room and bath in the Alhambra, a fortress palace built in the middle of the thirteenth century for Moorish kings. (There is another picture of this palace on page 203.)

A room (Sala de Camas) in the Alhambra in Granada, Spain

The two tessellations in the center of the Alhambra photo have figures that are curved. In the following paragraphs, however, we will concern ourselves only with polygons that tessellate. The triangle is an easy case to consider first. Any triangle will tessellate by simply putting together two copies of the triangle to form a parallelogram (see shaded region). Copies of the parallelogram can then be moved horizontally and vertically. The points of tessellation at which the vertices of the triangle

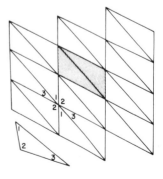

meet are called the *vertex points* of the tessellation. Since the sum of the angles in a triangle is 180°, the 360° about each vertex point of the tessellation will be filled in by using each angle of the triangle twice. In this tessellation of triangles, angles 1, 2, and 3 occur twice about each vertex point.

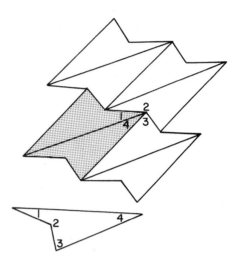

The size of the angles in a polygon and the sums of these angles will determine whether or not the polygon will tessellate. The fact that the sum of the angles in a quadrilateral is 360° suggests that quadrilaterals have the right combinations of angles to fit around each vertex point of a tessellation. In this tessellation, each angle of the quadrilateral (angles 1, 2, 3, and 4) occurs once about each vertex point of the tessellation. As in the case of triangles, two copies of the quadrilateral were matched together (see shaded region) and then moved horizontally and vertically to fill out the plane.

The quadrilateral in the preceding tessellation is nonconvex. It is quite surprising that every quadrilateral, convex or nonconvex, will tessellate. The case for polygons with more than four sides is not as strong. While there are some pentagons that will tessellate, there are also others that will not. Similarly, some hexagons will tessellate (for example, the regular hexagon), but not all of them.

If we consider only convex polygons, it can be proven that no polygon with more than six sides will tessellate. However, there are countless possibilities for tessellations of nonconvex polygons of more than six sides. The polygon in this tessellation is 12-sided.

SUPPLEMENT *(Activity Book)*

Activity Set 4.2 Regular and Semiregular Tessellations

Just for Fun: Escher-type Tessellations

Additional Sources

Association of Teachers of Mathematics. *Notes on Mathematics in Primary Schools.* London: Cambridge University Press, 1967.

Maletsky, E. "Designs with Tessellations." *The Mathematics Teacher,* **67** No. 4 (April 1974), 335–38.

Mann, J.E. "Polygon Sequences—an Example of a Mathematical Exploration Starting with an Elementary Theorem." *The Mathematics Teacher,* **63** No. 5 (May 1970), 421–28.

Mold, J. *Tessellations.* London: Cambridge University Press, 1971.

Ranucci, Ernest R. "Space Filling in Two Dimensions." *The Mathematics Teacher,* **64** No. 7 (November 1971), 587–93.

EXERCISE SET 4.2

1. This drawing of algae (sea life) is one of the unusual life forms studied by the German biologist, Ernst Haeckel (1834–1919). The polygons in the surface of this algae form the beginning of a tessellation.

   ★ a. There is a regular pentagon at the center. What are the polygons adjacent to this pentagon? Are they regular?

   b. There is a second ring of polygons surrounding the inner six. What kind of polygons are these?

★ 2. This regular pentagon has been inscribed inside a circle. Angle *BAC* is a central angle of the pentagon.

   ★ a. How many degrees are in each central angle of a regular pentagon?

   ★ b. Write the number of degrees for the central angles of the regular polygons in the following table.

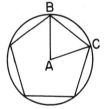

	Triangle	Quadrilateral	Pentagon	Hexagon	Heptagon	Octagon	Nonagon	Decagon		
Number of sides	3	4	5	6	7	8	9	10	20	100
Central angle	120°	90°								

★ 3. Fill in the table for these regular polygons.

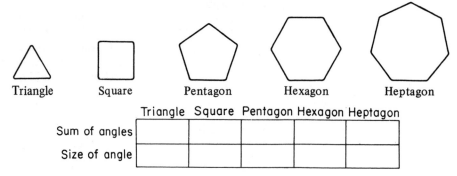

	Triangle	Square	Pentagon	Hexagon	Heptagon
Sum of angles					
Size of angle					

4. This Canadian nickel is a dodecagon (12-sided). Assume that you are to design a 12-sided one-dollar coin. Use your knowledge of central angles to inscribe a dodecagon in the given circle.

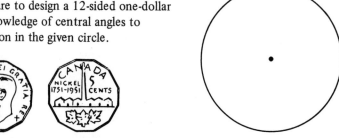

5. Use a protractor to construct a regular octagon having $\overline{AB}$ as a side.

A   B

6. This regular heptagon has been *circumscribed* about a circle.

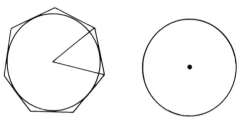

★   a. How many degrees are in the central angle of this heptagon?

    b. Construct a decagon which is circumscribed about the second circle.

7. What regular polygons will be formed by the following methods?

★ a. Tie a long rectangular strip of paper into a knot and smooth it down.

  b. Cut out an equilateral triangle and fold each vertex into the center.

  c. Draw a circle with a compass. With the opening of the compass equal to the radius of the circle, "pace off" this distance on the circumference and connect adjacent points.

8. Draw some figures to make conjectures about the truth or falsity of the following statements. Which three of these properties will not hold?

★ a. In any quadrilateral the sum of the opposite angles is 180°.

  b. A line segment from a vertex of a triangle to the midpoint of the opposite side is called a *median*. The three medians of a triangle meet in a point.

★ c. The diagonals of a quadrilateral bisect each other.

  d. If the midpoints of the sides of any quadrilateral are connected, they form a parallelogram.

★ e. A line segment from a vertex of a triangle which is perpendicular to the opposite side is called an *altitude*. The three lines containing the altitudes of a triangle meet in a point.

  f. If the midpoints of the sides of any triangle are connected, they form an equilateral triangle.

9. Some regular polygons will tessellate and others will not.

  a. Which of the five regular polygons in Exercise 3 will tessellate?

  b. What is very special about the sizes of the angles in regular polygons that will tessellate?

★ c. Will a regular octagon tessellate? Give a reason for your answer.

10. All triangles and quadrilaterals will tessellate, but this is not true for polygons with more than four sides. Which pentagon and which hexagon in parts **a** through **d** will tessellate? Draw a portion of the tessellation.

a. b. c. d.

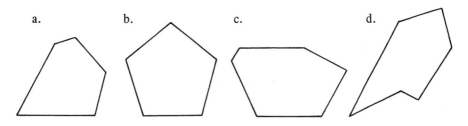

★ 11. Which of the following letters will tessellate? Draw portions of the tessellations.

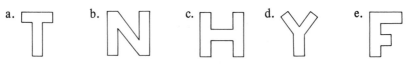

a.   b.   c.   d.   e.

12. A polygon with more than six sides will not tessellate if it is convex. The following polygons have more than six sides but they are nonconvex. Sketch a portion of a tessellation for each of these polygons.

a.   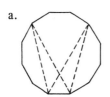   b.   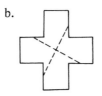   c.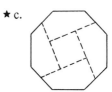

13. *Dissection Puzzles:*   Trace and cut out the pieces of each of these figures. Reassemble each set of pieces into a square.

a.   b.   ★ c.

★ 14. *Mirror Experiment:* Draw a line (*L*) and place two hinged mirrors on the line as shown here. By varying the angle between the mirrors, different regular polygons can be formed.

★   a. Adjust the mirrors to form a regular pentagon, as shown in this figure. Hold the mirrors firmly in place and trace the angle where the mirrors rest on the paper. How many degrees are in this angle?

★   b. Form each of the following regular polygons and measure the angle between the mirrors for each one.

   Triangle   Square   Pentagon   Hexagon   Heptagon   Octagon

15. *A Gestalt Puzzle:* Gestalt psychology was developed in Germany in the 1930s and is concerned primarily with the laws of perception. What is represented by these polygons and their background?

Cubic Space Division by M. C. Escher

"Space is an infinite sphere whose center is everywhere and whose surface is nowhere."
Cassius Jackson Keyser

4.3 SPACE FIGURES

In this lithograph by the Dutch artist Maurits C. Escher (1898–1972), the girders intersect
at right angles to form the edges of large cubes. In this way Escher represents space as being
filled with cubes of the same size. The Canadian mathematician H.S.M. Coxeter calls it
the cubic honeycomb and says its perspective gives a wonderful sense of infinite space.

The notion of space in geometry is an undefined term, just as the ideas of point, line,
and plane. We intuitively think of space as being 3D or three-dimensional and a plane as

only 2D or two-dimensional. In his Theory of Relativity, Einstein tied together the three dimensions of space with the fourth dimension of time. He showed that space and time affect each other and give us a "four-dimensional" universe.

Polyhedra   The three-dimensional figures in photos (a) and (b) are crystals. Their flat sides and straight edges were not cut by people but were shaped by Nature. The crystal in (a), which is shown embedded in rock, is pyrite. It has 12 pentagonal faces, 4 of which can be seen. The crystal in (b) is zircon. Its faces are rectangles and triangles. The surfaces of space figures such as these, whose sides are polygons, are called *polyhedra.* The polygons are called *faces* and they intersect in the *edges* and *vertices* of the polyhedron. The union of a polyhedron and its interior is called a *solid region or solid.*

(a) Crystal of pyrite

(b) Crystal of zircon

Regular Polyhedra—The most famous of all the polyhedra are known as *regular polyhedra* or *Platonic solids.* Their faces are congruent regular polygons and there is the same arrangement of polygons at each vertex. It can be proven that there are just five such polyhedra. The *tetrahedron* has 4 triangular faces; the *cube* has 6 square faces; and the *octahedron* has 8 triangular faces. These three polyhedra are found as crystals. Two of these, the cube and the octahedron, are pictured in (c). These are forms of the common mineral pyrite. The cube, which is embedded in rock, was found in Vermont, and the octahedron is

(c) Crystals of pyrite

from Peru, South America. The two other regular polyhedra are the *dodecahedron,* with 12 pentagonal faces, and the *icosahedron,* with 20 triangular faces. These cannot occur as crystals but have been found as skeletons of microscopic sea animals, called radiolarians. Posterboard models of the five regular polyhedra are shown next.

From left to right: tetrahedron, cube (hexahedron),
octahedron, dodecahedron, icosahedron

Semiregular Polyhedra—There is a greater
variety of polyhedra once we allow two or
more different types of regular polygons
for faces. The faces of the boracite crystal
are squares and equilateral triangles. Strange
as it may seem, this crystal developed these
flat, regularly shaped faces on its own, without the help of machines or people. There
are only 13 polyhedra that have two or
more regular polygons as faces and for
which each vertex is surrounded by the
same arrangement of polygons. They are
called *semiregular polyhedra*. The boracite
crystal is one of these. Each of its vertices
is surrounded by three squares and one
equilateral triangle.

Crystal of boracite

The regular and semiregular polyhedra are convex. A polyhedron is *convex*, if for any
two points on its surface, the line segment connecting these points is on the surface or in
the interior of the polyhedron. The polyhedra pictured next are nonconvex. They are
called the Kepler-Poinsot solids. Like those of the regular and semiregular polyhedra, the
faces of these polyhedra are regular polygons.

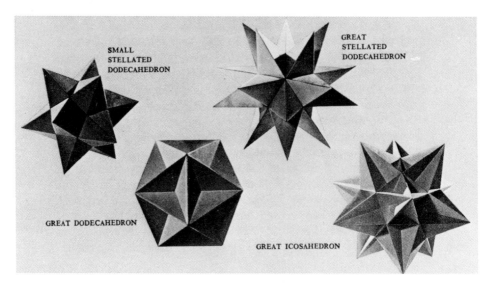

Kepler-Poinsot solids

Pyramids and Prisms   Chances are that when you hear the word "pyramid" you think, that's what the Egyptians built. Each of the Egyptian pyramids has a square base and triangular sides rising up to the vertex. This is just one type of pyramid. In general, the base of a pyramid can be any polygon but its sides are always triangular. Pyramids are named according to the shape of their base. Several pyramids with different bases are shown below. Church spires are familiar examples of pyramids. These are usually square, hexagonal, or octagonal pyramids. The spire in the photo on the right is a hexagonal pyramid.

The Bruton Steeple, Williamsburg, Virginia

Triangular pyramid
(also called tetrahedron)

Square pyramid

Pentagonal pyramid

Hexagonal pyramid

Prisms are another common type of polyhedra. You probably remember from your science classes that a prism is used to produce the spectrum of colors ranging from violet to red. Due to the angle between the vertical faces of the prism, a light directed into one face will be bent when it passes out through the other face.

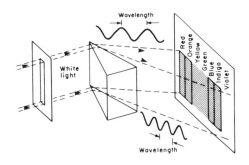

A *prism* has two parallel bases, an upper and a lower, which are congruent polygons. Like pyramids, prisms get their names from the shape of their bases. If the sides of a prism are perpendicular to the bases, as in the case of the triangular, quadrilateral, and hexagonal prisms, shown next, they are rectangles. Such prisms are called *right prisms,* or simply *prisms.* If some of the vertical faces are parallelograms, as in the pentagonal prism, the prism is called an *oblique prism.*

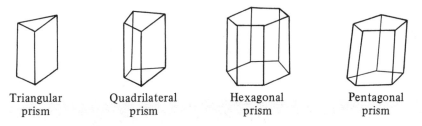

| Triangular prism | Quadrilateral prism | Hexagonal prism | Pentagonal prism |

The following two hexagonal prisms are crystals that grew with these flat, smooth faces and straight edges. These are oblique prisms whose faces are parallelograms.

Prisms of the crystal orthoclase feldspar

Cones and Cylinders   Cones and cylinders are the circular counterparts to pyramids and prisms. Ice-cream cones, paper cups, and New Year's Eve hats are common examples of cones. A *cone* has a circular region (disc) for a base and a lateral surface that slopes to the vertex point. If the vertex lies directly above the center of the base, the cone is called a *right cone* or usually just a *cone*. Otherwise, it is an *oblique cone*.

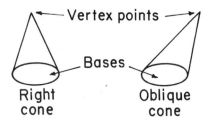

Right cone   Oblique cone

Ordinary cans are models of cylinders. A *cylinder* has two circular bases (discs) of the same size and a lateral surface that rises from one circle to the other. If the center of the upper base lies directly above the center of the lower base, the cylinder is called a *right cylinder,* or simply a *cylinder,* and if it does not, it is an *oblique cylinder*. Almost without exception, the cones and cylinders we use are right cones and right cylinders.

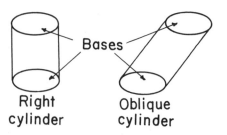

Right cylinder   Oblique cylinder

Spheres and Maps   The easiest way to create a sphere is to blow a bubble. Of all the different three-dimensional figures that will hold a given amount of air, the sphere is the most economically shaped in terms of the amount of material needed.

The same laws of Nature which govern the shape of bubbles are also responsible for the spherical shape of the sun, its planets, and the moons around the planets. The accompanying photo shows a fantastic view of the earth's spherical shape. It was photographed from the *Apollo 17* spacecraft during its 1972 lunar mission. The dark areas are water. The Red Sea and the Gulf of Aden are near the top center, and the Arabian Sea and Indian Ocean are on the right. Nearly the entire coastlines of Africa and the Arabian Peninsula are visible.

Earth, as seen from *Apollo 17* during 1972 lunar mission

Definition: A *sphere* is the set of points in space which are the same distance from a fixed point called the *center*. The union of a sphere and its interior is called a *solid sphere*.

The geometry of the sphere is especially important for navigating on the surface of the earth. The shortest distance between two points on a sphere is along an arc of a great circle. In this drawing of a sphere, G is a great circle, because its center is also the center of the sphere, and B is not a great circle. The distance between points X and Y along an arc of circle G is less than the distance between these points along an arc of circle B.

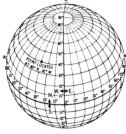

Locations on the earth's surface are often given by naming cities, streets, and buildings. A more general method of describing location is accomplished by two systems of circles. The circles that are parallel to the equator are called *parallels of latitude* [see Figure (a) shown next]. Except for the equator, these circles are not great circles. Each parallel of latitude is specified by an angle from 0° to 90°, both north and south of the equator. For example, New York City is at a northern latitude of 41°, and Sydney, Australia, is at a southern latitude of 34°. The second system of circles pass through the North and South Poles [see Figure (b) below] and are called *meridians of longitude*. These are great circles, and each is perpendicular to the equator. Since there is no natural point at which to begin numbering the meridians of longitude, the meridian that passes through Greenwich, England, is chosen as the zero meridian. Each meridian of longitude is given by an angle from 0° to 180°, both east and west of the zero meridian. The longitude of New York City is 74° West, and that for Sydney, Australia, is 151° East. These two systems of circles provide a grid or coordinate system for locating any point on earth [see Figure (c)].

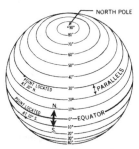

(a) Parallels of latitude

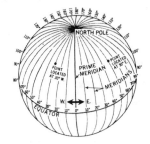

(b) Meridians of longitude

(c) Grid formed by both types of circles

The sphere cannot be placed flat on a plane without separating or overlapping some of its surface. This creates difficulties in making maps of the earth. There are three basic solutions to this problem: copying the earth's surface onto a cylinder, a cone, or a plane. These methods of copying are called *map projections*. In each case, some distortions of shapes and distances occur.

For a *cylindrical projection*, a cylinder is placed around a sphere (as shown in the next photo) and the surface of the sphere is copied onto the cylinder. Regions close to the equator are reproduced most accurately. The closer to the poles, the greater the map is distorted. The cylinder is then cut, and we have a flat map.

With a *conical projection*, a portion of the surface of a sphere is copied onto the cone. The cone can then be cut and laid flat. This type of map construction is commonly used for countries and other local regions of the earth's surface. The maps of the United States which are issued by the American Automobile Association are constructed by a conical projection.

Plane projections are made by placing a plane next to any point on a sphere and projecting the surface onto the plane. To visualize this, think of a light at the center of the sphere and the boundary of a country as being pierced with small holes. The light shining through these holes (see dotted lines) forms an image of the country on the plane. Roughly, half of the sphere's surface can be copied onto a plane by a plane projection, with the greatest distortion taking place at the outer edges of the plane.

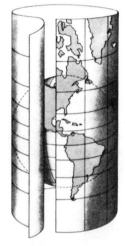

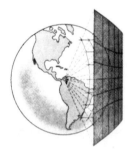

| Cylindrical projection | Conic projection | Plane projection |

SUPPLEMENT *(Activity Book)*

Activity Set 4.3 Models for Regular and Semiregular Polyhedra

Just for Fun: Penetrated Tetrahedron

Additional Sources

Beck, A., M.N. Bleicher, and D.W. Crowe. *Excursions into Mathematics.* New York: Worth, 1969. pp. 3–20 (proof of existence of only five regular polyhedra).

Cundy, H.M. and A.P. Rollett. *Mathematical Models.* New York: Oxford University Press, 1961. pp. 76–160.

Gardner, M. "On map projections (with special reference to some inspired ones)." *Scientific American,* **233** No. 5 (November 1975), 120–25.

Holden, A. *Shapes, Space and Symmetry.* New York: Columbia University Press, 1971.

Ridout, T.C. "The Mathematical Behavior of Minerals." *The Mathematics Teacher,* **42** No. 2 (February 1949), 88–90.

Smith, L.R. "Discovery in One, Two, and Three Dimensions." *The Mathematics Teacher,* **70** No. 8 (December 1977), 733–38.

Wenninger, M.J. *Polyhedron Models.* Reston, Virginia: NCTM, 1975 (instruction for making regular and semiregular polyhedra).

EXERCISE SET 4.3

1. There are six categories of illusions.* One category is called "impossible objects," because the illusions are produced by drawing three-dimensional figures on two-dimensional surfaces. Find the impossible feature in each of these pictures.

a.

Waterfall by M. C. Escher

b.

c.

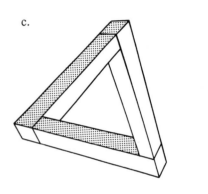

★ 2. A second type of illusion involves depth perception. We have accustomed our eyes to see depth in objects that are drawn on two-dimensional surfaces. Answer questions **a** and **b** by *disregarding the depth illusions.*

★   a. Is one of these cylinders on the right larger than the others?

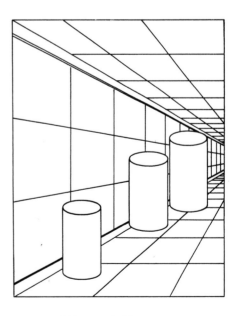

*P.A. Rainey, *Illusions* (Hamden, Conn: The Shoe String Press, 1973), pp. 18–43).

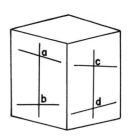

★ b. Which of the four lettered angles, *a, b, c,* or *d,* has the most degrees? Which are right angles?

★ 3. A cube can be dissected into triangular pyramids in several ways. Pyramid *FHCA* partitions the cube into five triangular pyramids. Name the four vertices of each of the other four pyramids.

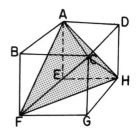

4. *F, E, G, H,* and *C* are the vertices of a square pyramid inside this cube. Name the five vertices of two more square pyramids such that the cube is partitioned into three pyramids.

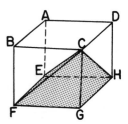

★ 5. There is a remarkable formula that relates the numbers of vertices, edges, and faces of a polyhedron. This formula was first stated by René Descartes (1596–1650) about 1635. In 1752 it was discovered again by Leonhard Euler (1707–1783) and is now referred to as Euler's formula. Count the number of vertices (v), edges (e), and faces (f), for each of these four polyhedra and find Euler's formula. (Don't forget to count the bottom face of each figure.)

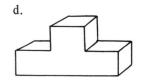

Figure	v	f	e
a			
b			
c			
d			

6. What are the names of each of these figures?

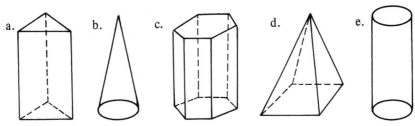

a.   b.   c.   d.   e.

★ 7. Count the vertices, faces, and edges of the regular polyhedra and complete the table. The cube and the octahedron are illustrated in Exercise 8. These numbers can be checked by Euler's formula (Exercise 5).

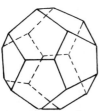 

Dodecahedron   **Icosahedron**

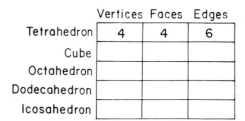

	Vertices	Faces	Edges
Tetrahedron	4	4	6
Cube			
Octahedron			
Dodecahedron			
Icosahedron			

8. The centers of the faces of a cube can be connected to form a regular octahedron. Also, the centers of the faces of an octahedron can be connected to form a cube. These polyhedra are called *duals.* Can you see how this dual relationship is suggested by the table in Exercise 7? Find two other Platonic solids that are duals. Which Platonic solid is its own dual?

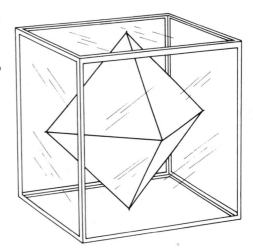

★ 9. Use your knowledge of the earth's coordinate system to match each of the following cities with its approximate longitude and latitude.

	Latitude	and	Longitude
Tokyo	38°N		120°W
San Francisco	56°N		4°W
Melbourne	35°N		140°E
Glasgow	35°S		20°E
Capetown	38°S		145°E

10. This collection of polyhedra shows some of the limited number of forms which crystals may take. These shapes were not invented by people but were revealed to us as products of Nature. The polygons at the tops of the marked off columns are the horizontal cross sections of the polyhedra in these columns.

★   a. List the numbers of the polyhedra which are pyramids.

    b. List the numbers of the polyhedra which are prisms.

★   c. Which one of these polyhedra is most like a dodecahedron?

    d. Which one of these polyhedra is most like an octahedron?

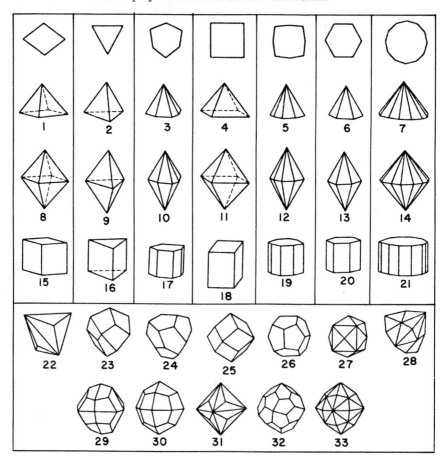

11. Two points on the earth's surface which are on opposite ends of a line segment through the center of the earth are called *antipodal points*. The coordinates of such points are nicely related. The latitude of one point is as far above the equator as the other is below, and the longitudes are supplementary angles (in opposite hemispheres). For example, (30°N, 40°E) is in Saudi Arabia, and its antipodal point (30°S,140°W) is in the South Pacific.

Babson College globe.
Diameter 28 feet, weight 21 tons.

★ a. The globe in the accompanying photo shows that (20°N,120°W) is a point in the Pacific Ocean just west of Mexico. Its antipodal point is just east of Madagascar. What are its coordinates?

b. The point (30°S,80°E) is in the Indian Ocean. What are the coordinates of its antipodal point? What country is it in?

★ 12. China is bounded by latitudes of 20°N and 55°N, and by longitudes of 75°E and 135°E. It is playfully assumed that if you could dig a hole straight through the center of the earth you would come out in China. For which of the following locations is this true?
a. Panama (9°N,80°W) b. Buenos Aires (35°S,58°W) c. New York (41°N,74°W)

★ 13. Which of the three types of projections for making maps is most appropriate for obtaining flat maps of the following regions?
★ a. Australia. ★ b. North, Central, and South America.
★ c. The entire equatorial region within the latitudes of 30° North and 30° South.

14. Hurricane Ginger was christened on September 10, 1971, and became the longest-lived Atlantic hurricane on record. This tropical storm formed approximately 275 miles south of Bermuda and reached the U.S. mainland 20 days later.

Hurricane Ginger

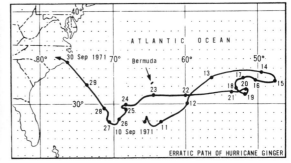

★   a. The storm's coordinates on September 10 were (28°N,66°W). What were the coordinates for the following dates: September 15; September 23; and September 30?

       b. At this latitude on the earth's surface each degree of longitude spans a distance of approximately 60 miles. About how many miles did this hurricane travel from September 10 to September 30? (*Hint:* Use a piece of string.)

★ 15. *Matchstick Puzzle:* Make 1 square and 4 triangles using 8 matchsticks of equal length which touch only at their end points.

★ 16. *Riddle:* Some sportspeople, having pitched camp, set out to go bear hunting. They walked 15 miles due south, then 15 miles due east, where they shot a bear. Walking 15 miles due north, they returned to their camp. What was the color of the bear? Name at least two locations on earth where the three 15-mile legs of this trip will end at the starting point.

Taj Mahal, built (1630–1652) on the banks of the Jumna in Agra, India

4.4 SYMMETRIC FIGURES

The Taj Mahal is considered by many to be the most beautiful building in the world. It is made entirely of white marble and surrounded by a landscaped walled garden on the banks of the Jumna River in Agra, India. The Taj Mahal was built by the emperor Shah Jahan in memory of his wife, Mumtaz Mahal. It is an octagonal building with four of its eight faces containing massive arches rising to a height of 33 meters (108 feet). The center is an octagonal chamber containing the tombs of the emperor and his wife.

The Taj Mahal has a form and balance which can be described by saying it is symmetrical. The concept of symmetry was one of the earliest attempts of the human race to find order and harmony in a world of hostility and confusion. Perhaps the most influential factor in our desire for symmetry is the shape of the body. Even children in their earliest drawings show an awareness of body symmetry.

Reflectional Symmetry for Plane Figures   Many years before the recent trend to teach geometric ideas in the elementary school, cutting out symmetric figures was a routine activity. The procedure is to fold a piece of paper and draw a figure which encloses part or all of the crease. When the figure is cut out and unfolded it is symmetrical. The crease is called a *line of reflection* or *line of symmetry,* and the figure is said to have a *reflectional symmetry.*

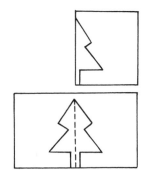

The intuitive idea of reflectional symmetry is that the two halves of the figure are "the same" or would coincide if one were placed on the other. The word "reflection" is a natural one to use because of the mirror test for symmetry. If the edge of a mirror is placed along a line of reflection, the "half-figure" and its image will look like the whole figure. For example, you can use this test on the front of the following building by placing the edge of a mirror along the vertical center line of the photo. With the mirror in this position, half of the building and its reflection will look like the whole building. Since this is the only way the mirror can be placed so that this will happen, the front of this building has only one line of symmetry.

Putnam Hall, University of New Hampshire

Some figures have more than one line of reflection. To produce a figure with two such lines, fold a sheet of paper in half and then in half again. Then cut out a figure whose end points are on the creases. When the paper is opened, the two perpendicular creases will be lines of reflection for the figure.

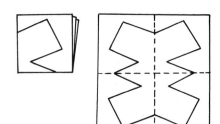

The idea of symmetry can be made more precise by adopting the terminology of "image," which is suggested by mirrors. If a line can be drawn through a figure so that each point on one side of the line has a matching point on the other side at the same perpendicular distance from the line, it is a *line of reflection.* For any two matching points, one is called the *image* of the other. A few points and their images on the house figure have been labelled. *A* corresponds to *A'*, *B* to *B'*, *C* to *C'*, and *D* to *D'*. The line segment connecting a point and its image is perpendicular to the line of reflection.

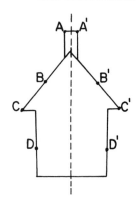

Each of the following polygons has two or more lines of reflection. The equilateral triangle has three such lines, along its medians. The square has four lines of reflection: one horizontal, one vertical, and two diagonal. The rectangle has only two lines of reflection, one horizontal and one vertical. The diagonals of the rectangle are not lines of reflection.

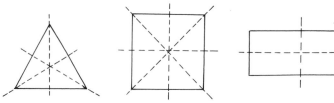

Rotational Symmetry for Plane Figures

This is not a picture of a plant, as you might expect, but a type of jellyfish called Aurelia. While it seems to have the form and balance of a symmetric figure, it has no lines of reflection. It does, however, have what we call *rotational symmetry,* because it can be turned about its center so that the figure coincides with itself. For example, if it is rotated 90° clockwise, the top "arm" will move to the 3 o'clock position, the bottom "arm" will move to the 9 o'clock position, etc.

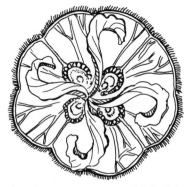

Aurelia, common coastal jellyfish

Let's consider another example of rotational symmetry. Trace the figure shown here and mark the center X and the arms A, B, and C. Cut it out and place it on the page so both figures coincide. By holding a pencil at point X, the top figure can be rotated clockwise so that A goes to B, B to C, and C to A. This is called a *rotational symmetry*, and X is called the *center of rotation*. Since the figure is rotated 120° (1/3 of a full turn) it has a *120° rotational symmetry*. From its original position this figure can also be made to coincide with itself after a 240° clockwise rotation, with A going to C, B to A, and C to B. This is a *240° rotational symmetry*. Since every figure can be rotated back onto itself after a 360° rotation, every figure has a *360° rotational symmetry*.

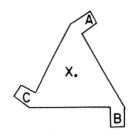

Some figures have both rotational and reflectional symmetries. The regular polygons have both types. The central angles of these polygons determine the angles for the rotational symmetries. In the regular hexagon the central angle is 360 ÷ 6, or 60°. Therefore, rotations of 60°, 120°, 180°, 240°, 300°, and 360° each move the hexagon back onto itself. Every regular hexagon has six rotational symmetries and six lines of symmetry.

Each of the hexagonal-shaped snowflakes possesses the rotational and reflectional symmetries of the hexagon. Beyond this similarity there is endless variety in their structure. These next photographs are from the more than 2200 pictures of snowflakes in the book, *Snow Crystals.* *

*W. A. Bentley and W. J. Humphreys, *Snow Crystals* (New York: McGraw-Hill, 1931).

Reflectional Symmetry for Space Figures   The intuitive idea of reflectional symmetry for three-dimensional objects is similar to that for plane figures. With plane figures we found lines such that one-half of the figure was the reflection of the other. With space figures there are planes of symmetry such that the two halves "look the same." Consider, for example, this antique chair. By cutting down the center of the back and across the seat to the front of the chair, it can be divided into left and right halves, both of which will be mirror images of each other. We can think of this cut as being done by a plane, called the *plane of reflection* or *plane of symmetry*. The chair is said to have *reflectional symmetry*.

Ebonized walnut armchair, 1865–1875

This type of symmetry can be mathematically defined by requiring that for each point on the left-hand side of the chair, there is a corresponding point on the right-hand side, such that both points are the same perpendicular distance from the plane of reflection. For the antique chair, A corresponds to A' and B corresponds to B'. These points are called *images* of each other, and the segments AA' and BB' are perpendicular to the plane of reflection.

"Two-sided symmetry," such as for the antique chair, the Long-Horn Beetle, and the Zebra Butterfly, is sometimes called *bilateral symmetry* or *vertical symmetry,* because the plane of reflection is perpendicular to the ground. Martin Gardner offers the explanation on the following page for the common occurrence of this type of symmetry.*

Long-Horn beetle

Zebra butterfly

*M. Gardner, The *Unexpected Hanging* (New York: Simon and Schuster, 1969), pp. 114–15.

"Because the earth is a sphere toward the center of which all objects are drawn by gravity, living forms have found it efficient to evolve shapes that possess strong *vertical symmetry* combined with an obvious lack of horizontal or rotational symmetry. In making objects for his use man has followed a similar pattern. Look around and you will be struck by the number of things you see that are essentially unchanged in a vertical mirror: chairs, tables, lamps, dishes, automobiles, airplanes, office buildings—the list is endless."

Each of the objects shown above and on the preceding page has one or more planes of reflection. The chair, beetle, and butterfly each have one plane of reflection. The small square-shaped table has four planes of reflection: one from front to back; one from side to side; and one through each diagonal of the surface. The lamp has six vertical planes of reflection because its shade has six sections. The wastebasket has eight vertical planes of reflection, since its base has the shape of an octagon. If the wastebasket had a top cover to match its bottom, it would have a horizontal plane of symmetry. A sphere mounted on a pentagonal base, similar to this 1940 beechwood engraving by M. C. Escher, would have five planes of reflection because of the shape of its base.

Sphere with Fish
by M. C. Escher

Rotational Symmetry for Space Figures

Some three-dimensional objects, such as the table shown here, have *rotational symmetries.* By rotating this table 120° the legs will change places and the table will be moved back into the same location or position. That is, leg *A* will go to leg *B, B* to *C,* and *C* to *A.* In this example the table is being rotated about line *L* which passes through the center of the tabletop and its base. Line *L* is called the *axis of rotation,* and the table is said to have a *rotational symmetry.* Since the angles formed by pairs of legs of this table have 120°, the table has 120°, 240°, and 360° rotational symmetries.

The three-legged table also has three vertical planes of reflection, one passing through each leg. It is not difficult to find objects with both planes of reflection and axes of rotation. The small table, the lamp, and the wastebasket pictured on page 201 all have both types of symmetry. Occasionally, however, you will see objects such as this paper windmill which have rotational symmetries but no plane of reflection. There are rotational symmetries of 90°, 180°, 270°, and 360° about its axis, which is the line perpendicular to this page and through the center of the windmill.

Paper windmill

SUPPLEMENT *(Activity Book)*

Activity Set 4.4 Creating Symmetric Figures by Paper Folding

Additional Sources

Birkhoff, G.D. *Aesthetic Measure.* Cambridge, Massachusetts: Harvard University Press, 1933. "Polygonal Forms," pp. 16–48.

Gardner, M. *The Unexpected Hanging.* New York: Simon and Schuster, 1969. "Rotations and Reflections," pp. 114–21.

Lockwood, E.H. "Symmetry in Wallpaper Patterns." *Mathematics in Schools,* 2 No. 4 (July 1973), 14–15.

Weyl, H. *Symmetry.* Princeton: Princeton University Press, 1952. pp. 3–38.

The Alhambra, built in the thirteenth century for Moorish kings, Granada, Spain

EXERCISE SET 4.4

1. The pool, building, and fortress in this picture have a vertical plane of symmetry, about which their left sides are the reflections of their right sides.

    a. List five objects in this picture which have images about the plane of reflection.

    b. Physical objects can never be perfectly symmetric. In this scene, for example, there are several deviations from perfect symmetry. List three objects which do not have an image for the vertical plane of symmetry.

★   c. There are several individual items in this photo which have vertical lines of symmetry. Name an object in this picture which has a horizontal line of symmetry.

2. In 1850 gold was so plentiful that dozens of different banks and business firms minted their own coins. Some were square and others were eight-sided, such as this octagonal fifty-dollar gold piece.

Panama-Pacific octagonal
fifty-dollar gold piece

★   a. How many rotational symmetries are there for a regular octagon?

b. How many degrees are there in the smallest rotational symmetry?

c. How many lines of reflection are there for a regular octagon?

3. The following organisms have lines of reflection and rotational symmetries. Determine the number of lines of reflection and rotational symmetries for each one.

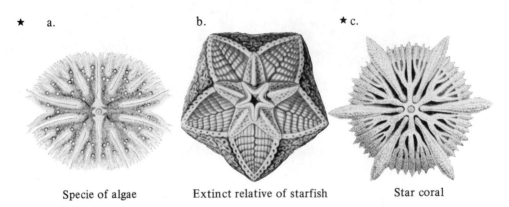

★   a.   b.   ★ c.

Specie of algae   Extinct relative of starfish   Star coral

★ 4. Which two of these polygons have no lines of reflection? Draw all of the lines of reflection for the remaining polygons. Find the number of rotational symmetries for each figure.

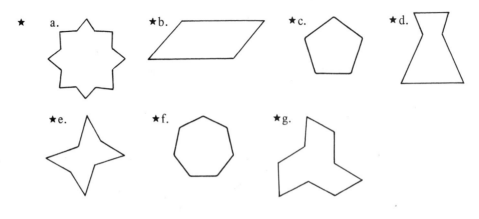

★   a.   ★ b.   ★ c.   ★ d.

★ e.   ★ f.   ★ g.

★ 5. The mirror test for symmetry is very effective when the reflecting is done with a piece of plexiglass.* In addition to reflecting, plexiglass allows you to see through it, so the whole figure can be seen. To find a line of symmetry, it is necessary only to move the plexiglass until the reflected image on the plexiglass coincides with the portion of the figure behind it. This cannot be done for several of the following figures. Which ones?

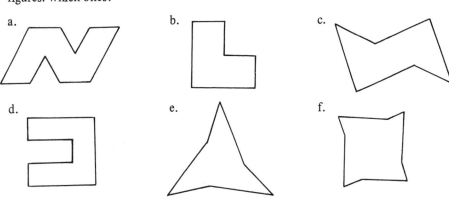

a.　　　　b.　　　　c.

d.　　　　e.　　　　f.

★ 6. Which letters have at least two lines of reflection? Which letters have two rotational symmetries but no lines of reflection?

ABCDEFGHIJKLMNOPQRSTUVWXYZ

7. If you write the letter P on a piece of paper and hold it in front of a mirror, it will look reversed.

★　a. Which capital letters will not appear reversed if you repeat this experiment? What type of symmetry do these letters have?

　b. Use the letters from part **a** to write a word such that its reflection in a mirror is also a word.

*E. Woodward, "Geometry with a Mira," *The Arithmetic Teacher*, **25** No. 2 (November 1977), 117–18.

8. The following figures have been formed on a circular geoboard. Three of these figures have no lines of reflection. Sketch in all lines of reflection for the remaining figures. Find the number of rotational symmetries and give the number of degrees for each.

★ a.   b.   ★c.   d.

★ e.   f.   g.   h.

9. Sketch figures on these geoboards which have the given numbers of symmetries.

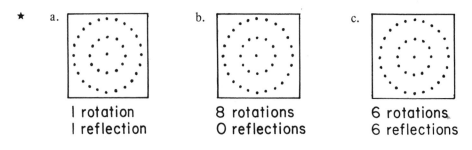

★ a.   1 rotation   1 reflection

b.   8 rotations   0 reflections

c.   6 rotations   6 reflections

10. Complete the sketches of these figures so that they are symmetric about the given line. You might want to first find the image with a mirror or Mira, as in Exercise 5.

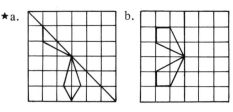

★a.   b.

11. Finish these figures so that they will be symmetric about the two perpendicular lines.

a.   ★b.

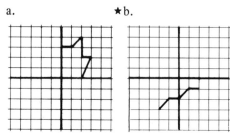

12. Shade in the smallest number of squares so that the pattern of shaded squares will have the given rotational symmetries.

a.

180° rotational symmetry

★b.

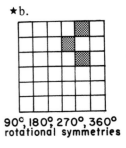

90°, 180°, 270°, 360° rotational symmetries

13. How many planes of reflection are there for each of these objects?

14. These geometric solids are highly symmetric.

From left to right: cone, cylinder, prism, sphere, cube, pyramid (square base)

★ a. Which of these solids has a horizontal plane of reflection?

b. Does each of these solids have at least one vertical plane of reflection?

★ c. For each vertical axis of symmetry give the number of rotational symmetries for each solid.

15. *Mirror Cards:* * Mirror Cards were developed by Marion Walter to informally teach spatial relationships and symmetry in the early grades. The problem posed by Mirror Cards is one of using a mirror to match a pattern on one card to that on another. Place a mirror on the card shown here and try to obtain the patterns in parts **a** through **d**. Which ones can be obtained from this card?

*M. Walter, "An Example of Informal Geometry: Mirror Cards" *The Arithmetic Teacher,* **13** (October 1966), 448-52.

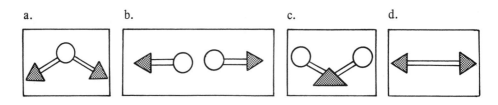

16. These metalwork designs have many pleasing symmetries. How many rotational symmetries and lines of symmetry are there for each design?

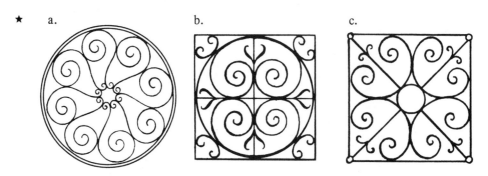

17. How many rotational symmetries are there for each of the Japanese crests?

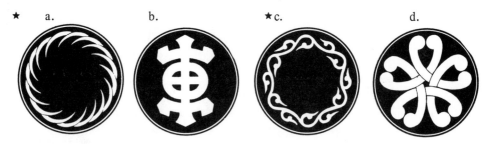

★ 18. The subject of beauty has been discussed for thousands of years. Aristotle felt that the main elements of beauty are order and symmetry. The American mathematician George Birkhoff (1884–1944) developed a formula for rating the beauty of objects.* Part of his formula involved counting symmetries. If only symmetry is used to rate the beauty of the polygons in Exercise 4, which would have the highest rating (counting all lines of reflection and rotational symmetries)? Which is the least beautiful?

*G. D. Birkhoff, *Aesthetic Measure* (Cambridge, Mass.: Harvard University Press, 1933), pp. 33–46.

19. *Words with Symmetry:* The word "HIDE" has a
    horizontal line of symmetry, and "ATOM," written
    vertically, has a vertical line of symmetry. Write
    a few more words which have horizontal and
    vertical lines of symmetry.

    A
    T
    O
    M

20. *Dates with Symmetry:* The date 1881 has a horizontal and a vertical line of symmetry.

    a. Find another date which has horizontal and vertical lines of symmetry.

    b. The date 1691 has a 180° rotational symmetry. What year in this century has this
       type of symmetry?

CHAPTER FIVE

MEASUREMENT

When you can measure what you are speaking about and express it in numbers, you know something about it; but when you cannot measure it, when you cannot express it in numbers, your knowledge is of a meagre and unsatisfactory kind.

Lord Kelvin

Stonehenge (1900–1700 B.C.), Salisbury Plain, England

"The man who undertakes to solve a scientific question without the help of mathematics undertakes the impossible. We must measure what is measurable and make measurable what cannot be measured."

Galileo

5.1 SYSTEMS OF MEASURE

The daily rotations of the earth, the monthly changes of the moon, and our planet's yearly orbits about the sun provided some of the first units of measure. The day was divided into parts by sunrise, midday, sunset, and the tides, while the year was divided into seasons. One of the early attempts to measure the length of a year and its seasons was the prehistoric Stonehenge in southern England. This monument and some 40 or 50 others like it

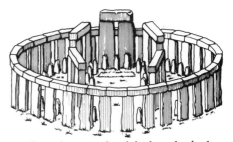

Stonehenge as it might have looked 4000 years ago

served as astronomical observatories as well as temples for conducting religious and agricultural rituals. By adjusting movable stones it was possible to predict the arrival of the summer solstice, winter solstice, and occurrence of eclipses. Eventually, the sundial was invented to measure smaller periods of time, and it remained the principal method of measuring time until the nineteenth century. Today, atomic clocks measure time to within one ten-millionth of a second. (See Exercise 1, Exercise Set 7.1.)

Historical Development   It took many centuries before accurate units of measure were needed or developed. At first, rough-and-ready methods were sufficient. A distance might be described by a *day's walk* or *a number of paces*. The Chinese had an *uphill mile* and a *downhill mile,* with the uphill mile being the shorter distance. The traditional *Roman mile* was 1000 (mille) double paces of 5 feet each.

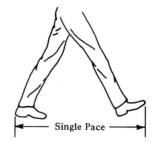

— Single Pace —

Many of the first units of measure were parts of the body. The early Babylonian and Egyptian records indicate that the *hand,* the *span,* the *foot,* and the *cubit* were all units of measure. The hand was used as a basic unit of measure by nearly all ancient civilizations and is still used to measure the heights of horses. The height of a horse is measured by the number of hand breadths from the ground to the horse's shoulders.

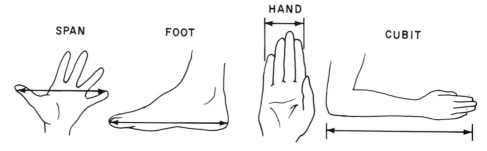

SPAN   FOOT   HAND   CUBIT

Evidence of other early units of measure still exists today. Seeds and stones were common units for measuring weight. The word "carat," which is a unit of weight for precious stones, was derived from the carob seeds of Mediterranean evergreen trees. Carat also expresses the fineness of a gold alloy. *Fourteen carat* means 14 parts of gold to 10 parts of alloy or that 14 out of 24 parts are pure gold. The *grain,* which was obtained from the average weight of grains of wheat, is another unit of weight used by jewelers. Until the past few years the *stone* was a common unit of weight in England and Canada. A newborn baby would weigh about one-half a stone.

In the English system of measure, shoe sizes are measured by an old unit called the *barleycorn.* The length of three barleycorns placed end to end is an inch, so shoe sizes vary by thirds of an inch. A size 7 shoe is 1 inch longer than a size 4 shoe. The *yard* and *fathom* are two more English system units for length which have their origins in

body measurements. The yard was established by royal decree in the twelfth century by King Henry I, as the distance from his nose to his thumb. The fathom, a common unit for depths of water, is 2 yards or approximately the distance from fingertip to fingertip of a person's outstretched arms.

The *mouthful* is an ancient unit of measure for volume which was used by the Egyptians. It was also part of a unique English doubling system: 2 mouthfuls equals 1 jigger; 2 jiggers equals 1 jack; 2 jacks equals 1 jill; etc. The familiar "Jack and Jill went up the hill" nursery rhyme contains three units of volume, the *jack, jill,* and *pail,* and was a protest against the King of England for his taxation of the *jack* or *jackpot* as it was called. The origin of the common expression, "hit the jackpot," which is used by gamblers, is due to the king's mishandling of the jackpot tax. The phrase, "broke his crown," in the nursery rhyme, refers to King Charles I and is prophetic. Not only was his crown broken but he lost his head in Britain's bourgeois revolution, not many years after the taxation of the jackpot.*

English System—Units of Volume	
2 mouthfuls	= 1 jigger
2 jiggers	= 1 jack (jackpot)
2 jacks	= 1 jill
2 jills	= 1 cup
2 cups	= 1 pint
2 pints	= 1 quart
2 quarts	= 1 pottle
2 pottles	= 1 gallon
2 gallons	= 1 pail

As societies evolved, weights and measures became more complex. Since most units of measure had developed independently of one another, it was difficult to change from one unit to another. As examples, in the English system there are 12 inches in a *foot;* 16 ounces in a *pound;* 2150.42 cubic inches in a *bushel;* and 5280 feet in a *mile.* By the fifteenth and sixteenth centuries the need for a simplified worldwide system of measurement had become apparent, and several proposals for a new system were made during this period.

© 1972 United Feature Syndicate, Inc.

Metric System In 1790, in the midst of the French Revolution, the metric system was developed by the French Academy of Sciences. To create a system of "natural standards" they subdivided the distance from the equator to the North Pole to obtain the basic unit of length, the *meter.* The basic

Greek prefixes	*kilo	 1000
	hecto	 100
	deca	 10
Latin prefixes	*deci	. . . . 1/10
	*centi	. . . . 1/100
	*milli	. . . . 1/1000

units for weight, the *gram,* and for volume, the *liter,* are obtained from the meter. Smaller measures are obtained by dividing these basic units into 10, 100, and 1000 parts. Larger measures are 10, 100, and 1000 times greater than the basic units. These new measures are named by attaching prefixes to the names of the three basic units (see table above). In everyday nonscientific needs, only the above four prefixes that are marked with an asterisk are commonly used.

*A. Kline, *The World of Measurement* (New York: Simon and Schuster, 1975), pp. 32–39.

According to these prefixes, a *kilometer* is 1000 meters, a *centimeter* is one-hundredth of a meter, a *millimeter* is one-thousandth of a meter, and so on. There are more prefixes continuing in both directions for naming larger and smaller measures, with each new measure being 10 times greater or one-tenth as great as the previous one. This relationship between measures is a major advantage of the metric system. Converting from one measurement to another can be accomplished by multiplying or dividing by powers of 10, which, since we use base ten in our numeration system, is just a matter of moving decimal points.

The metric system was a radical change for the French people and met widespread resistance. Finally in 1837, the French government passed a law forbidding the use of any measures other than those of the new system. Steadily, other nations adopted the metric system. In 1866 the Congress of the United States enacted a law stating that it was lawful to employ the weights and measures of the metric system and that no contract or dealing could be found invalid because of the use of metric units. Today, less than 200 years after its creation, the metric system has been adopted by almost every country, with the last major nonmetric country, the United States, moving toward the metric system as its standard of measure. The Metric Conversion Act of 1975 set a national policy of voluntary conversion with no overall timetable.

Length

A meter is roughly the distance from the floor to the waist of an adult. If 10,000,000 meters were placed end to end, they would reach one-quarter of the distance around the world. This was the origin of the meter, but once it was established, an official meter bar of platinum was produced as the standard length. Today every country that uses the metric system has a reproduction of the French meter bar as its standard unit of length. In the United States a copy of the meter bar is maintained by the Bureau of Standards in Washington, D.C.

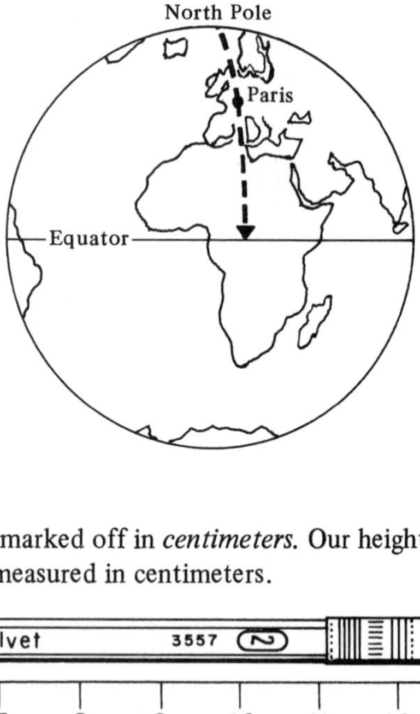

For everyday requirements, the centimeter, meter, and kilometer are sufficient for lengths and distances. The ruler shown here is marked off in *centimeters*. Our height, waist, hat size, and other such measurements are measured in centimeters.

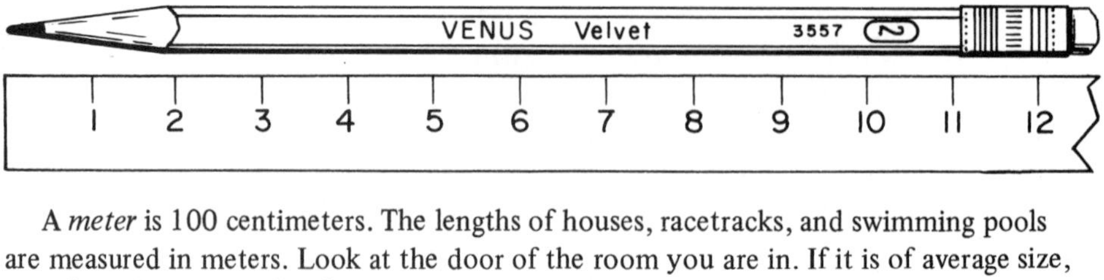

A *meter* is 100 centimeters. The lengths of houses, racetracks, and swimming pools are measured in meters. Look at the door of the room you are in. If it is of average size, it is about 2 meters in height.

Distances between cities are measured by *kilometers*. A kilometer is shorter than a mile, approximately 3/5 as long. A person who walks 3 miles per hour will walk about 5 kilometers in an hour; and a speed limit of 30 miles per hour is equal to approximately 50 kilometers per hour.

Occasionally, you will need to measure something smaller than a centimeter. One-tenth of a centimeter is a *millimeter*. The thickness of a pencil lead is about 2 millimeters, and the width of a pencil is about 7 millimeters or 7/10 of a centimeter.

Length	
kilometer	km
hectometer	hm
decameter	dkm
meter	m
decimeter	dm
centimeter	cm
millimeter	mm

Volume   The most familiar unit of measure for volume or capacity is the *liter*. Compared to a quart the liter is just a "little bit" bigger. The capacities of fuel tanks, aquariums, and milk containers are measured in liters. For volumes that are less than a liter, such as some grocery store and drugstore items, the *milliliter* (1/1000 of a liter) is the common measure. Larger volumes, such as a community's reserve water supply, are measured by *kiloliters* (1000 liters).

Volume	
kiloliter	kℓ
hectoliter	hℓ
decaliter	dkℓ
liter	ℓ
deciliter	dℓ
centiliter	cℓ
milliliter	mℓ

One *milliliter* is 1/1000 of a liter and is the amount contained in a cube having a length, width, and height of 1 centimeter. A cube with these dimensions is called a *cubic centimeter.* Therefore, a liter is the amount contained in 1000 such cubes or in a container that has a length, width, and height of 10 centimeters. Converting from milliliters to liters, or from cubic centimeters to liters, is easily accomplished by dividing by 1000.

One cubic centimeter (1 cm by 1 cm by 1 cm) equals one millileter.

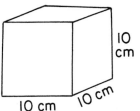

One cubic decimeter (10 cm by 10 cm by 10 cm) equals one liter.

10 cm   10 cm   10 cm

Weight

The basic unit of weight is the *gram.* This is a relatively small weight, approximately the weight of a medium-size paper clip or a dollar bill. The majority of grocery store items are measured in grams.

Heavier objects are measured in kilograms. A kilogram (about 2.2 pounds) is the common unit for people's weight. An average size woman will weigh from 50 to 60 kilograms; and a weight-lifting record of 255 kilograms was set by Russia's Vasily Alexeyev in the 1976 Olympics.

Scientists distinguish between "weight" and "mass," with the difference in these two concepts being due to the effects of gravity. An object will weigh more at sea level than on top of a mountain, because

the earth's gravity exerts a greater force on it. The same object in a spaceship would weigh practically nothing. Yet, in each of these three locations the amount of material in these objects hasn't changed! Because of this situation, it is necessary for scientific purposes to refer to the *mass* of an object as a measurement which doesn't change as it is moved further from the center of the earth. We can think of mass intuitively as the amount of matter which makes up the object. *Weight,* on the other hand, is the force that gravity exerts on the object; it varies with different locations. At sea level the mass and weight of an object are essentially equal. Since the variation in weight between sea level and our highest mountains is very small (.1% difference), the concept of weight is accurate enough for everyday measurements.

A *gram* is the weight of 1 cubic centimeter of water. (Technically, the water must be at a temperature of 4° Celsius because temperature affects volume.) This means that 1 milliliter of water weighs 1 gram. This simple relationship between weight, length, and volume gives the metric system its superiority over other systems of measure.

Weight	
kilogram	kg
hectogram	hg
decagram	dkg
gram	g
decigram	dg
centigram	cg
milligram	mg

One cubic centimeter of water equals one milliliter of water and weighs one gram.

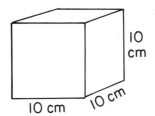

One cubic decimeter of water equals one liter of water and weighs one kilogram.

Temperature   In 1714 Gabriel Fahrenheit, a German instrument maker, invented the first mercury thermometer. The lowest temperature he was able to attain with a mixture of ice and salt he called 0 degrees. He used the normal temperature of the human body, which he selected to be 96 degrees, for the upper point of his scale. (We now know it is about 98.6 on the Fahrenheit scale.) On this scale of temperatures, water freezes at 32 degrees and boils at 212 degrees. In 1742, before the development of the metric system, the Swedish astronomer Anders Celsius devised a temperature scale by selecting 0 as the freezing point of water and 100 as the boiling point. He called his system the *centigrade* (100 grades) thermometer, but in recent years it has been changed to *Celsius* in his honor.

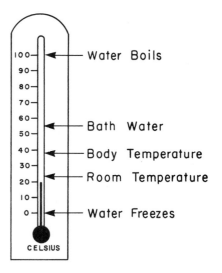

Heat is related to the motion of molecules, and the faster their motion, the greater the heat. All movement stops at ⁻273.15° Celsius. Lord Kelvin called this temperature "absolute zero" and devised the *Kelvin scale* which increases 1 degree for each increase of 1 degree Celsius. Thus, 273.15° on the Kelvin scale is 0° on the Celsius scale. Both the Celsius and Kelvin scales are part of the metric system. The Celsius scale is used for weather reports, cooking temperatures, and other day-to-day needs, and the Kelvin scale is used for particular scientific purposes.

Precision and Small Measurements   The objects in this picture are DNA and ribosome molecules in an active chromosome. Each DNA molecule is so small that 100,000 of them lined up side by side would fit into the thickness of this page. Measurements to this type of precision are possible with the transmission electron microscope. More recently, with the development of the field ion microscope, scientists have been able to view atoms that are ten times smaller than DNA molecules. The following scale shows eight decreasing measures from a centimeter down to an angstrom, each being one-tenth the size of the preceding one.

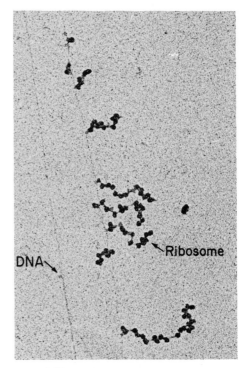

DNA and ribosome molecules

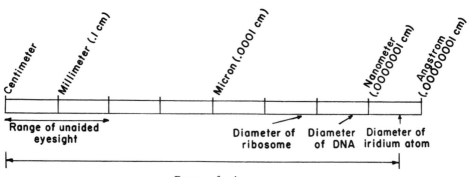

Range of microscopes

The amount of precision that is possible in taking measurements depends on the smallest unit of the measuring instrument. Using a centimeter ruler this paper clip has a length of 3 centimeters, or we might say, just over 3 centimeters. With a ruler marked off in millimeters the length of the paper

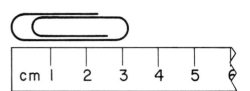

clip can be measured as 32 millimeters or 3.2 centimeters. With finer and finer measuring instruments we might measure the length of this paper clip to be 3.24, 3.241, or 3.2412 centimeters. It would never be possible, however, to measure its length or the length of any other object exactly.

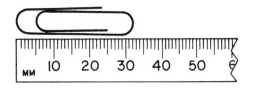

If the smallest unit on the measuring instrument is a millimeter, then the measurement can be taken to the nearest millimeter. This means that the measurement could be off by one-half of a millimeter, either too much or too little. In general, the *amount of precision* is to within one-half of the smallest unit of measure being used.

Conversely, if a measurement is given as 7.6 centimeters, we can assume that it was measured to the nearest tenth of a centimeter and that it is closer to 7.6 centimeters than to 7.5 centimeters or 7.7 centimeters. In other words, it is 7.6 ± .5 centimeters. A measurement of 7.62 centimeters means that it has been obtained to the nearest one-hundredth of a centimeter and may be off by as much as .005 centimeter. Sometimes you will see a measurement such as 15.0 centimeters. This means that the measurement is accurate to the nearest tenth of a centimeter and implies more precision than if it had been given as 15 centimeters.

International System of Units   The International System of Units is a modern version of the metric system which has been established by international agreement. Officially abbreviated as SI, this system is built on the metric units discussed previously, plus units for: time *(second)*; electric current *(ampere)*; light intensity *(candela)*; and the molecular weight of a substance *(mole)*. This system provides a logical and interconnected framework for all measurements. To enable the type of precision needed in science today the meter is now defined as 1,650,763.73 wavelengths in a vacuum of the orange-red line of the spectrum of krypton 86. A second of time is defined as the deviation of 9,192,631,770 cycles of the radiation associated with the cesium 133 atom (see Exercise 1, Exercise Set 7.1, page 324).

SUPPLEMENT *(Activity Book)*

Activity Set 5.1 Measuring with Metric Units
Just for Fun: Metric Games

Additional Sources

Hopkins, R.A. *The International (SI) Metric System and How It Works.* Tarzana, Ca: Polymetric Services, 1973.

National Council of Teachers of Mathematics. *The Arithmetic Teacher,* **20** No. 4 (April 1973). Six articles devoted to the metric system.

National Council of Teachers of Mathematics. *Measurement in School Mathematics—1976 Yearbook.* Reston, Virginia: NCTM.

Perry, J. *The Story of Standards.* New York: Funk and Wagnalls, 1955.

Willerding, M. *Mathematical Concepts, An Historical Approach.* Boston: Prindle, Weber and Schmidt, 1967. "A Brief History of Measurement," pp. 79–89.

EXERCISE SET 5.1

1. In 1976 New Jersey started replacing damaged or worn-out road signs with new signs showing both miles and kilometers. While almost all educators agree that we should not teach the metric system by converting back and forth from English units to metric units, it is sometimes helpful to make rough comparisons between the two systems.

★   a. Approximately what does the 55 mph speed limit equal in kilometers per hour (kph)?

      b. Jersey City to New York is 24 miles. Approximately what is this distance in kilometers?

★   c. It has been suggested that the speed limit be increased so that the new limit will be 100 kph. Approximately how many miles per hour is this equal to?

"ALL RIGHT—NOW CONVERT THE WHOLE THING TO METRIC."

2. Write the most appropriate metric units for determining the following measurements. Fill in the crossword puzzle with your answers.

Across

2. Weight of a truck
5. Length of a building
7. Volume of a city water supply
8. Length of a river

Down

1. Volume of a gasoline tank
3. Volume of a perfume bottle
4. Width of a television screen
6. Weight of a 50-cent coin

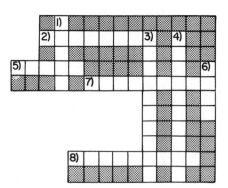

3. Complete the statements and write your answers in the cross-number puzzle, placing one digit in each square. (Statements 4 and 9 Across and 6 and 7 Down involve volumes of water.)

★ *Across*

1. 16.5 cm = ___ mm
4. 3.15 ℓ weighs ___ g
5. .12 ℓ = ___ mℓ
8. ___ kg = 92,000 g
9. ___ mℓ weighs 7.920 kg
10. ___ m = 5.55 km

Down

2. 632,000 ℓ = ___ kℓ
3. 4.5 km = ___ m
6. ___ g is the weight of 432 mℓ
7. ___ kg is the weight of 190 ℓ
8. ___ mg = .9 g
9. .75 m = ___ cm

★ 4. Each of the following grocery store items is measured either by weight or by volume. Connect each item to the appropriate measurement.

946 mℓ   4.536 kg   567 g   384 mℓ   40 g   59 mℓ

5. A measurement taken to within a certain unit may be off by as much as plus or minus one-half of that unit. If a can of pineapple juice, which is labelled 1.32 ℓ, was measured to the nearest hundredth of a liter, its volume is greater than the minimum of 1.315 ℓ and less than the maximum of 1.325 ℓ. Find the minimum and maximum numbers associated with each of the following measurements.

★ a. A two-speed heavy-duty washing machine weighed 112 kg, to the nearest kg.

b. The patient's temperature at 6:00 P.M. was 36.2°C, to the nearest tenth of a degree.

★ c. One of the speakers with a stereo 8-track player/recorder was found to have a width of 48.3 cm, to the nearest tenth of a centimeter.

d. A three-day-old baby weighed 3.46 kg, to the nearest hundredth of a kilogram.

6. *Consumer Mathematics*

★ a. A shopper purchased the following items: tomatoes, 754 g; soup, 772 g; potatoes, 3.45 kg; sugar, 4.62 kg; raisins, 425 g; vegetable shortening, 1.361 kg; and baking powder, 218 g. What is the total weight of this purchase in kilograms?

b. Material that is 1 m wide and 120 cm long is required to make a single window curtain. There are two curtains per window and six windows. If the curtain material comes in rolls of 1-m widths, how many meters of length will be needed to make curtains for all six windows?

★ c. The following amounts of gasoline have been charged on a credit card: 38.2 ℓ; 26.8 ℓ; 54.3 ℓ; 44.7 ℓ; and 34 ℓ. At a price of 18 cents per liter, what is the cost of this gasoline?

d. A car owner has his/her tank filled and records the odometer reading as 14368.7 (km). After a trip in the country, the tank takes 34.5 ℓ to fill and the odometer reads 14651.6. How many kilometers to the liter is this car getting?

★ e. A 24-kg bag of birdseed is priced at $16.88. If 75 g of this feed are put in a bird-feeder each day, how many days before the bag of seed will run out? Rounded off to the nearest penny, how much does it cost to feed the birds each day?

7. A fruit punch calls for these ingredients: 3.5 ℓ of unsweetened pineapple juice; 400 mℓ of orange juice; 300 mℓ of lemon juice; 4 ℓ of ginger ale; 2.5 ℓ of soda water; 500 mℓ of mashed strawberries; and a base of sugar, mint leaves and water which has a total volume of 800 mℓ.

★ a. What is the total amount of this punch in liters?

b. For a party of 30 people, how many milliliters of punch will there be per person?

c. This punch was used at a fair and each drink of 80 mℓ cost 25 cents. What was the profit on the sale of this punch if the ingredients cost $12.50?

8. Prescriptions for the antibiotic garamycin vary from 20 mg for a child to 80 mg for an adult. Garamycin is contained in a vial which has a volume of 2 cm³ (2 mℓ). The garamycin in each vial weighs 80 mg.

★ a. How many cubic centimeters of garamycin are needed for 12 injections of 24 mg each?

b. How many injections of 60 mg each can be obtained from 24 vials?

9. Roof de-icers prevent ice dams on roofs and gutter pipes. An electric heating cable is clipped to the edge of the roof in a sawtooth pattern. In the following questions, assume that this pattern is to run along the edges of a roof with a total length of 28 m.

★   a. How many meters of heating cable will be needed for the edges of the roof, if each 1 m of roof-edge requires 2 m of cable?

      b. There are two gutter pipes which run along the edges of the roof. Each has a length of 14 m. There are two downspouts, one from each gutter pipe, to the ground. Each downspout has a length of 3.2 m. How many meters of cable will be required to go along the gutter pipes and the downspouts?

★   c. The heating cable sells for $1.20 per meter. What is the cost of the total cable?

★ 10. The National Assessment of Educational Progress (NAEP) administered the following question to 17-year-olds in both 1969 and 1973. In 1969, 36% of 17-year-olds attending high school chose the correct answer, but only 27% could do so in 1973. Select the correct answer from 1 to 5.

Suppose that a rubber balloon filled with air does not leak and that it is taken from earth to the moon. One can be sure that, on the moon, the balloon will have the same

     1. size as on earth.   *4. rate of fall as on earth.*
     2. mass as on earth.   *5. ability to float as on earth.*
     3. weight as on earth.   *6. I don't know.*

11. The first unit of measure that was recorded by history is the *cubit*. This is the distance from elbow to fingertips and was used more than 4000 years ago by the Egyptians and Babylonians. An ancient Egyptian cubit measuring 52.5 cm is preserved in the Louvre in Paris. Here is a picture of this cubit and its subdivisions.

Ancient Egyptian cubit

a. How does the length of the Egyptian cubit compare with your cubit?

★   b. A large dark object under the ice on Mount Ararat in Turkey may be Noah's Ark. The biblical dimensions of Noah's Ark from the sixth chapter of Genesis are contained in this table. Convert these measures to the nearest meter using the length of the Egyptian cubit.

	Cubits	Meters
Length	300	
Breadth	50	
Height	30	

12. Track events traditionally recorded in yards and miles are being changed to meters as shown in the next table. This change is the result of an international rule which went into effect on June 1, 1975. Using the fact that a yard is about 91.5 cm, determine whether the metric event is longer or shorter, and compute the difference in meters and centimeters.

	Old Race	New Race	Difference
★ a.	100 yd	100 m	
b.	220 yd	200 m	
★ c.	440 yd	400 m	
d.	880 yd	800 m	
e.	1 mi	1500 m	

Ready or not—metric system is coming.

LOS ANGELES (AP)—The mile run and the 100-yard dash, two of track's glamor races, may soon join the horse-drawn carriage and the five-cent beer as relics of days gone by.

The United States soon will be forced to switch from measuring track meets in yards to measuring them in meters. The Amateur Athletic Union and the National Collegiate Athletic Association have long fought such a switch, but both agree it is becoming mandatory.

Under an international rule which went into effect on June 1, an athlete who runs a race in yards may not qualify for the Olympics, whether he sets a record or not.

The International Amateur Athletic Fed-

13. The Celsius and Fahrenheit temperature scales are related by the following formulas:

$$C = 5(F - 32)/9 \quad \text{and} \quad F = 9C/5 + 32$$

Write in the missing two temperatures for each thermometer to the nearest tenth of a degree.

★   a. Highest recorded temperature, Libya, 1922.

b. At this body temperature, see a doctor.

c. At this temperature, check your car's antifreeze.

d. Lowest recorded temperature, Antarctica, 1960.

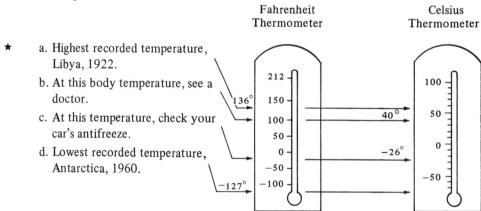

14. Explain how this standard set of 11 brass metric weights can be used with a balance scale to weigh objects with the given weights. *(Hint:* The weights can be placed on both sides of the scale.)

    a. 147 g   ★b. 900 g   c. 128 g

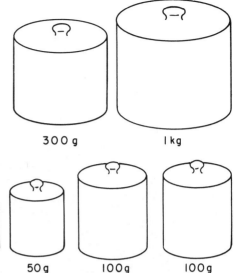

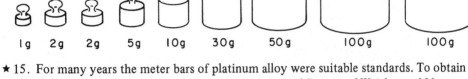

★ 15. For many years the meter bars of platinum alloy were suitable standards. To obtain greater precision and permanence, the International Bureau of Weights and Measures has adopted wavelengths of light as the ultimate standard of length. In 1960 the meter was defined as 1,650,763.73 wavelengths of the orange-red line in the spectrum of the krypton 86 atom.

★   a. How many of these wavelengths are there in 1 mm (to the nearest wavelength)?

★   b. How many of these wavelengths are there in one micron (to the nearest wavelength)?

16. *Challenge:* Find a collection of just 10 weights that can be used to weigh any object from 1 g to 1.023 kg, to the nearest gram. (It is necessary only to place the weights on one side of the scale.)

★ 17. *Bookworm Puzzle:* There are four books which are numbered 1 through 4 from left to right on a shelf. Each book has 300 pages. The thickness of 300 pages is two centimeters and the thickness of each cover is three millimeters. If a worm begins at page 1 of book 1, and eats to page 300 of book 4, what is the shortest distance he could have travelled?

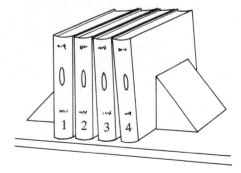

Federal Reserve Bank in Minneapolis

5.2 AREA AND PERIMETER

The design of the Federal Reserve Bank in Minneapolis is based on the mathematical curve called a *catenary*. This is the curve commonly used in the construction of bridges and is formed by suspending cables from two towers. In fact, this building can be thought of as a bridge that is 10 stories deep. There was no prototype for its structural design. Each floor has an unobstructed area of 60 by 275 feet. No occupied floors without internal columns have ever spanned such a length before.

This bank was designed to fulfill some unusual zoning restrictions. One of these was that the "coverage," or ground area occupied by the building, could be only 2.5% of the area of the city block which the bank was to be built on. To satisfy this condition, the bank is supported by two towers and its lower floor is 20 feet above the plaza. The coverage is the small amount of area which the two towers are built on. A 2.5-acre plaza runs under the building and is entirely for the public.

There are two catenaries of steel cable, one on the front side and one on the back side of the building. These cables support the weight of the bank's floors. One of these catenaries and the windows above and below can be seen in the above photograph. By counting the rectangles and parts of a rectangle which are formed by the windows, the areas above and below the catenary can be estimated (see Exercise 1, Exercise Set 5.2).

Relating Area and Perimeter   To measure the sizes of plots of
land, panes of glass, floors, walls, and other such surfaces,
we need a new type of unit—one which can be used to
cover a surface. Theoretically, this unit can have any
shape: rectangular, triangular, etc. The number of units
needed to cover a surface is called the *area.* Squares
have been found to be the most convenient shape for

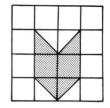

measuring area. The area of this figure is 4 square units because it can be covered by
2 squares and 4 half-squares. Covering a region or surface with unit squares is the basic
concept of area, and yet, it is often poorly understood. A Michigan State assessment
found that fewer than half of the seventh graders knew the area of this figure. About 20
percent of them thought the area was 6.

Another measure associated with a region
is its *perimeter,* the length of its boundary.
The perimeter of the figure on the right is
23 centimeters.

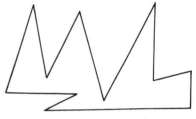

Intuitively, it seems as though the area of a region
should depend on its perimeter. For example, if one
person requires more fence to close in a piece of land
than another person, it is tempting to assume the first
person has the greater amount of land. However, this
is not necessarily true. Figure (a) has an area of 4 square
units and a perimeter of 8 centimeters, while Figure (b)
has the same area but a perimeter of 10 centimeters.

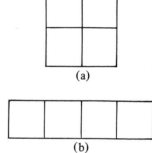

(a)

(b)

It is also possible for two figures to have the same perimeter but different areas. The
perimeters of Figures (c) and (d) are each 10 centimeters, but their areas differ by 1
square.

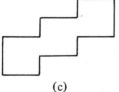

(c)

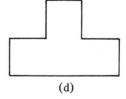

(d)

Units for Area   The earliest units for measuring area were associated with agriculture. The amount
of land that could be plowed in a day with the aid of a team of oxen was called an *acre.* In
Germany, a scheffel was a volume of seed, and the amount of land that could be sown
with this volume of seed became known as a *scheffel of land.* Just as for units of length,
volume, and weight, more carefully defined units of area were eventually adopted.

In the metric system there is a square unit for each unit of length. The common area units are the square kilometer (km^2), square meter (m^2), square centimeter (cm^2), and square millimeter (mm^2). These units are related to one another. *The square centimeter,* for example, is equal to *100 square millimeters.*

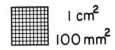

1 cm^2
100 mm^2

The *are* (pronounced "air") is a metric unit for measuring the areas of house lots, gardens, and other such "medium-size" regions. The *are* is equal to the area of a square whose sides measure 10 meters each. The *hectare* is the area of a square whose sides measure 100 meters each. This is approximately the area of two football fields, including end zones.

Rectangles   Rectangles have right angles and pairs of opposite parallel sides, so that unit squares fit onto them quite easily. This rectangle can be covered by 24 whole squares and 6 half-squares. Its area is 27 square centimeters. This area can be obtained from the product 6 × 4.5, because there are four and one-half squares in each of six columns. In general, the *area of a rectangle* is the product of its length times its width.

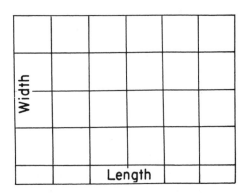

Area = length × width   ($A = lw$)

For a given perimeter, the shape of the rectangle affects its area. The same knotted piece of string in the following photos has been formed into three different rectangular shapes. By using the lengths and widths of each rectangle, you will see that their perimeters are 36 centimeters. Yet, the areas decrease from 80 square centimeters (8 × 10), to 72 square centimeters (6 × 12), to 32 square centimeters (16 × 2), as the shape of the rectangle is changed. If we continue to decrease the width of the rectangle, its area can be made as small as we please, even though the perimeter remains 36 centimeters.

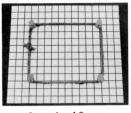

8 cm by 10 cm

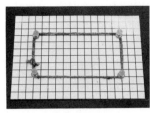

6 cm by 12 cm

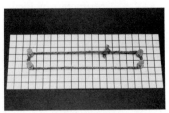

16 cm by 2 cm

Parallelograms, Triangles, and Polygons

Fitting unit squares onto a figure is a good way to acquire an understanding of the concept of area. However, actually placing squares on a region is usually difficult because of the boundary, as in the case of the slanted sides of the parallelogram shown here.

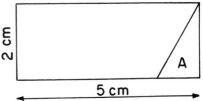

One of the basic principles in finding area is that a region can be cut into parts and reassembled without changing its area. This principle is useful in developing a formula for the area of a parallelogram. In the figure shown here, part *A* has been moved from the left end of the paral-

lelogram in the previous figure to the right end in order to form this rectangle. The parallelogram and the rectangle both have an area of 10 square centimeters. The length (5 centimeters) and width (2 centimeters) of the rectangle are equal to the *base* and *height* (or *altitude*) of the parallelogram, respectively. This suggests the following formula for the *area of a parallelogram*. The height in this formula is the perpendicular distance between the upper and lower bases of the parallelogram.

$$\text{Area} = \text{base} \times \text{height} \quad (A = bh)$$

As in the case of a rectangle, for a given perimeter the area of a parallelogram depends on its shape. The two parallelograms in the following photo are constructed from linkages. Both have the same perimeter, but as the parallelogram is skewed more to the right, its height and area decrease. The area of the first parallelogram is approximately 66 square centimeters and the area of the second one is 40 square centimeters. The height, and consequently the area, of the parallelogram can be made arbitrarily small by further skewing the linkage, while the perimeter will stay constant.

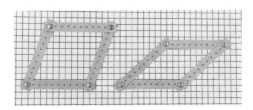

The area of the triangle in this sketch is at least 4 square centimeters. Since the unit squares do not conveniently fit inside its boundary, we will use a different approach for finding its area. By placing two copies of the triangle side by side, they form a parallelogram, as shown in the next figure.

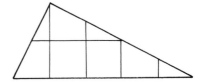

If the length of the *base* of the triangle is b and its *height* (or *altitude*) is h, the area of the parallelogram is b × h. Since the parallelogram has twice the area of the triangle, the triangle's area is one-half of the base times the height. In general, this is the formula for the area of any triangle.

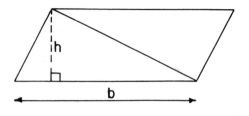

$$\text{Area} = \frac{1}{2} \times \text{base} \times \text{height} \qquad (A = \frac{1}{2} \times bh)$$

Since the preceding triangle has a base of 5 centimeters and a height of 2 centimeters, its area is 5 square centimeters. Each of the three sides of a triangle may be considered as the base, and each side has its corresponding altitude. Triangle *ABC* is drawn below in two positions showing two sets of bases and altitudes. One of these altitudes falls outside the triangle. The area of this triangle can be found by computing 1/2 × 5.8 × 2.0, or 1/2 × 4 × 2.9. In both cases, the area is the same.

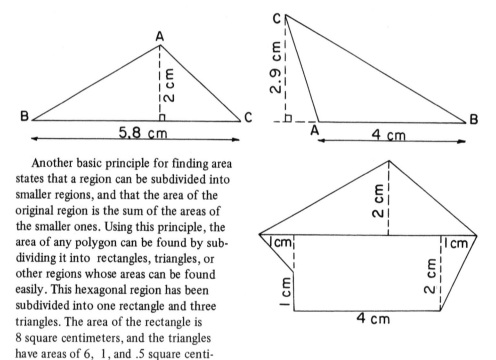

Another basic principle for finding area states that a region can be subdivided into smaller regions, and that the area of the original region is the sum of the areas of the smaller ones. Using this principle, the area of any polygon can be found by subdividing it into rectangles, triangles, or other regions whose areas can be found easily. This hexagonal region has been subdivided into one rectangle and three triangles. The area of the rectangle is 8 square centimeters, and the triangles have areas of 6, 1, and .5 square centimeters. The sum of these areas is 15.5, the approximate area of the hexagon.

Irregular Shapes   Sometimes it is necessary to find the areas of irregular or nonpolygonal shapes. Botanists, for example, compute the areas of leaves to determine the amount of water they lose for each square centimeter of surface area. The water you see on a leaf in the early morning is often not dew condensed from the air, but water which is given off by the leaf. Some of the water which is supplied to the leaf by its system of tiny veins is lost through small openings called stomates. The number of stomates may range from a few thousand to over a hundred thousand for each square centimeter of surface.

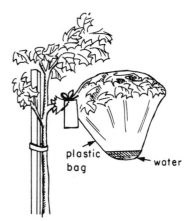

plastic bag — water

Collecting water from leaves

One method of finding the area of a leaf is to trace it on graph paper or place it under a transparent grid. In this example there are 18 large squares (1 centimeter by 1 centimeter) which fall completely inside the boundary of the leaf. Thus, the leaf has a *lower bound* of at least 18 square centimeters. The total number of large squares which contain any part of the leaf is 37. Thus, the leaf has an *upper bound* of at most 37 square centimeters. For each square which intersects the boundary, the area of the square which falls inside the boundary can be estimated. These estimates can be made more accurate by subdividing the boundary squares into smaller parts. In this figure each boundary square has been subdivided into 4 smaller squares. There are 20 of these smaller squares which fall completely inside the boundary of the leaf. These 20 squares

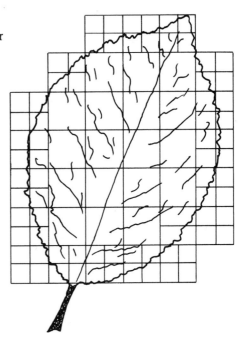

have an area of 5 square centimeters. Thus, the leaf has a new lower bound of 18 + 5, or 23 square centimeters. There are 32 of these small squares which contain some part of the boundary. These squares have an area of 8 square centimeters. Thus, the upper bound of the area of the leaf is not more than 23 + 8 or 31 square centimeters. This process of subdividing squares into smaller squares can be continued to give lower and upper bounds which are closer to the actual area of the leaf.

We have found that the area of this leaf is between 23 and 31 square centimeters. Let's approximate its area by counting the small squares on the boundary that are partially covered by the leaf. If the leaf covers half or more than half of a small square, we will count the square. There appear to be 16 such squares. The area of these squares is 4 square centimeters. Therefore, the area of this leaf is approximately 23 + 4, or 27 square centimeters. If this leaf loses 2 milliliters of water in 24 hours, it will lose 2/27 or .074 milliliter of water per square centimeter.

Engineers and scientists sometimes use a *planimeter* for measuring area. To find the area of a figure, its boundary is traced by one arm of the planimeter while the other stays fixed. As the arm traces the boundary, a disc moves along a scale whose readings are converted into area. It is surprising that a planimeter will determine area by tracing out the perimeter of a

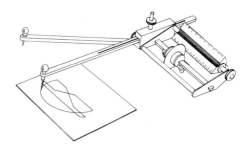

figure, since we have seen that two figures with the same perimeter may have different areas. There is also an instrument, called a *map tracer,* for measuring the length of a curve. This device can be adjusted for different map scales.

Approximating the areas of regions is often easier than approximating the lengths of curves or the perimeters of figures. For a given region, no matter how irregular its boundary, we can always estimate its area fairly accurately by a grid. Two lines, however, can lie fairly close to each other and still differ considerably in length. For example, the following straight line is a poor approximation for the zigzag line. The line has a length of 9 centimeters. The zigzag path has a length of 12.6 centimeters, which is 40 percent longer.

One method of approximating the length of a simple closed curve is to inscribe a polygon in its interior. This method can also be used to approximate the area of the interior by finding the area of the polygon. More accuracy can be obtained by increasing the number of sides of the polygon and drawing the lines closer to the curve.

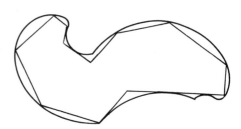

Circumferences of Circles

Circles are part of our natural environment. The sun, moon, flowers, whirlpools, and cross sections of trees all contain circular shapes. The nearly perfect concentric circles on this thin section of a natural pearl were formed by many layers of growth. The circular beach stones shown below were tumbled and shaped by the ocean's tides.

Concentric circles of a section
of natural pearl

The circle was among the most powerful of the early symbols or signs of magic. It had neither beginning nor end and represented eternity. Making a circle with the thumb and index finger is an old good luck gesture that is still used today to mean "OK."

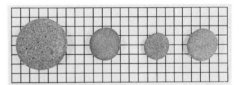

There is something deceptive about trying to estimate the *circumference* (perimeter) of a circle. For example, you may be surprised to know that if you place a piece of string around the edge of the pearl in the photo, the length of the string will be greater than the distance from the top of the pearl photo to the bottom of the tennis can photo.

The tendency to underestimate the circumference of a circle can be dramatized by using objects such as a drinking glass, jar, or can and asking someone to compare the circumference with its height. A tennis ball can is excellent for this experiment because its height appears to be greater than its circumference. How would you answer this question?

If a piece of string just fits around a tennis ball can, as shown in this picture, how would the length of the string compare to the height of the can?

A very common but wrong answer (see photo on the right) is that the string would reach about two-thirds of the way up the can, which shows that some people tend to estimate the circumference of a circle by doubling its diameter. Actually, the distance around a circle is a little more than three times greater than the distance across.

To illustrate this relationship, cut a strip of paper or a piece of string which will reach around a circular object and then fold the paper (or string) into three equal parts. The folded strip will be a little longer than the diameter of the circle. This relationship between the diameter (*d*) or radius (*r*) of a circle, and its circumference (*C*), is expressed by the following formulas:

$$C = \pi \times d \quad \text{or} \quad C = \pi \times 2r,$$

where pi (π) is approximately 3.14 or 22/7. For many estimation purposes, a value of pi equal to 3 is accurate enough.

Areas of Circles   The area of a circle can be approximated reasonably well by the area of an inscribed polygon. As you can see from this drawing, the area of the dode-cagon is just a little less than the area of the circle. The triangle formed by one side of the dodecagon and two radii of the circle has an area of $1/2 \times bh$. Since the dodecagon can be subdivided into 12 such triangles, its area is $1/2 \times 12bh$.

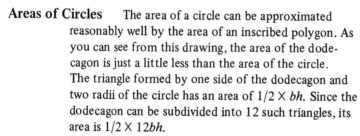

If we were to inscribe a polygon with a greater number of sides, its area would be even closer to the area of the circle. Notice that the perimeter of the dodecagon, $12b$, is very close to the circumference, *C*, of the circle. Also the altitude, *h*, of the triangle is close to the radius, *r*, of the circle. If $12b$ is replaced by *C*, and *h* is replaced by *r*, then $1/2 \times 12bh$ will be approximately equal to $1/2 \times Cr$. Since the area of the inscribed polygon can be made arbitrarily close to the area of the circle, this suggests that the area of the circle is $1/2 \times Cr$. Since the circumference is equal to $2\pi r$,

$$\frac{1}{2} \times Cr = \frac{1}{2} \times 2\pi r \times r = \pi r^2$$

In general, for any circle of radius *r*, its area *A* is π times the square of the radius.

$$\text{Area} = \pi \times (\text{radius})^2 \quad (A = \pi r^2)$$

Precision The sides of this pentagon have been measured to within one-tenth of a centimeter. This means that the side whose length is labelled 4.6 cm is accurate to within .05 centimeter. That is, its true length is between 4.55 and 4.65 centimeters. The side labelled 2.0 cm could have an actual length as small as 1.95 centimeters or as big as 2.05 centimeters. Using minimum and maximum numbers, such as these, an interval can be deter-

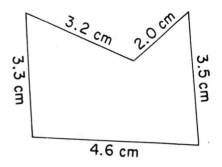

mined within which the perimeter of this pentagon is contained. The minimum and maximum numbers for the five measurements are listed below. The perimeter of the pentagon is greater than the sum of the minimum numbers and less than the sum of the maximum numbers.

Minimum	Given Measurements	Maximum
3.25 cm	3.3 cm	3.35 cm
4.55 cm	4.6 cm	4.65 cm
3.45 cm	3.5 cm	3.55 cm
1.95 cm	2.0 cm	2.05 cm
+ 3.15 cm	+ 3.2 cm	+ 3.25 cm
16.35 cm	16.6 cm	16.85 cm

The sum of the given lengths of the pentagon is 16.6 centimeters. The actual length of the perimeter could be as small as 16.35 centimeters or as big as 16.85 centimeters.

The amount of precision in computing with measurements depends on the precision of the original measurements. The previous example shows that precision cannot be increased by computation. The sides of the pentagon were measured to within one-tenth of a centimeter, but the perimeter, 16.6 centimeters, is accurate to within only .25 centimeter.

A range or interval of error can also be computed for products of measurements. The length and width of this rectangle were measured to within one-tenth of a centimeter. These dimensions could be as small as 2.75 and 5.25 centimeters, respectively, or as big as 2.85 and 5.35 centimeters, respectively. The interval within which the area of this rectangle falls is determined by computing the area for these minimum numbers and then the area for the maximum numbers.

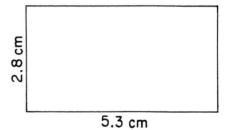

Minimum	Given Measurements	Maximum
2.7 5	2.8	2.8 5
× 5.2 5	× 5.3	× 5.3 5
1 3 7 5	8 4	1 4 2 5
5 5 0	1 4 0	8 5 5
1 3 7 5	1 4.8 4	1 4 2 5
1 4.4 3 7 5		1 5.2 4 7 5

The area of the rectangle, according to the given measurements (length = 5.3 centimeters and width = 2.8 centimeters), is 14.84 square centimeters. The actual area of this rectangle could be as small as 14.4375 square centimeters or as big as 15.2475 square centimeters.

SUPPLEMENT *(Activity Book)*

Activity Set 5.2 Areas on Geoboards

Just for Fun: Pentominoes

Additional Sources

Beck, A., M.N. Bleicher and D.W. Crowe. *Excursions into Mathematics.* New York: Worth, 1969. "What Is Area," pp. 147–209.

Bruni, J.V. *Experiencing Geometry.* Belmont, California: Wadsworth, 1977. "Perimeter and Area," pp. 76–110.

Hirstein, J.J., C.E. Lamb and A. Osborne. "Some misconceptions about area measure." *The Arithmetic Teacher,* 25 No. 6 (March 1978), 10–16.

Holmes, P. "The Area of a Circle Lies Between $3r^2$ and $4r^2$." *Mathematics in Schools,* 2 No. 5 (September 1973), 25–28.

Menninger, K.W. *Mathematics in Your World.* New York: The Viking Press, 1962. "Peach Trees, Honeycombs and Queen Dido" pp. 41–50.

School Mathematics Study Group. *Mathematics and Living Things.* Student Text. Palo Alto: Stanford University, 1965. pp. 49–53 (leaf surface area and water loss).

Walter, M. "A common misconception about area." *The Arithmetic Teacher,* 17 No. 4 (April 1970), 286–89.

Willerding, M. *Mathematical Concepts, A Historical Approach.* Boston: Prindle, Weber and Schmidt, 1967. pp. 113–21 (Archimedes' method for evaluating π).

EXERCISE SET 5.2

1. This is the skeletal structure of the Minneapolis Federal Reserve Bank. Its ten floors and the vertical beams running down the front of the bank, partition the area above and below the catenary into rectangles. The dimensions of these rectangles are approximately 2 m by 4 m.

   ★ a. Use this diagram of the bank to approximate the area of the front of the bank which lies under the catenary by counting rectangles and parts of rectangles.

   ★ b. Use the rectangles to find the width and height of the bank's ten stories. What is the area enclosed by these dimensions?

   c. Use the areas in parts **a** and **b** to approximate the area above the catenary (not including the portion above the ten stories).

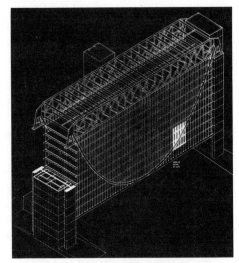

Skeletal structure of the Minneapolis Federal Reserve Bank

2. The wheel has been called the most important invention of all time. Consider the diameter of each wheel in this cartoon to be 75 cm and the distance between opposite pairs of wheels to be 300 cm. How many revolutions of the wheels will it take to turn this contraption in one complete circle?

"Just because you invented the *wheel*, it doesn't follow logically that you can go on to the *wagon*."

3. The following question is taken from a mathematics test which was given to 9-year-olds by the National Assessment of Educational Progress (NAEP).

a. Which of the following figures has the same area as the 4 by 4 square?

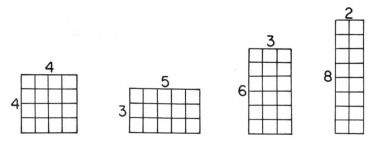

★ b. Only 44 percent chose the 8 by 2 rectangle, and almost as many selected the 3 by 5 rectangle. The selection of the 3 by 5 rectangle may indicate confusion over which two concepts of measurement? Explain why they might have made this choice.

★ 4. There are seven rectangles with a perimeter of 28 cm, whose lengths and widths are whole numbers of centimeters. One of these has a width of 3 cm, a length of 11 cm, and an area of 33 cm². Write the lengths and areas of the remaining six rectangles in the table, and plot these areas on the graph. For any given perimeter, what is the shape of the rectangle that will have the greatest area?

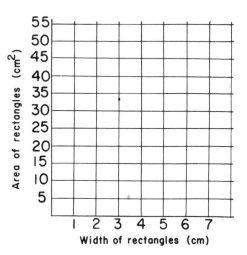

Width of rectangles	Length of rectangles	Area of rectangles
I cm		
2 cm		
3 cm	II cm	33 cm²
4 cm		
5 cm		
6 cm		
7 cm		

5. *Consumer Mathematics*

★ a. A store sells two types of Christmas paper: Type A has four rolls per package and costs $2.99, and each roll is 75 cm by 150 cm; Type B has a single roll, costs $3.19, and is 88 cm by 500 cm. In which case will you get more paper for your money?

b. An all-purpose rug costs $27.50 per square meter. What is the cost of a rug whose dimensions are 360 cm by 400 cm?

★ c. Tru-site glass for picture frames sells at the rate of $20.00 per square meter. What is the total cost for the glass in these two picture frames: Frame A, 58 cm by 30 cm; and Frame B, 40 cm by 60 cm?

d. The length of a kitchen cupboard is 3.5 m. There are three shelves in the cupboard, each of width 30 cm. How many rolls of shelf paper will be needed to cover these shelves, if each roll is 30 cm by 3 m?

★ e. The instructions on a bag of lawn fertilizer recommend that 35 g of fertilizer be used for each square meter of lawn. How many square meters of lawn can be fertilized with a 50-kg bag?

6. Each dimension of a 30 cm by 58 cm pane of antique stained glass was measured to the nearest centimeter.

★ a. According to these measurements, what is the area in square centimeters?

b. Assuming that these measurements could be off by as much as plus or minus one-half centimeter, what are the minimum numbers for the length and width of this glass?

★ c. What is the area of this glass for the minimum numbers in part **b**?

d. If this glass sells for 15 cents per square centimeter, how much will it cost for the area in part **a**?

e. If this glass has the minimum dimensions which were found in part **b**, how much money would you lose on this purchase?

7. Of all simple closed curves of equal length, the circle encloses the largest area. James R. Newman in *The World of Mathematics* tells how this fact was used by the ancients.* "The Phoenician princess Dido obtained from a native North African chief a grant of as much land as she could enclose with an ox-hide. A clever girl, she cut the ox-hide into long thin strips, tied the ends together, and staked out a large and valuable territory on which she built Carthage. Horatio—he who made his reputation defending the bridge—was rewarded by a gift of as much land as he could plow round in a day . . . ." Each of the following figures has a perimeter of 120 mm. Compute their areas in square millimeters and note the differences between them.

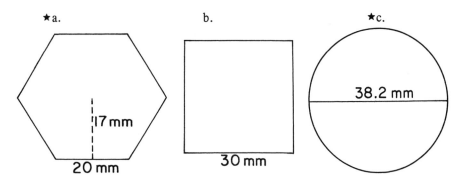

★a.

b.

★c.

17 mm

20 mm

30 mm

38.2 mm

*J.R. Newman, *The World of Mathematics*, 2 (New York: Simon and Schuster, 1956), p. 882.

8. A trundle wheel is a convenient device for measuring distances along the ground. Every time the wheel makes one complete revolution it has moved forward 1 m. If you were to cut this wheel from a square piece of plywood, what would be the dimensions of the smallest square you could use? (*Hint:* Find the diameter of the wheel.)

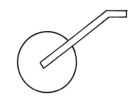

9. Some humidity in homes is necessary for comfort, but too much can cause mold and peeling paint. Paint-destroying moisture can come from walls, crawl spaces, and attics.

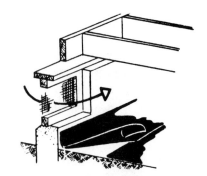

★   a. An attic should have 900 cm² of ventilation for each 27 m² of area. How many square centimeters of ventilation is needed for a 4 m by 12 m attic area?

   b. A crawl space should have 900 cm² of ventilation for each 27 m² of area, plus 1800 cm² for each 30 m of perimeter around the crawl space. How many square centimeters of ventilation is needed for a 10 m by 15 m crawl space?

10. Many of the problems found on ancient Egyptian scrolls (1850–1650 B.C.) contain formulas for computing land areas. Two of these formulas are given in parts **a** and **c**.

★   a. Find the area of this circle by the Egyptian formula: "The area of a circle is equal to the area of a square whose width is 8/9 the diameter of the circle."

   b. Compute the area of this circle using

   $$A = \pi r^2 \quad (\pi = 3.14)$$

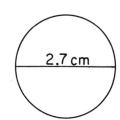

2.7 cm

★   c. The Egyptians used the following formula to find the area of a quadrilateral: "For successive sides of lengths *a*, *b*, *c*, and *d*, the area is $(a + c)(b + d)/4$." Use this formula to compute the area of the quadrilateral shown here.

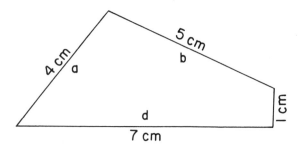

d. Find the area of this quadrilateral by subdividing it into smaller regions and finding their areas. (This can be done by two triangles and one rectangle. Measure lengths to the closest millimeter.)

e. Compare your answers in parts **b** and **d** to the areas obtained in parts **a** and **c**. Did the Egyptian formulas give areas that are too big or too small?

11. Compute the areas of each of these figures in square millimeters and square centimeters.

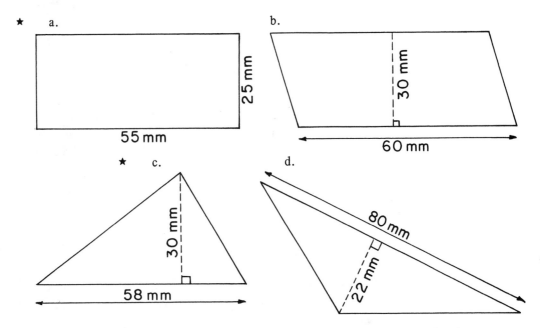

★ a.

25 mm

55 mm

b.

30 mm

60 mm

★ c.

30 mm

58 mm

d.

80 mm

22 mm

12. The 70-story cylindrical building shown at the top of the next page is the Peachtree Plaza Hotel in Atlanta, Georgia. It contains a 7-story central court with a half-acre lake and over 100 trees. The diameter of this building is 35.36 m (116 ft). According to its architect, John Portman, cylindrical walls were chosen rather than the more common rectangular walls because a circle encloses more area with less perimeter than any other shape.

★ a. What is the area of its cross section (that is, the area of any floor)?

b. What is the perimeter to the nearest meter of a square that encloses the same area as you found in part **a**? (*Hint:* First find the square root of the area in part **a**.)

★ c. What is the perimeter to the nearest meter of the Peachtree Plaza Hotel?

d. How many more meters is the perimeter of the square in part **b** than the perimeter of the hotel?

★ e. The Peachtree Plaza Hotel is 230 m (754 ft) tall. This number multiplied by your answer in part **d** will give the additional wall area that would be needed to enclose the same space if the hotel had a square base rather than a circular one. What is this area?

Peachtree Plaza Hotel in Atlanta, Georgia

13. This wall has a length of 540 cm and a height of 240 cm. A few dimensions are also given around the window and fireplace.

★ a. How many square centimeters of wallpaper will this wall require?

b. A standard roll of wallpaper is 12.8 m by 53 cm, and partial rolls are not sold. How many rolls must be purchased to cover this wall?

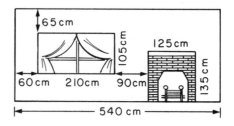

14. The common starfish has five arms. Most species grow as large as 20 to 30 cm in diameter (greatest distance between tips of arms), but some species reach only 1 cm. The body surface contains hairlike projections capable of producing a waving motion, which carries food to the mouth and removes unwanted particles from the body. This starfish was photographed on a centimeter grid and has a diameter of approximately 9.5 cm.

★ a. Compute an upper bound for the area of the underside of this starfish by counting every square that is covered or partially covered. (*Hint:* Count the squares which are not covered or not partially covered.)

b. Compute a lower bound for the area of this starfish by counting only those squares which are completely covered. (*Hint:* Use the region outside the boundary of the starfish.)

c. Approximate the area of this starfish by counting only those squares which are at least half covered. (*Note:* Your answer should be between the two areas from parts **a** and **b**.)

15. There are many factors that enter into the assessment of a house for determining property taxes. Once the proper category is determined, the assessment rate is per square foot or square meter. Compute the value of the following one-floor dwellings for an assessment rate of $189 per square meter.

★ a. Ranch Style: 8 m by 13.75 m

b. L-shaped: 8.7 m by 11 m plus 8 m by 9.5 m

★ c. How much in taxes must be paid on these houses if the tax rate is $52 on every $1000 worth of assessment?

★ 16. The sides of the square on this circle have a length equal to the radius of the circle. The area of the circle is obviously less than four of these squares. Continue placing the pie-shaped sectors of the circle on the squares (as shown) to see what part of the four squares they will cover. Estimate your answer to the nearest one-tenth of a square. How would your answer be affected if the circle were divided into smaller sectors? How should your answer be related to π?

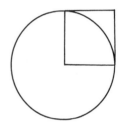

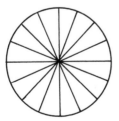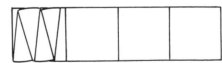

17. Physicists study cosmic radiation to learn about properties of our galaxy and levels of sun activity. This proton histogram contains information on the intensity level of cosmic rays. Region A (shaded) under the histogram, which is called the *background area,* is compared with the total area of regions A and B. (Use a millimeter ruler for parts **a** and **b**.)

★ a. What is the approximate area of A in square millimeters?

   b. What is the approximate area of B?

★ c. Region A is what percent of the total area of A and B?

18. *Earth Belt Puzzle:* Suppose there were a belt which fit tightly around the earth and you wished to cut it and splice in an additional piece so that the belt would be 1 m from the earth. About how much new belt would need to be spliced in?

19. *Demonstration:* The vertical height of a tennis ball can is approximately equal to the circumference of a tennis ball. This can be illustrated by rolling a tennis ball along the edge of a can. The ball will make one complete revolution. Since a can holds three tennis balls, what does this demonstration imply about the diameter of a ball as compared to its circumference?

★ 20. *Twelve-Penny Conjecture*:* Here are two methods of packing 12 discs or pennies into a rectangle. One of these rectangles has less wasted space than the other. Which one? It is an unproven conjecture that this rectangle has the smallest amount of wasted space of all rectangles containing 12 pennies.

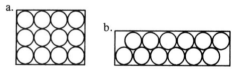

*C.S. Ogilvy, *Tomorrow's Math,* 2nd ed. (New York: Oxford University Press, 1972), pp. 32–33.

Liquefied natural gas tanker, *Aquarius*

5.3 VOLUME AND SURFACE AREA

This is the *Aquarius*, one of twelve liquefied natural gas tankers being built by the Quincy Shipbuilding Division of General Dynamics. The *Aquarius* is longer than three football fields (285 meters) and carries five spherical aluminum tanks, each having a diameter of 36.58 meters. The space inside these tanks is measured by the number of unit cubes that are required to fill it. The number of such cubes is the measure which we call *volume*. The appropriate unit cube for these tanks is cubic meters (m^3), the volume of a cube whose edges have a length of 1 m. Each of these spheres holds 25,000 cubic meters of liquefied gas at ⁻165° Celsius (⁻265° Fahrenheit).

Relating Volume and Surface Area   The cubic centimeter is the common unit for measuring small volumes. The first box shown here has 12 cubes on its base and will hold 12 more above these. Its volume is 24 cubic centimeters (24 cm³).

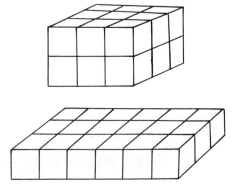

Another important measure for space objects is their amount of surface. For example, the effects of the wind and temperature changes on a building depend on the size of the building's outside surface. Just as in the case of two-dimensional figures, surface area is the number of unit squares needed to cover the surface. The top of this box, for example, has a surface area of 12 square centimeters and the front face has an area of 8 square centimeters. The total surface area, including the base, is 52 square centimeters.

The size of the surface area of an object cannot be predicted by knowing its volume, any more than the perimeter of a figure is determined by its area. These two boxes on the right, for example, both have a volume of 18 cubic centimeters, but there is a good deal of difference in the areas of their surfaces. The top box has a surface area of 42 square centimeters, and the lower box has a surface area of 54 square centimeters. If you were going to build a box that had a volume of 18 cubic centimeters, building the lower box would require about 30 percent more material than the top box. In general, for a given volume, the box that is closer to the shape of a cube will have the smaller surface area.

Prisms   In this drawing the length (20) times the width (10) gives the number of cubes (200) on the floor of the box (or base of the rectangular prism). Since the box can be filled with six such floors of cubes, it will hold 1200 cubes. This volume of 1200 cubic centimeters can be obtained by multiplying the three dimensions of the box. In general, a rectangular prism with length l, width w, and height h has the following volume:

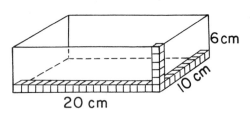

6 cm

10 cm

20 cm

$$\text{Volume} = \text{length} \times \text{width} \times \text{height} \quad (V = lwh)$$

For any prism the volume can be found in a similar way. The base of this prism is a right triangle which is covered by four and one-half cubes. Since six different floors or levels of four and one-half cubes each fill the prism, its volume is 6 × 4.5, or 27 cubic centimeters.

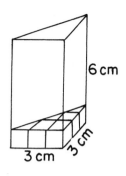

The number of cubes which cover the base of this prism is the same as the area of the base. Therefore, the volume of the prism can be computed by multiplying the area of the base by the height (or altitude) of the prism. In general, for any right prism having a base of area B and a height of h, its volume can be computed by the following formula:

Volume = area of base × height $(V = B \times h)$

A formula for the volume of an oblique prism can be suggested by beginning with a stack of cards, and then pushing them sideways to form an oblique prism. If each file card is 12.5 centimeters by 7.5 centimeters and the stack is 5 centimeters high, its volume will be 12.5 × 7.5 × 5, or 468.75 cubic centimeters. The base of the oblique prism is also 12.5 centimeters by 7.5 centimeters, and its *height* (or *altitude*), the perpendic-

ular distance between its upper and lower bases, is 5 centimeters. Since both stacks contain the same number of cards, their volumes are both 468.75 cubic centimeters. This means that the volume of the oblique prism can be computed by multiplying the area of its base by its height. In general, the area of any prism, right or oblique, is the area of its base times its height.

Sometimes a prism can have more than one base, as shown by the following sketches. The first prism is a right prism with a base that is a parallelogram. By turning the prism as shown in the two remaining positions, it can be classified as oblique with a rectangular base.

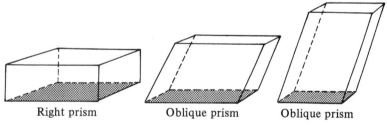

Right prism   Oblique prism   Oblique prism

The surface area of a prism is the sum of the areas of its bases and faces. For right prisms these faces are rectangles, and for oblique prisms the faces are rectangles and parallelograms.

Cylinders   These cylindrical buildings are part of the Renaissance Center in Detroit. Just as with the conventional rectangular-shaped buildings, architects need to know the volumes and surface areas of these glass-walled cylinders.

To compute the volumes of cylinders, we continue to use unit cubes even though they do not conveniently fit into a cylinder. The cubes in the cylinder pictured below to the right show that more than 33 cubes are needed to cover its base. Furthermore, since the cylinder has a height of 12 centimeters, it will take at least 12 × 33, or 396, cubes to fill the cylinder.

If we were to use smaller cubes they could be packed closer to the boundary of the base, and a better approximation could be obtained to the volume of the cylinder. This suggests that the formula for the volume of a cylinder is the same as that for the volume of a prism. For a cylinder with a base area of *B* and a height of *h*, its volume is computed by this formula:

Renaissance Center in Detroit

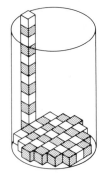

Volume = area of base × height   ($V = B \times h$)

A cylinder without bases can be formed by joining the opposite edges of a rectangular sheet of paper. The circumference of the base of the cylinder is the length of the rectangle, and the height of the cylinder is the height of the rectangle. Therefore, the surface area of the sides of a cylinder is the circumference of the base of the cylinder times its height.

For any cylinder whose base has a radius r and whose height is h, the base has a circumference of $2\pi r$, and the surface area of the sides of the cylinder is $2\pi rh$. Adding the area of each base, πr^2, the total surface area is $2\pi rh + 2\pi r^2$.

Pyramids   The Egyptian Great Pyramid, or Pyramid of Cheops, as it is called, has a height of 148 meters and a square base with a perimeter of 930 meters. The Transamerica Pyramid in San Francisco has a height of 260 meters and a square base with a perimeter of 140 meters. In spite of the height differences in these giant pyramids, the Egyptian pyramid has several times more volume.

Transamerica Pyramid in San Francisco

The shaded pyramid inscribed in the cube on the right has a square base, *EFGH,* and a height *GC.* This pyramid, together with the two pyramids having bases *EABF* and *EADH,* and vertices *C,* partition the cube into three congruent pyramids. Therefore, the volume of the shaded pyramid is one-third of the volume of the cube. That is, the volume of the pyramid is one-third of the area of the base of the cube times the height of the cube.

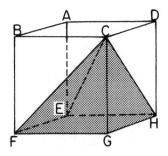

In general, if *B* is the area of the base of a pyramid and *h* is the perpendicular distance from the upper vertex to the base, the volume of the pyramid is computed by this formula:

$$\text{Volume} = \frac{1}{3} \times \text{area of base} \times \text{height} \qquad (V = \frac{1}{3}Bh)$$

Let's apply this formula to find the volume of the Great Pyramid. Its base has an area of 54057 square meters, and its height is 148 meters. Therefore, its volume is 1/3 X 54057 X 148, or 2,666,812 cubic meters.

Cones

This pile of crude salt has the shape of a cone with a circular base, and it has been formed in the same way as you might build a sand castle by letting sand run through your hands. The salt has been solar-evaporated from ocean water and is being stored to await further purification. Cone-shaped piles of sand, gravel, and stone are common sights at construction companies.

The volume of a cone can be approximated by the volume of a pyramid which is inscribed in the cone. The hexagonal pyramid in this figure has a volume of 430 cubic centimeters, and the volume of the cone is approximately 523 cubic centimeters. As the number of sides of the base of the pyramid is increased, its volume becomes closer to the volume of the cone. Since the volume of the pyramid is one-third the area of its base times its height, this suggests that the same formula will give the volume of a cone. In general, the volume of any cone whose base has area *B* and whose height is *h,* is given by the following formula:

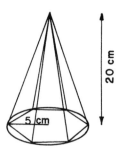

20 cm

5 cm

$$\text{Volume} = \frac{1}{3} \times \text{area of base} \times \text{height} \quad (V = \frac{1}{3}Bh)$$

Let's apply the formula for the volume of a cone to compute the amount of salt in the photo on the previous page. The height of this cone is 12 meters, and the diameter of its base is 32 meters. Therefore, its volume equals $1/3 \times \pi \times 16^2 \times 12$, or approximately 3215 cubic meters. This will fill about 40 railroad flat cars having a capacity of 80 cubic meters each.

Spheres

This is a view of the earth as seen from the *Apollo 10* spacecraft as it passed over the moon's surface. The earth, the planets, and their moons are all spheres, spinning and orbiting about a spherical-shaped sun. There is considerable variation in the volumes of these spheres. The earth has about 18 times more volume than the smallest planet, Mercury. The largest planet, Jupiter, is 10,900 times bigger than the earth. The sun is the largest body in our solar system with a volume which is more than a million times greater than the earth's.

Apollo 10 view of earth over moon's surface, 1969

Formulas for the volume and surface area of a sphere were known by Archimedes (287–212 B.C.) and described in his work *On the Sphere and Cylinder.* Archimedes discovered the following remarkable relationships between a sphere and the smallest cylinder containing it: The volume of the sphere is two-thirds the volume of the cylinder; and the surface area of the sphere is two-thirds the surface area of the cylinder. Archimedes rated these discoveries among his greatest accomplishments and requested that they be engraved on his tombstone.

For a sphere of radius r, the smallest cylinder to contain it has a height of $2r$. The volume of this cylinder is $\pi r^2 \times 2r$, or $2\pi r^3$. Two-thirds of this, $2/3 \times 2\pi r^3$, is $4/3 \times \pi r^3$, the formula for the volume of a sphere. If you think of π as approximately 3, the volume of the sphere is about $4r^3$, that is, four times the volume of the cube whose sides have the length of the radius of the sphere.

The surface area of this cylinder is $6\pi r^2$. Two-thirds of this, $2/3 \times 6\pi r^2$, is $4\pi r^2$, the surface area of the sphere. This shows that the surface area of a sphere is exactly four times greater than the area of a disc whose center is the center of the sphere (see shaded disc).

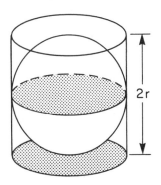

Irregular Shapes   The volume of an irregular-shaped figure can be determined quite easily by submerging it in water and measuring the volume of the water which is displaced. To illustrate this method we will find the volume in cubic centimeters of this miniature statue. Before submerging the statue, the cylinder has been filled to a height of 700 milliliters. When the statue is placed in the cylinder, the water level rises to the 800-milliliter level. This means that the volume of the statue is equal to the volume of 100 milliliters of water. Since each milliliter of water has a volume of 1 cubic centimeter, the volume of the statue is 100 cubic centimeters.

Creating Surface Area   A potato can be cooked in a shorter time if it is cut into pieces. Ice will melt faster if it is crushed, and coffee beans will provide a better coffee if they are ground before they are boiled. The purpose of crushing, grinding, cutting, or, in general, subdividing, is to increase the surface area of a substance. While you may be aware of this, it is the rate and amount of increase of surface produced that is surprising.

To illustrate how rapidly surface area can be created, consider the cube that is 2 centimeters on each edge. Its volume is 8 cubic centimeters, and its surface area is 24 square centimeters. If this cube is cut into 8 smaller cubes, the total volume is still 8 cubic centimeters, but the surface area is doubled to 48 square centimeters. This can be easily seen by looking at the small cube in the upper right-hand corner. Faces a, b, and c contributed 3 square centimters to the area of the original cube, and after the cut, its remaining three faces were exposed, contributing 3 more square centimeters of area. Since this is true for each of the 8 smaller cubes, the total increase in surface area is 8 × 3 square centimeters, or 24 square centimeters.

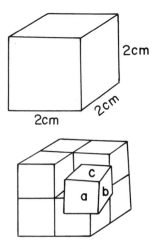

If we continue this process, by cutting each of the centimeter cubes into 8 smaller cubes, there will be 64 cubes whose edges have lengths of 1/2 centimeter. The total volume of these cubes is still 8 cubic centimeters, but the surface area is doubled to 96 square centimeters.

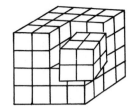

For the third such cut the surface area becomes $2^3 \times 24$, or 192 square centimeters, and after the 12th such cut it has increased to $2^{12} \times 24$, or 98,304 square centimeters! During this splitting process the volume has remained the same, 8 cubic centimeters.

This process of subdividing can also be used to double the surface area of a sphere. If for example, a sphere of radius 2 centimeters is formed into 8 smaller spheres, each with a radius of 1 centimeter, the total volume will remain the same but the surface area will be doubled. As in the case of the cubes, if this subdividing is carried out far enough, a surface area can be increased many times. Consider this effect when water is sprayed into the air in a fine mist, as happens with snow-making machines. The surface area of each drop of water is increased hundreds of times, and the minute particles of water freeze into snowflakes.

Nature also divides and splits substances into smaller parts to increase surface area. Our lungs contain about 300 million tiny cavities to provide the necessary amount of surface area for oxygen. Their area is about 120 square meters, which is more than the area of the walls of a large size room! If the internal surfaces of lungs were smooth and flat, their area would be only about 1.5 square meters.

SUPPLEMENT *(Activity Book)*

Activity Set 5.3 Models for Volume and Surface Area (Prisms, Pyramids, Cones, Cylinders, and Spheres)

Just for Fun: Soma Cubes

Additional Sources

Aaboe, A. *Episodes from the Early History of Mathematics.* New York: Random House, 1964. pp. 73–99.

Dodd, W.A. "History of Mensuration." *Mathematics in Schools,* **6** No. 4 (September 1977), 8–10.

Loeb, A.L. "Remarks on some elementary volume relations between familiar solids."

The Mathematics Teacher, **58** No. 5 (May 1965), 417–19.

Menninger, K.W. *Mathematics in Your World.* New York: The Viking Press, 1962. "The Strange Interplay of Volume and Surface," pp. 219–36.

"Galileo showed men of science that weighing and measuring are worthwhile. Newton convinced a large proportion of them that weighing and measuring are the only investigations that are worthwhile."

Charles Singer

EXERCISE SET 5.3*

1. Each of the mathematical objects listed below can be seen in the accompanying picture. Try to find the objects.

   a. Cone   ★b. Pyramid
   c. Cylinder   d. Sphere
   e. Circle   ★f. 30° angle
   g. Rectangle   h. Semicircle
   i. Square   ★j. 45° angle

2. The liter is the metric unit which will replace the quart. A 1-quart milk carton has a square base of 7 cm by 7 cm and a height of 19.3 cm.

   ★   a. What is its volume?
   b. Which is greater, a quart or a liter?

Thompson Hall, University of New Hampshire

3. Thor has invented a "new wheel" but B.C. doesn't seem to be overly impressed.

*Use π = 3.14 in this Exercise Set.

★   a. What is the volume of this wheel to the nearest cubic centimeter, if it has a length and height of 1 m, a thickness of 20 cm, and an inner diameter of 46 cm?

   b. If the stone this wheel is made of weighs 7 g per cubic centimeter, what is the weight of the wheel in kilograms?

4. *Consumer Mathematics*

★   a. A house with ceilings that are 2.4 m high has five rooms with the following dimensions: 4 m by 5 m; 4 m by 4 m; 6 m by 4 m; 6 m by 6 m; and 6 m by 5.5 m. Which of the following air conditioners will be adequate to cool this house: an 18,000 Btu unit for 280 m^3; a 21,000 Btu unit for 340 m^3; or a 24,000 Btu unit for 400 m^3?

   b. A woodshed has a length, width, and height of 3 m by 2 m by 2 m. If each 1.5 m^3 of firewood sells for $25, how much will it cost to completely fill the shed with wood?

★   c. A catalog describes two types of upright freezers: Type A has a 60 cm by 60 cm by 150 cm storage capacity and costs $339; Type B has a 55 cm by 72 cm by 160 cm storage capacity and costs $379. Which freezer has the most cubic centimeters for each dollar?

   d. A drugstore sells the same brand of talcum powder in two types of cylindrical cans: Can A has a diameter of 5.4 cm, a height of 9 cm, and sells for $1.59; Can B has a diameter of 6.2 cm, a height of 12.4 cm, and sells for $2.99. Which can is the better buy?

5. The number of fish that can be put in an aquarium depends on the amount of water, the size of the fish, and the capacity of the pump and filter system.

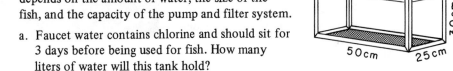

★   a. Faucet water contains chlorine and should sit for 3 days before being used for fish. How many liters of water will this tank hold?

   b. The recommended number of tropical fish for this tank is 30. How many cubic centimeters does this allow for each fish?

   c. Goldfish need more space and oxygen than tropical fish. Those about 5 cm in length require 3000 cm^3 of water. How many of these fish can adequately live in this tank?

6. In an NAEP (National Assessment of Educational Progress) question on volume, students were shown a sketch of a box divided into cubes, similar to this one, and asked how many cubes the box contained. Only 6% of the 9-year-olds, 21% of the 13-year-olds, and 43% of the 17-year-olds correctly answered the question.

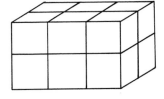

★ a. The most common wrong answer was 16. Explain how the students might have obtained this answer.

b. What two concepts of measure might they have confused?

7. Swimming pools require daily testing to determine the pH factor and the chlorine content. Pumps and filters are also necessary, and some pools have heating systems.

★ a. What is the depth of a 6 m by 12 m pool that contains 193 kℓ of water?

b. If this pool requires 112 g of chlorine every 2 days, how many kilograms of chlorine should be purchased for a 3-month period?

★ c. The Alcoa Solar Heating System for pools has 32 square panels which are 120 cm by 120 cm. Will this heating system fit onto a 5 m by 8 m roof?

d. Each panel for this system holds 5.68 ℓ of water. What is the total weight of water for 32 panels?

8. This art form is Alex Lieberman's *Argo* which was placed on Newport, Rhode Island's, Ocean Drive for the 1974 Monumenta Exhibition. It is now at the Walker Art Center in Minneapolis. The entire display weighs about 4535 kg.

a. The cylinder shown in front is approximately 2 m tall and 1 m in diameter. What is its volume?

★ b. What is the weight of a volume of water which is equal to the volume of the front cylinder? (One cubic centimeter of water weighs approximately 1 g.)

9. Each aluminum sphere for the liquefied natural gas tankers which are being built by General Dynamics has a diameter of approximately 36.6 m and a weight of 725,750 kg (800 tons). These spheres are constructed in Charleston, South Carolina, and then towed by tug to Quincy, Massachusetts.

★ a. Compute the volume of one of these spheres in cubic meters.

   b. A heavy external coating of insulation enables the sphere to maintain the liquefied natural gas at ⁻165° Celsius. What is the surface area of one of these spheres?

10. The concrete foundation for the new office building on the corner of Congress and State streets in Boston, required 496 truck loads of concrete

Spheres for liquefied natural gas tankers

and was formed in one continuous pouring over a 30-hour period.

   a. The concrete was poured to a depth of 1.8 m and covered an area of 2420 m². How many cubic meters of concrete did this require?

   b. If each truck load was the same size, what was the volume of a load of concrete?

11. One of the silos pictured here holds corn and the other holds hay. Chopped corn and hay are blown into the top of the silos through the pipes which can be seen. Each silo has a height of 18 m and a diameter of 6 m.

★ a. What is the volume in cubic meters of one of these silos?

   b. Each cubic meter of corn weighs approximately 780 kg. What is the weight of a full silo of corn?

★ c. A full silo of hay weighs 208,650 kg. What is the weight to the nearest kilogram of 1 m³ of hay?

   d. A blower that is fed continuously can blow in a cubic meter of hay in 3 minutes. At this rate how long will it take to fill a silo with hay?

12. Egypt's Great Pyramid, or Pyramid of Cheops, was built about 2600 B.C. and is one of the 7 wonders of the ancient world. It is estimated that 100,000 men required 20 years to complete it.

★ a. How many times greater is the volume of the Great Pyramid than the volume of the Transamerica Pyramid in San Francisco?

   b. The heights (altitudes) of the triangular faces of the Great Pyramid and Transamerica Pyramid are 188 m and 261 m, respectively; and the bases of these triangles are 232 m and 35 m, respectively. About how many times greater is the surface area of the Egyptian pyramid (not counting the areas of their bases)?

13. Each of the following open top boxes can be formed by cutting out the corners of a 16 by 16 grid of centimeter paper. For each box write the dimensions of its base and the volume and surface area in the table. For the given heights plot the volumes of the boxes on the graph. Which box has the greatest volume?

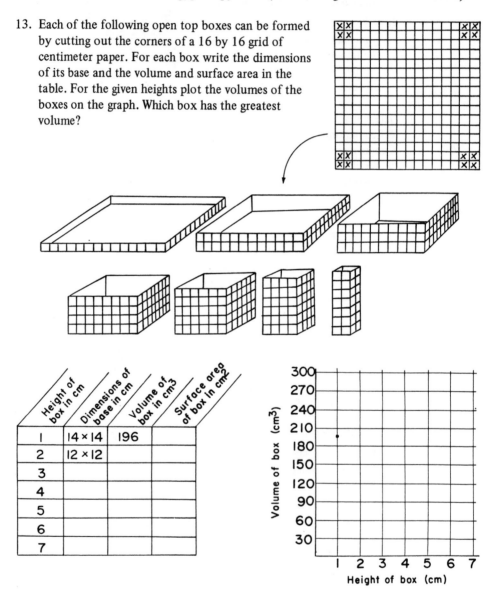

Height of box in cm	Dimensions of base in cm	Volume of box in cm³	Surface area of box in cm²
1	14 × 14	196	
2	12 × 12		
3			
4			
5			
6			
7			

14. Assume that a drop of unvaporized gasoline is a sphere with a diameter of 4 mm.

★   a. If this drop is divided into 8 smaller drops, each with a diameter of 2 mm, how many times greater is the total surface area of these 8 drops, compared to the surface area of the original drop?

   b. If each drop with a diameter of 2 mm is divided into 8 smaller drops, each with a diameter of 1 mm, how many times greater is the total surface area of the 64 drops, as compared to the surface area of the original drop?

★   c. If the vaporizing mechanism in a car's engine carries out this process of splitting 20 times, how many times is the surface area of the original drop increased?

15. Compute the volumes of the following figures in cubic centimeters. Figures **b** and **c** have a square base and figures **a**, **f**, and **g** have rectangular bases. (*Hint for Figure* **a**: Think of the front face as the base of a prism.)

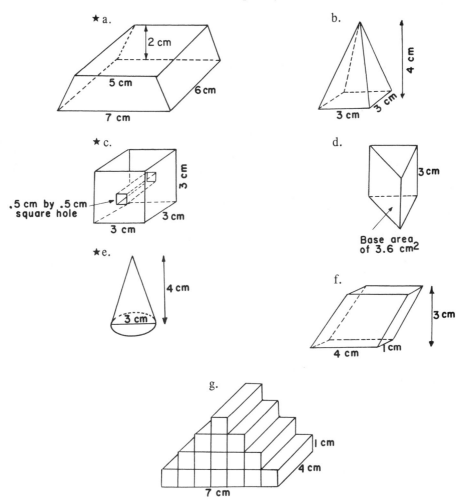

★ a.

2 cm
5 cm
6 cm
7 cm

b.

4 cm
3 cm
3 cm

★ c.

3 cm
.5 cm by .5 cm square hole
3 cm
3 cm

d.

3 cm
Base area of 3.6 cm²

★ e.

4 cm
3 cm

f.

3 cm
4 cm
1 cm

g.

1 cm
4 cm
7 cm

★ 16. *Riddle:* What is a 1.6-mm compacted graphite marking instrument encased in the exact center of a wood fiber handle, with a metal-bonded aft-mounted synthetic rubber tip?

★   a. In case you didn't recognize the description, this instrument is called a pencil. If the graphite cylinder in a new pencil has a diameter of 1.6 mm and is 17.5 cm long, what is its volume in cubic millimeters?

★   b. How many cubic centimeters of graphite are needed to make 20 million pencils? How many cubic meters are needed?

17. *Legend:* There is a story about an ancient cubical monument that sits on a square plaza. Both the cube and the square plaza were constructed from the same number of smaller cubes. The plaza is twice as wide as the cube. How many cubes were needed to build the monument and the plaza?

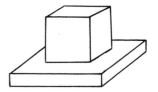

18. *Puzzle:* A cylindrical can such as the one shown here is full of water. If you were to pour the water from the can, how would you know when half the water was gone if you had no measuring device? (*Hint:* Try it. Watch the water inside the can.)

CHAPTER SIX

FRACTIONS AND INTEGERS

The whole of mathematics consists in the organization of a series of aids to the imagination in the process of reasoning.

A.N. Whitehead

New York Stock Exchange trading floor

6.1 EQUALITY AND INEQUALITY OF FRACTIONS

The sale of stocks and bonds on the New York Stock Exchange is carried out on this three-story trading floor. Some 2000 private telephone lines connect brokers on the floor with their offices, which in turn are linked to hundreds of branch offices throughout the United States. The annunciator board in the upper left of this picture is a calling system of flashing numbers to brokers on the floor.

The prices of stocks are stated in dollars, as well as halves, fourths, eighths, and six-teenths of a dollar. Historically, when a smaller measure or unit was needed, half of it was used. For a still smaller amount, a half of a half produced a fourth. Similarly, eighths, sixteenths, and thirty-seconds resulted from repeatedly halving the original unit. The inch with its halves, fourths, eighths, etc., is a familiar example of this process. Another example is the speed of shutter openings of a camera, which is calibrated in fractions of a second: 1/2, 1/4, 1/8, 1/15, 1/30, 1/60, etc. (see Exercise Set 6.1, Exercise 1). On the other hand, when larger measures were needed, the common units were doubled (see page 213). The custom of doubling and halving seems to be a natural tendency. Karl Menninger, in *Number Words and Number Symbols,* calls them "primitive operations."*

Historical Development   The use of fractions is ancient. The Egyptians were using them before 2500 B.C. With the exception of 2/3, all Egyptian fractions were *unit fractions,* that is, fractions with a numerator of 1 (1/3, 1/4, etc.).

These fractions were represented by placing the symbol ○, which meant "part," above their hieroglyphic numerals for whole numbers.

$$\overset{\text{O}}{|\,|\,|} = \frac{1}{3} \qquad \overset{\text{O}}{|\,|\,|\,|\,|} = \frac{1}{5} \qquad \overset{\text{O}}{\cap|\,|} = \frac{1}{12}$$

When they used their hieratic (sacred) numerals, as shown in this scroll, a dot was placed above the numeral to represent a unit fraction. For example, ∧ represents the number 30, and ∧̇ represents the fraction 1/30. It is interesting to note that as late as the eighteenth century, over 3000 years later, the symbols $\dot{\overline{2}}$ and $\dot{\overline{4}}$ are found in English books for the fractions 1/2 and 1/4.

Egyptian leather scroll containing simple relations between fractions (ca. 1700 B.C.)

While the Egyptians used fractions with fixed numerators, the Babylonians (ca. 2000 B.C.) used only fractions with denominators of 60 and 60^2. The fraction 1/60 was referred to as the "first little part" and $1/60^2$ as the "second little part." These fixed denominators were extensions of the Babylonian base sixty and were used in their study of astronomy. Our use of minutes, 1/60 of an hour, and seconds, $1/60^2$ of an hour, has been handed down from the Babylonians.

Gradually, fractions in which the numerator and denominator could be any number gained acceptance. These more general fractions were being used by the Hindus in the seventh century. There have been many notations for writing fractions. Our present form

*K. Menninger, *Number Words and Number Symbols* (Cambridge, Massachusetts: MIT Press, 1969), pp. 359–60.

of writing one number above the other was used the Hindus. The Arabs introduced the bar between the two vertically placed numbers. The resulting notation, $\frac{a}{b}$, was in general use by the sixteenth century.

Fraction Terminology   "Fraction" is from the French "frangere," which means *to break.* The words "broken number" and "fragment" have been frequently used in the past for "fraction." These numbers were first needed for measurements that were less than a whole unit. One of the problems which occurs in the Egyptian writings from 1650 B.C. requires that

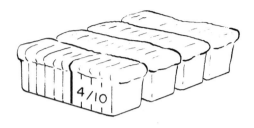

4 loaves of bread be divided equally among 10 men. If a loaf of bread is divided into 10 equal parts, the share for each man is 1/10 of a loaf. Since there are 4 loaves of bread, each man will receive 1/10 of each loaf or 4/10 of a loaf. This "broken number" tells *how much* each man is to receive.

The word "fraction" is used to refer both to a number written in the form *a/b* and to the numeral, *"a/b."* The *a* and *b* can be any numbers, as long as *b* is not equal to zero. (See Sections 7.3 and 7.4 for fractions involving decimals and irrational numbers.) When we speak of adding two fractions, we are talking about adding numbers, not numerals (symbols). On the other hand, when we say the denominator of the fraction 5/12 has a factor of 3, we are referring to the bottom half of the numeral "5/12." You will not have to be concerned about whether or not the word "fraction" is being used as a number or as a numeral; it will be clear from its use in any particular instance.

In the fraction *a/b* the bottom number, *b,* is called the *denominator* and can be thought of as the number of equal parts in a whole. The top number in a fraction is called the *numerator* and indicates the number of parts being considered. Here are some examples: If we think of an iceberg as having nine parts of equal volume, then eight of these parts are underwater; The surface of the earth can be divided into three parts of equal area such that two of these parts are water; and California's coastal region can be divided into four parts of equal length so that the San Andreas Fault runs the length of three of these parts.

8/9 of the volume of an iceberg is underwater.

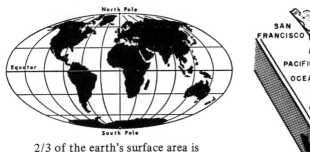

2/3 of the earth's surface area is
covered by water.

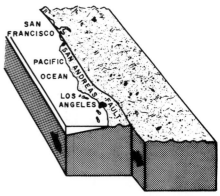

The San Andreas Fault runs 3/4 of the
length of California's coastal region.

Fractions and Division   One of the major influences in the early development of fractions was the need to solve problems involving division of whole numbers. The Egyptians used fractions in cases where the quotient was not a whole number. In the example cited previously, the Egyptians needed a fraction in order to compute 4 divided by 10 ($4 \div 10 = 4/10$).

BUT FIFTY IS MORE THAN TWENTY-FIVE!

YOU SIMPLY DON'T UNDERSTAND DIVISION...NO WONDER YOU'VE BEEN GETTING SUCH POOR GRADES...

YOU CAN'T MAKE FIFTY GO INTO TWENTY-FIVE!

YOU CAN IF YOU PUSH IT!

© 1967 United Feature Syndicate, Inc.

Charlie Brown's little sister Sally has a similar problem. She is trying to divide 25 by 50. Charlie Brown's comment shows that he thinks of division only in terms of *measurement,* that is, "how many 50s in 25?" However, there is another approach to division. Remember from page 124 that division has two meanings, measurement and partitive. With the *partitive* concept, dividing by 50 means there will be 50 parts. If we picture dividing 25 objects of equal size, such as sticks of gum, into 50 equal parts, each part will be one-half of a stick. Stated in equation form, $25 \div 50 = 1/2$.

This close relationship between fractions and division is often used to define a fraction as a quotient of two numbers.

Definition: For any numbers *a* and *b*, with $b \neq 0$,

$$\frac{a}{b} = a \div b$$

Mixed Numbers and Improper Fractions

Historically, a fraction stood for part of a whole and represented numbers less than 1. The idea of a fraction such as 4/4 or 5/4 having a numerator greater than or equal to the denominator was uncommon even as late as the sixteenth century. Such fractions are sometimes called *improper fractions.*

When these fractions are written as a combination of whole numbers and fractions, they are called *mixed numbers.* The numbers 1 1/5, 2 3/4, and 4 2/3 are examples of mixed numbers. Placing a whole number and a fraction side by side, as in mixed numbers, indicates the sum of the two numbers. For example, 1 1/5 means 1 + 1/5. The following number lines are labelled with fractions, mixed numbers, and whole numbers.

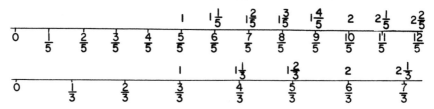

Models for Fractions

Generations of students have been introduced to fractions by "pies," which are divided into 2, 3, 4, or more equal parts. A pizza divided into eight equal parts by four cuts of a roller-bladed knife is a well-known example. The pie shown here is a type of optical illusion. About 1/8 of the pie appears to be missing. Turn this page upside down and see what part of the pie is missing.

Squares and rectangles are also frequently used to illustrate fractions, because they can be conveniently divided into parts of equal size. The two commercially produced models which are described in the following paragraphs have rectangular shapes.

Fraction Bars—Fraction bars are rectangular and are divided into 2, 3, 4, 6, and 12 parts. The denominator of a fraction is represented by the number of parts to a bar and the numerator by the number of shaded parts. The bars are colored depending on their numbers of parts: Halves are green; thirds are yellow; fourths are blue; sixths are red; and twelfths are orange. A bar that is all shaded is called a *whole bar* and represents the unit. A bar with no shading is called a *zero bar* and represents zero.

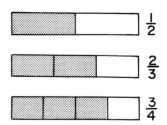

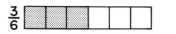   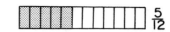

Cuisenaire Rods—Cuisenaire rods are used as models for fractions by comparing the lengths of 2 rods. It takes 3 red rods to equal the length of 1 dark green rod, so the length of the red rod is 1/3 the length of the dark green rod. If the length of the dark green rod is chosen as the unit, then the red rod represents 1/3. If a different unit is selected, the red rod will represent a different fraction. For example, if the yellow rod is the unit length, then the red rod represents 2/5, because the length of the red rod is 2/5 the length of the yellow

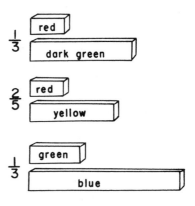

rod. Depending on the choice of unit there are several different rods which can represent the same fraction. One-third was represented above by a red rod, but if the unit is the length of a blue rod then the green rod represents 1/3.

Number Line—Fractions are represented on the number line by selecting a unit length and dividing the interval from 0 to 1 into parts of equal length. To locate the fraction *r/s,* subdivide the interval into *s* equal parts, and, beginning at 0, count off *r* of these parts. Here are some examples showing the sixths and tenths between 0 and 1.

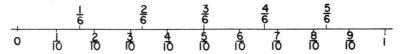

Equality   Equality of fractions can be illustrated by comparing the shaded amounts of regions. This chart shows that 1/3, 2/6, and 4/12 are equal fractions. Another combination of regions shows that

$$\frac{2}{3} = \frac{4}{6} = \frac{8}{12}$$

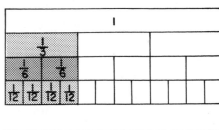

Different equalities can be illustrated by dividing the unit regions into different numbers of parts. In this chart we see that

$$\frac{1}{2} = \frac{2}{4} = \frac{4}{8}$$

What other equalities can you see?

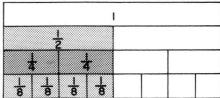

Equality of fractions can also be illustrated by sets of points. Three out of 12 or 3/12 of the points shown here are circled. Viewed in another way, 1/4 of the points are circled because there are four rows containing the same number of points each, and one row is circled. So, 3/12 and 1/4 are equivalent fractions representing the same amount.

For every fraction there are an infinite number of other fractions which represent the same number. The bars on the right show how to obtain two fractions equal to 3/4. Each part of the first bar has been split into two equal parts, to show that 3/4 = 6/8. We see that doubling the number of parts in a bar also doubles the number of shaded parts. This is equivalent to multiplying the numerator and denominator of 3/4 by 2. Similarly, splitting each part of a 3/4 bar into three equal parts triples the number of parts in the bar and triples the number of shaded parts. This has the effect of multiplying the numerator and denominator of 3/4 by 3 and shows that 3/4 is equal to 9/12.

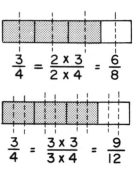

$$\frac{3}{4} = \frac{2 \times 3}{2 \times 4} = \frac{6}{8}$$

$$\frac{3}{4} = \frac{3 \times 3}{3 \times 4} = \frac{9}{12}$$

In general, two fractions are equal if one can be obtained from the other by multiplying its numerator and denominator by a whole number greater than 0.

Definition: For any fraction $\frac{a}{b}$ and whole number $k > 0$,

$$\frac{a}{b} = \frac{ka}{kb}$$

Lowest Terms—When the numerator and denominator of a fraction have a common factor greater than 1, the fraction can be written in simplified form. This is done by dividing the numerator and denominator by their greatest common factor. For example, both 6 and 9 in the fraction 6/9 can be divided by 3 to get the fraction 2/3. When the only common factor of the numerator and denominator is 1, the fraction is said to be in *lowest terms*.

$$\frac{6}{9} = \frac{6 \div 3}{9 \div 3} = \frac{2}{3}$$

$$\frac{2}{3} \quad \frac{7}{12} \quad \frac{5}{8} \quad \frac{4}{9}$$

Fractions in lowest terms

Common Denominators   One of the more important skills in the use of fractions is that of replacing two fractions with different denominators by two fractions having equal denominators. The fractions 1/6 and 1/4 have different denominators, and the bars representing these fractions have different numbers of parts (see top of next page). If each part of the 1/6 bar is split into 2 equal parts and each part of the 1/4 bar is split into 3 equal parts, both bars will have 12 equal parts. The fractions for these new bars, 2/12 and 3/12, have a common denominator of 12.

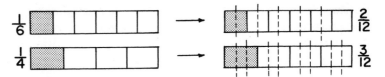

Obtaining the same number of equal parts for two bars is a visual way of illustrating common denominators. Another method for finding common denominators of two fractions is to list the multiples of their denominators. The arrows shown here point to the common multiples of 6 and 4. The least common multiple (l.c.m.) of 6 and 4 is 12. This is also the smallest common deonominator of 1/6 and 1/4. In general, the *smallest common denominator* of two fractions is the least common multiple of their denominators.

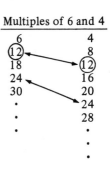

Multiples of 6 and 4

Once a common denominator is found, two fractions can be replaced by fractions having the same denominator.

$$\frac{1}{6} = \frac{2 \times 1}{2 \times 6} = \frac{2}{12} \qquad \frac{1}{4} = \frac{3 \times 1}{3 \times 4} = \frac{3}{12}$$

Sometimes only one fraction has to be replaced in order to obtain two fractions with the same denominator. For the fractions 2/3 and 11/18, 18 is the least common multiple of the two denominators. Therefore, it is necessary only to replace 2/3 by 12/18 to obtain two fractions with equal denominators.

Inequality   Charts, such as the one pictured here, show many different inequalities between fractions. Place the edge of a piece of paper on this chart to compare the regions for 4/5 and 7/9. You will see that 7/9 is just a little less than 4/5.

We can also see the pattern of decreasing inequalities,

$$\frac{1}{2} > \frac{1}{3} > \frac{1}{4} > \frac{1}{5} > \frac{1}{6} > \frac{1}{7} > \frac{1}{8} > \frac{1}{9} > \frac{1}{10}$$

and that of the increasing inequalities,

$$\frac{1}{2} < \frac{2}{3} < \frac{3}{4} < \frac{4}{5} < \frac{5}{6} < \frac{6}{7} < \frac{7}{8} < \frac{8}{9} < \frac{9}{10}$$

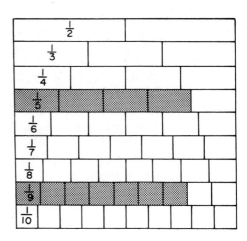

One of the reasons for finding a common denominator for two fractions is to be able to determine the greater fraction. For the fractions 5/8 and 3/5, it is difficult to know which is greater without first replacing them by fractions having a common denominator. The least common multiple of 8 and 5 is 40. Replacing both 5/8 and 3/5 by fractions having a denominator of 40, we see that 5/8 is the greater fraction, but that they differ by only 1/40.

$$\frac{5}{8} = \frac{5 \times 5}{5 \times 8} = \frac{25}{40} \qquad \frac{3}{5} = \frac{8 \times 3}{8 \times 5} = \frac{24}{40}$$

Any two fractions, a/b and c/d, can be compared by replacing them with fractions having a common denominator, ad/bd and bc/bd. An equality or inequality for these fractions can then be determined by comparing the numerators, ad and bc, of these fractions.

Definition: For any fractions a/b and c/d,
(1) $a/b < c/d$ if and only if $ad < bc$, and
(2) $a/b = c/d$ if and only if $ad = bc$.

Density of Fractions The whole numbers are evenly spaced on the number line, and for any whole number there is "the next" whole number. However, for fractions this is not possible. There is no fraction which is the next one greater than 1/2. We know that 6/10 is greater than 1/2 and differs from it by 1/10. But 11/20 is also greater than 1/2 and differs from it by only 1/20. We could continue with the fractions 16/30, 21/40, 26/50, etc., all greater than 1/2 and each one closer to 1/2.

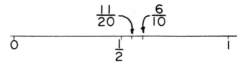

Similarly, there is no fraction next to 0. Stated in another way, there is no smallest fraction greater than 0. The following sequence of fractions gets closer and closer to 0, but no matter how far we go in this sequence, these fractions will always be greater than 0.

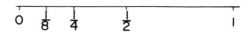

$$\frac{1}{2} \quad \frac{1}{4} \quad \frac{1}{8} \quad \frac{1}{16} \quad \frac{1}{32} \quad \frac{1}{64} \quad \frac{1}{128} \cdots$$

These examples are special cases of the more general fact that between any two fractions there is always another fraction. We refer to this by saying that the fractions are *dense*. As a result of this fact, there are an infinite number of fractions between any two fractions. Thus, it appears that the number line is just about filled up with fractions. However, in Section 7.4 you will see that there are lots of points on the number line which do not have corresponding fractions.

Fractions and Calculators   Every fraction, *a/b*, is equal to the quotient of two numbers, $a \div b$. One method of comparing two fractions to determine whether or not they are equal, or which is the greater, is to find their decimal representations by dividing the numerator by the denominator. For example, 7/12 and 9/16 are both very close to 1/2, but which is the greater fraction? The top calculator display was obtained by dividing 7 by 12 and shows that 7/12 is about 58 hundredths. The second display is from dividing 9 by 16 and shows that 9/16 is about 56 hundredths. Therefore, 7/12 is greater than 9/16 by about 2 hundredths.

7/12 | 0.58333333 |

9/16 | 0.56250000 |

For those fractions which are represented by *infinite repeating decimals,* such as 7/12 = .5833333333 . . . , the number in the calculator display is an approximation because it shows only a few of the digits. For most applications this is sufficient accuracy, and we will not need to be concerned over the fact that the decimal is not exactly equal to the fraction. The decimal .5625, on the other hand, is exactly equal to 9/16 because it is a *terminating decimal* whose digits are contained on the calculator display. For more details on infinite repeating decimals and terminating decimals see page 317.

SUPPLEMENT *(Activity Book)*

Activity Set 6.1   Models for Equality and Inequality (Fraction Bars and Cuisenaire Rods)

Just for Fun: Fraction Games

Additional Sources

Aaboe, A. *Episodes from the Early History of Mathematics.* New York: Random House, 1964. pp. 1–21.

Beardslee, E.C., G.E. Gau and R.T. Heimer. "Teaching for generalization: An array approach to equivalent fractions." *The Arithmetic Teacher,* **20** No. 7 (November 1973), 591–99.

Bell, K.M. and D.D. Rucker. "An algorithm for reducing fractions." *The Arithmetic Teacher,* **21** No. 4 (April 1974), 299–300.

Gillings, R.J. "The Remarkable Mental Arithmetic of the Egyptian Scribes (Part I)." *The Mathematics Teacher,* **59** No. 4 (April 1966), 372–81.

National Council of Teachers of Mathematics. *The Rational Numbers.* Reston, Virginia: NCTM, 1972. "Fractions and Rational Numbers," pp. 25–54.

Smith, D.E. *History of Mathematics,* **2.** Lexington, Mass: Ginn, 1925. pp. 208–23.

EXERCISE SET 6.1

1. This is a shutter speed knob on top of
a 35 mm camera. The numerals on this
knob determine the amount of time
the camera's shutter stays open. The
settings of "4" and "2," following
"B," open the shutter for 4 and 2
seconds, respectively. The remaining numerals 1, 2, 4, 8, etc., represent
1, 1/2, 1/4, 1/8, etc., of a second. The
fastest opening on this camera is 1/1000
of a second.

35 mm camera shutter speed knob

★ a. The less light which is available, the longer the shutter must stay open for the film
to be properly exposed. Will a shutter setting of 15 allow more or less light than a
setting of 60?

b. If a shutter setting of 250 doesn't allow quite enough light, what number should
the dial be set on?

2. There are only a few fractions with different denominators which occur frequently
in newspapers, magazines, sales ads, etc.

a. Name the different denominators for the fractions in the following collage.

b. There are many mixed numbers in the collage. Name ten of them.

c. Write your mixed numbers in part **b** as improper fractions.

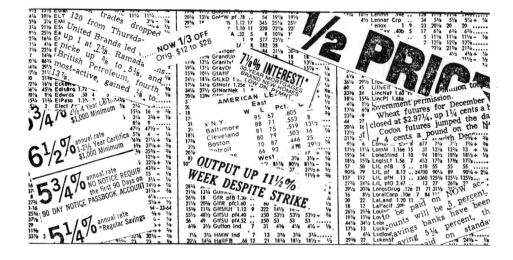

3. This news clipping shows the 5 most active and the 20 top stocks in dollars and fractions of a dollar for a given day. The first column of numbers contains the daily closing prices. The second column shows the fraction of a dollar which the stock was up (u), or down (d), or unchanged (unc.) from the previous day's price.

★ a. Find the cost in dollars and cents for these stocks.

    Dow Ch   Gen Elec   Gulf Oil

★ b. Find the daily price change in cents for these stocks.

    RCA   Xerox   Tyco Lb

5 MOST ACTIVE			
Occid. Pt	17 3/8	d	1/8
Dow Ch	44 3/8	d	1/4
Gn Mot	67 1/4	u	5/8
Aetna Lf	31 1/2	unc.	
Brist My	74	u	3/8
20 TOP STOCKS			
Am T&T	59	u	1/4
RCA	27 1/8	u	1/8
Data Genl	47 5/8	u	3/8
Nat Gyp	14 3/8	unc.	
P Sv Eg	20 3/4	u	1/8
Con Foods	25 1/4	d	1/8
Reyn Ind	60	u	1/4
Pub S.N.H.	20 1/2	unc.	
Xerox	63 3/8	u	3/16
U.S. Steel	47 3/4	d	1/2
Exxon	51 7/8	u	1/2
Whel Fry	22 1/8	d	1/4
Gen Elec	52 3/4	u	1/4
Gulf Oil	25 7/8	d	1/4
Polaroid	37 7/8	u	1/4
Unit Tech	33 1/4	u	1/16
Con Edis	18 7/8	d	1/8
Tyco Lb	12 5/8	d	1/4
Ca Pw Pf	27 7/8	unc.	
IBM	271 3/4	u	1/2

4. Here are five daily reports of stock prices with the lowest and highest costs per share for the day.

★ a. Which of these five stocks had the lowest cost for the day?

    b. Which stock had the highest cost for the day?

★ c. Alaska Airlines finished the day at 8 and 3/16 dollars. Is this greater than or less than the day's high price for Canadian Homestead?

Company	Low Price	High Price
Alaska Airlines	$7\frac{5}{8}$	$8\frac{3}{4}$
Canadian Homestead	$7\frac{15}{16}$	$8\frac{1}{4}$
Mobile Home Indiana	$23\frac{5}{8}$	$24\frac{1}{2}$
Ranger O Can	$19\frac{1}{2}$	$20\frac{3}{8}$
Vintage Enterprise	$28\frac{3}{8}$	$29\frac{3}{4}$

5. Write each of these fractions as mixed numbers or as whole numbers.

★ a. $\frac{5}{3}$ b. $\frac{8}{8}$ ★c. $\frac{25}{6}$ d. $\frac{21}{7}$ ★e. $\frac{17}{5}$

6. Write each of these mixed numbers as fractions.

★   a. $1\frac{3}{4}$ b. $2\frac{1}{5}$ ★c. $4\frac{2}{3}$ d. $2\frac{5}{6}$ e. $1\frac{3}{7}$

7. *Calculator Exercise:* As a result of the density of fractions on the number line, there is no single fraction greater than zero which is closest to zero. Use a calculator to convert these fractions to decimals. Explain why a calculator will eventually show all zeros if we continue converting fractions in this sequence to decimals.

$$\frac{1}{50} \qquad \frac{1}{500} \qquad \frac{1}{5000} \qquad \frac{1}{50000} \qquad \frac{1}{500000} \qquad \frac{1}{5000000}$$

8. Cuisenaire rods represent various fractions, depending on the choice of the unit rod. If the unit rod is brown, then the purple rod represents 1/2.

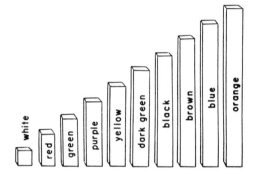

★ a. What is the unit rod if the purple rod represents 2/3?

    b. If the unit rod is the orange rod, what fraction is represented by the black rod?

★ c. If the dark green rod represents 3/4, what is the unit rod?

9. Split the parts of these fraction bars to illustrate the given equalities.

★ a.

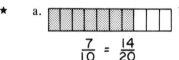

$$\frac{7}{10} = \frac{14}{20}$$

    b.

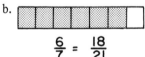

$$\frac{6}{7} = \frac{18}{21}$$

    c.

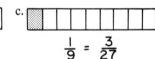

$$\frac{1}{9} = \frac{3}{27}$$

10. Find a way to split each of these figures to illustrate the equality of the two given fractions.

★ a. $\dfrac{2}{5} = \dfrac{8}{20}$ b. $\dfrac{2}{9} = \dfrac{6}{27}$

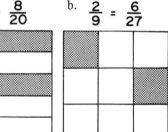

11. Write the missing numbers for these fractions.

★ a. $\dfrac{7}{8} = \dfrac{}{32}$　　b. $\dfrac{}{3} = \dfrac{40}{24}$　　★c. $\dfrac{2}{3} = \dfrac{12}{}$　　d. $\dfrac{5}{} = \dfrac{20}{24}$

12. Write each fraction as a fraction with a denominator of 9. (If at first it doesn't seem possible, replace the fraction by a fraction in lowest terms.)

★ a. $\dfrac{4}{18}$　　b. $\dfrac{12}{27}$　　★c. $\dfrac{4}{12}$　　d. $\dfrac{16}{24}$

13. Complete the equations so that each pair of fractions has the smallest common denominator.

★ a. $\dfrac{2}{3} =$　　b. $\dfrac{1}{6} =$　　★c. $\dfrac{3}{15} =$　　d. $\dfrac{5}{8} =$

　$\dfrac{4}{5} =$　　　$\dfrac{7}{12} =$　　　$\dfrac{5}{6} =$　　　$\dfrac{3}{10} =$

14. This chart represents fractions with denominators of 2 through 12. By placing the edge of a piece of paper or a ruler on the vertical lines, equalities and inequalities can be determined.

★ a. Use this chart to find the greater fraction in each of these pairs (write < or > between each pair).

　　$\dfrac{2}{3}$　$\dfrac{4}{7}$

　　$\dfrac{5}{8}$　$\dfrac{6}{10}$

　　$\dfrac{8}{11}$　$\dfrac{3}{4}$

　　$\dfrac{1}{4}$　$\dfrac{3}{10}$

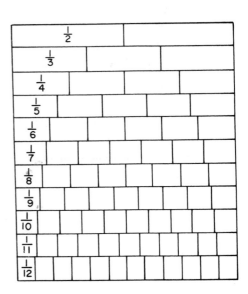

★   b. Check your inequalities in part **a** by replacing each pair of fractions by fractions having the same denominator.

$\dfrac{2}{3} =$

$\dfrac{4}{7} =$

$\dfrac{5}{8} =$

$\dfrac{6}{10} =$

$\dfrac{8}{11} =$

$\dfrac{3}{4} =$

$\dfrac{1}{4} =$

$\dfrac{3}{10} =$

15. *Calculator Exercise:* Use a calculator to convert each fraction to a decimal. Determine which fraction in each pair is the greater, and write the correct inequality symbol, $<$ or $>$.

★   a. $\dfrac{6}{32} \quad \dfrac{5}{24}$ b. $\dfrac{45}{7} \quad \dfrac{32}{5}$ c. $\dfrac{81}{246} \quad \dfrac{32}{94}$

16. *Calculator Exercise:* Use a calculator to rewrite these fractions in increasing order.

$$\dfrac{16}{20} \qquad \dfrac{19}{34} \qquad \dfrac{38}{52} \qquad \dfrac{21}{25} \qquad \dfrac{11}{17}$$

★ 17. Magnesium is lighter than iron but stronger. One-fiftieth of the earth's crust is magnesium, and 1/20 of the earth's crust is iron. Is there more iron or more magnesium?

18. Pure gold is quite soft. To make it more useful for such items as rings and jewelry it is mixed with other metals, such as copper and zinc. Pure gold is marked 24k (24 karat). What fraction of a ring is pure gold if it is marked 14k? Write your answer in lowest terms.

★ 19. Some health authorities say that we should have 1 gram of protein a day for each kilogram of our weight. There are about 40 grams of protein in a liter of milk.

★   a. If a person weighs 60 kilograms and his/her only source of protein is milk, how much milk will be needed to supply the daily requirement of protein?

★   b. A liter of fat-free milk weighs about 1040 grams. What fraction of milk's weight is protein?

20. *Fibonacci Numbers:* The Fibonacci numbers, 1, 1, 2, 3, 5, 8, 13, . . . , occur as the numerators and denominators of fractions that are associated with patterns of leaves on the stems of trees and plants. Begin with a leaf from a pear tree and count the number of leaves until you reach a leaf just above the first. This number of leaves, not counting the first (zero leaf), is the Fibonacci number 8. Furthermore, in passing around the branch from a leaf to the one directly above, you will make three complete turns. This means that each leaf is 3/8 of a turn from its adjacent leaves. This fraction, which is called *leaf divergence,* can be easily found by counting leaves and turns around stems. Find the leaf divergence (fractions) in the following cases.

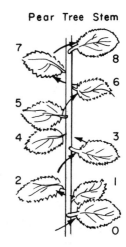

Pear Tree Stem

		Number of turns around branch	Number of leaves from a leaf to one directly above
★	a. Beech and hazel trees	1	3
	b. Elm and lime trees	1	2
★	c. Apple, oak, and cherry trees	2	5
	d. Weeping willow and pear trees	3	8
	e. Pussy willow and almond trees	5	13

21. *Poetry and Precision:* In Lord Tennyson's poem, "The Vision of Sin," there is a verse which reads,

> Every minute dies a man,
> Every minute one is born.

In response to these lines the English engineer Charles Babbage wrote a letter to Tennyson in which he noted that if this were true, the population of the world would be at a standstill. He suggested that the next edition of the poem should read:

> Every moment dies a man,
> Every moment 1 1/16 is born.

By Babbage's figure, there are 17 people born for every 16 that die. At the current rate (see population clock, page 147), there is a birth every 10 seconds and a death every 16 seconds. What mixed number should now be used in this poem in place of 1 1/16?

United States space shuttle, *Enterprise*

6.2 OPERATIONS WITH FRACTIONS

Fractions are often used to indicate how much smaller a diagram or picture is than the original object. A number which describes the relative size from an object to its representation is called a *scale factor*. Scale factors may be greater than 1 (see Section 9.2) or less than 1. If a model or drawing is bigger than the actual object, the scale factor is greater than 1. Models of buildings, planes, and ships have scale factors that are less than 1. The scale factor from the space shuttle orbiter *Enterprise* to the model of the space shuttle which is shown here is approximately 1/50. This model is in a wind tunnel which simulates space conditions at 4 1/2 times the speed of sound and up to 32 miles above the earth. The rocket engines in this model actually fire during tests.

In each of the following examples the scale factor from the object to its picture is a fraction less than 1. The picture of the foxhound is 32 times smaller than the actual dog, and so the scale factor from the dog to its picture is 1/32. The reductions in the size of the trumpeter and turkey buzzard are also indicated by fractions.

Foxhound

Trumpeter

Turkey buzzard

Addition The concept of addition of fractions is the same as that for whole numbers. The addition of whole numbers is illustrated by "putting together" or "combining" two sets of objects. Similarly, the addition of fractions can be illustrated by combining two regions. By placing two bars end to end, the total shaded amount represents the sum of the two fractions.

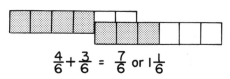

$$\frac{4}{6} + \frac{3}{6} = \frac{7}{6} \text{ or } 1\frac{1}{6}$$

Addition of fractions can also be illustrated on the number line by placing arrows for the fractions end to end.

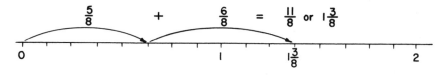

$$\frac{5}{8} \quad + \quad \frac{6}{8} \quad = \quad \frac{11}{8} \text{ or } 1\frac{3}{8}$$

Unlike Denominators—The difficulty in adding fractions occurs when the denominators are unequal. The 2/5 bar and the 1/3 bar show that 2/5 + 1/3 is greater than 3/5 but less than 4/5. To determine this sum exactly, the two fractions must be replaced by fractions having the same denominator.

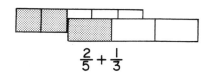

$$\frac{2}{5} + \frac{1}{3}$$

Since the smallest common denominator of 2/5 and 1/3 is 15, these two fractions can be replaced by 6/15 and 5/15, respectively. The sum of these two fractions is 11/15, so 2/5 + 1/3 = 11/15.

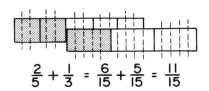

$$\frac{2}{5} + \frac{1}{3} = \frac{6}{15} + \frac{5}{15} = \frac{11}{15}$$

Multiplying the denominators of two fractions by each other will always yield a common denominator, but not necessarily the smallest one. Once two fractions have a common denominator, their sum is computed by adding the numerators and retaining the denominator.

Definition: For any fractions $\frac{a}{b}$ and $\frac{c}{d}$,

$$\frac{a}{b} + \frac{c}{d} = \frac{ad}{bd} + \frac{bc}{bd} = \frac{ad + bc}{bd}$$

Mixed Numbers—Mixed numbers are combinations of whole numbers and fractions. The sum of two such numbers can be found by adding the whole numbers and the fractions separately. If the denominators of the fractions are unequal, as in this example, a common denominator must be found before adding.

$$2\frac{1}{4} = 2\frac{3}{12}$$
$$+1\frac{2}{3} = +1\frac{8}{12}$$
$$\overline{\qquad\qquad}$$
$$3\frac{11}{12}$$

Subtraction   The concept of subtraction for fractions is the same as for subtraction of whole numbers. That is, we can think of subtraction as "take away" or in terms of "missing addends." These two bars, for example, show that 1/2 take away 1/6 is 2/6, and that 2/6 must be added to 1/6 to get 1/2.

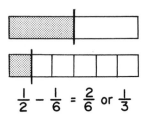

$$\frac{1}{2} - \frac{1}{6} = \frac{2}{6} \text{ or } \frac{1}{3}$$

To illustrate subtraction on the number line, the fraction to be subtracted is represented by an arrow from right to left, just as was done when subtracting whole numbers on the number line. In this example, 11/12 − 7/12 = 4/12, or 1/3.

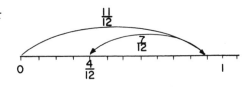

Unlike Denominators—These bars show that the difference between 5/6 and 1/4 is greater than 3/6 and less than 4/6. To compute this difference we can replace these fractions by fractions having a common denominator.

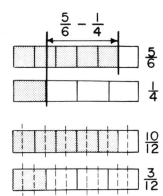

The smallest common denominator of 5/6 and 1/4 is 12, so these fractions can be replaced by 10/12 and 3/12, respectively. The difference between these two fractions is 7/12, so $5/6 - 1/4 = 7/12$.

$$\frac{5}{6} - \frac{1}{4} = \frac{10}{12} - \frac{3}{12} = \frac{7}{12}$$

The general rule for subtracting fractions is always stated by using a common denominator that is the product of the two denominators, even though this may not be the smallest common denominator. Once two fractions have a common denominator, their difference is computed by subtracting the numerators and retaining the denominators.

Definition: For any fractions $\frac{a}{b}$ and $\frac{c}{d}$,

$$\frac{a}{b} - \frac{c}{d} = \frac{ad}{bd} - \frac{bc}{bd} = \frac{ad - bc}{bd}$$

Mixed Numbers—The difference between two mixed numbers can be found by subtracting the whole number parts and the fractions separately. Sometimes before this can be done, borrowing or regrouping is necessary. In the accompanying example, the 4-1/5 is replaced by 3-6/5 before subtracting.

$$
\begin{aligned}
4\frac{1}{5} &= 3\frac{6}{5} \\
- 1\frac{2}{5} &= -1\frac{2}{5} \\
\hline
&2\frac{4}{5}
\end{aligned}
$$

$$4\frac{1}{5} = 4 + \frac{1}{5} = (3 + 1) + \frac{1}{5} = 3 + \left(1 + \frac{1}{5}\right) = 3 + \left(\frac{5}{5} + \frac{1}{5}\right) = 3 + \frac{6}{5} = 3\frac{6}{5}$$

If the denominators of the fractions in the mixed numbers are unequal, the fractions must be replaced by fractions having a common denominator. In this example, 3/4 and 1/3 were replaced by 9/12 and 4/12, respectively, before the numbers were subtracted. In some cases, both "borrowing" and changing denominators will be necessary before subtracting mixed numbers.

$$
\begin{aligned}
3\frac{3}{4} &= 3\frac{9}{12} \\
- 1\frac{1}{3} &= -1\frac{4}{12} \\
\hline
&2\frac{5}{12}
\end{aligned}
$$

Multiplication   In the product of two whole numbers, $m \times n$, we think of the first number m as indicating "how many" of the second number. For example, 2×3 means 2 "of the" 3s. Multiplication of fractions may be viewed in a similar manner. In the product $1/3 \times 6$, the first number tells us "how much" of the second, that is, $1/3 \times 6$ means $1/3$ "of" 6.

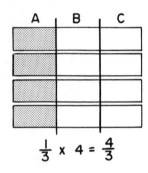

$\frac{1}{3} \times 6 = 2$

There is one major difference between multiplying by a whole number and multiplying by a fraction. When multiplying by a whole number greater than 1, the product is greater than the second number being multiplied. However, when multiplying by a fraction less than 1, the product is less than the second number being multiplied. This has always been a problem in explaining multiplication by fractions.*

Whole Number Times Fraction—In this case we can interpret multiplication to mean "repeated addition," just as we did for multiplication of whole numbers in Chapter 3. The whole number indicates the number of times the fraction is to be added to itself. For any whole number k, $k \times a/b$ is equal to a sum with a/b occurring k times. In this example, $k = 3$ and $a/b = 2/5$.

$$3 \times \frac{2}{5} = \frac{2}{5} + \frac{2}{5} + \frac{2}{5} = \frac{6}{5} \quad \text{or} \quad 1\frac{1}{5}$$

Fraction Times Whole Number—The product $1/3 \times 4$ means $1/3$ "of" 4. To illustrate this we will use 4 bars and divide them into 3 equal parts. The vertical lines partition these bars into equal parts A, B, and C. Part A, which is one-third of the 4 bars, consists of 4 one-thirds, or 4 thirds. That is,

$$\frac{1}{3} \times 4 = \frac{1}{3} + \frac{1}{3} + \frac{1}{3} + \frac{1}{3}$$

$$= \frac{4}{3}$$

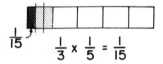

$\frac{1}{3} \times 4 = \frac{4}{3}$

Fraction Times Fraction—The product $1/3 \times 1/5$ means $1/3$ "of" $1/5$. This can be illustrated by using a $1/5$ bar and taking $1/3$ of its shaded amount. In order to do this, the shaded part of the bar has been split into three equal parts. The double-shaded part of the bar is $1/3$ of $1/5$. Each of these new parts is $1/15$ of a whole bar, so $1/3 \times 1/5 = 1/15$.

$\frac{1}{3} \times \frac{1}{5} = \frac{1}{15}$

*According to D.E. Smith, *History of Mathematics*, **2** (Lexington, Mass: Ginn, 1925), p. 225, even early writers (fifteenth and sixteenth centuries) on the subject of fractions expressed concern over this problem.

To illustrate 1/3 × 4/5, we will use a 4/5 bar and take 1/3 of each of the shaded regions. To do this, each shaded part of the 4/5 bar is split into three equal parts. Each of these new parts is 1/15 of a whole bar. The four double shaded parts of the bar represent 4/15, so, 1/3 × 4/5 = 4/15.*

The rule for computing the product of two fractions is the easiest of the algorithms for the four basic operations. This rule is suggested in the preceding illustrations of a fraction times a fraction. In each of these examples the product can be found by multiplying numerator by numerator and denominator by denominator.

Definition: For any fractions $\frac{a}{b} \times \frac{c}{d}$,

$$\frac{a}{b} \times \frac{c}{d} = \frac{ac}{bd}$$

Division   Division of fractions can be viewed in much the same way as division of whole numbers. One of the meanings of division of whole numbers is the "repeated subtraction" or "measurement" concept (see page 124). For example, to explain 15 ÷ 3, we often say, "How many times can we subtract 3 from 15?" Similarly, for 3/5 ÷ 1/10 we can ask, "How many times

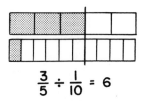

can we subtract 1/10 from 3/5?" The bars show that the shaded amount of a 1/10 bar can be subtracted from the shaded amount of a 3/5 bar six times. Or, viewed in terms of multiplication, the shaded amount of the 3/5 bar is six times greater than the shaded amount of the 1/10 bar.

This interpretation for division of fractions continues to hold even when the quotient is not a whole number. The bars in this sketch show that the shaded amount of the 1/3 bar can be subtracted from the shaded amount of the 5/6 bar two times, and there is a remainder. Just as in whole number division, the remainder is then compared to the divisor by a fraction. In this example, the remainder is 1/2 as big as the divisor, so the quotient is 2 1/2.

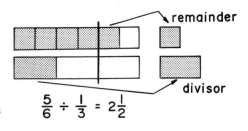

*Because of the distributive property, taking 1/3 of 4/5 of a bar is the same as taking 1/3 of each 1/5 of a bar: $\frac{1}{3} \times \frac{4}{5} = \frac{1}{3} \times \left(\frac{1}{5} + \frac{1}{5} + \frac{1}{5} + \frac{1}{5}\right) = \left(\frac{1}{3} \times \frac{1}{5}\right) + \left(\frac{1}{3} \times \frac{1}{5}\right) + \left(\frac{1}{3} \times \frac{1}{5}\right) + \left(\frac{1}{3} \times \frac{1}{5}\right) =$ $\frac{1}{15} + \frac{1}{15} + \frac{1}{15} + \frac{1}{15} = \frac{4}{15}$.

One of the early methods of dividing one fraction by another was to replace both fractions by fractions having a common denominator. When this is done the quotient can be obtained by disregarding the denominators and dividing the two numerators. For example,

$$\frac{3}{4} \div \frac{2}{3} = \frac{9}{12} \div \frac{8}{12} = 9 \div 8 = 1\frac{1}{8}$$

It may have been this approach, of getting a common denominator for dividing fractions by fractions, that eventually led to the present "invert and multiply" method. In this method of division, which came into general use in the seventeenth century, we invert the divisor and then multiply the two fractions. The Hindus and Arabs divided this way as early as the eleventh century, but it seems not to have been used for the next 400 years.*

Definition: For any fractions $\frac{a}{b}$ and $\frac{c}{d}$,

$$\frac{a}{b} \div \frac{c}{d} = \frac{a}{b} \times \frac{d}{c} = \frac{ad}{bc}$$

Number Properties   The commutative, associative, and distributive properties which hold for the operations of addition and multiplication with whole numbers also hold for these operations with fractions. The rules for adding and multiplying fractions will be used to verify each of these properties in the following examples.

Addition Is Commutative—Two fractions that are being added can be interchanged (commuted) without changing the sum: $7/8 + 3/5 = 3/5 + 7/8$.

$$\frac{7}{8} + \frac{3}{5} = \frac{35}{40} + \frac{24}{40} = \frac{59}{40} = 1\frac{19}{40} \qquad \frac{3}{5} + \frac{7}{8} = \frac{24}{40} + \frac{35}{40} = \frac{59}{40} = 1\frac{19}{40}$$

This property is illustrated next on the number lines by placing the lengths for the fractions end to end.

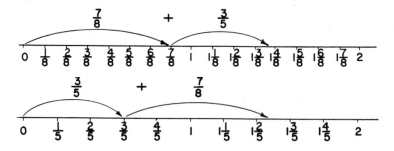

Addition Is Associative—In a sum of three fractions, the middle number may be grouped (associated) with either of the other two numbers. (The sums inside the parentheses are computed first.)

$$\left(\frac{1}{3} + \frac{1}{4}\right) + \frac{1}{6} = \frac{7}{12} + \frac{1}{6} = \frac{9}{12} \qquad \frac{1}{3} + \left(\frac{1}{4} + \frac{1}{6}\right) = \frac{1}{3} + \frac{5}{12} = \frac{9}{12}$$

*D.E. Smith, *History of Mathematics,* 2 (Lexington, Mass: Ginn, 1925), pp. 226–28.

Multiplication Is Commutative—Two fractions that are being multiplied can be interchanged (commuted) without changing the product: $1/2 \times 1/3 = 1/6$, and $1/3 \times 1/2 = 1/6$. A physical illustration of this property is interesting because the processes of taking 1/2 of something and taking 1/3 of something are quite different. To take 1/2 of 1/3 we begin with a 1/3 bar, and to take 1/3 of 1/2 we use a 1/2 bar.

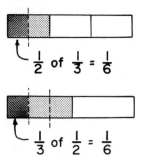

Multiplication Is Associative—In a product of three fractions the middle number may be grouped with either of the other two numbers. (The products inside the parentheses are computed first.)

$$\left(\frac{1}{2} \times \frac{3}{4}\right) \times \frac{1}{5} = \frac{3}{8} \times \frac{1}{5} = \frac{3}{40} \qquad \frac{1}{2} \times \left(\frac{3}{4} \times \frac{1}{5}\right) = \frac{1}{2} \times \frac{3}{20} = \frac{3}{40}$$

Multiplication Is Distributive over Addition—When a sum of two fractions is multiplied by a third number we can add the two fractions and then multiply, as in the first of the following equations; or we can multiply both fractions by the third number and then add, as in the second row of equations.

$$\frac{1}{2} \times \left(\frac{3}{4} + \frac{7}{10}\right) = \frac{1}{2} \times \left(\frac{15}{20} + \frac{14}{20}\right) = \frac{1}{2} \times \frac{29}{20} = \frac{29}{40}$$

$$\frac{1}{2} \times \left(\frac{3}{4} + \frac{7}{10}\right) = \left(\frac{1}{2} \times \frac{3}{4}\right) + \left(\frac{1}{2} \times \frac{7}{10}\right) = \frac{3}{8} + \frac{7}{20} = \frac{29}{40}$$

Inverses for Addition—For every fraction there is another fraction, called its *negative* or *inverse for addition,* such that the sum of the two fractions is 0. The fractions 3/4 and ⁻3/4 are inverses of each other for addition: $3/4 + {}^-3/4 = 0$.

$$\frac{3}{4} + \frac{{}^-3}{4} = \frac{3 + {}^-3}{4} = \frac{0}{4} = 0$$

Inverses for Multiplication—For every fraction not equal to 0 there is a nonzero fraction, called its *reciprocal* or *inverse for multiplication,* such that the product of the two numbers is 1. For the fraction 3/8, its reciprocal is 8/3, and $3/8 \times 8/3 = 1$.

$$\frac{3}{8} \times \frac{8}{3} = \frac{3 \times 8}{8 \times 3} = \frac{24}{24} = 1$$

Computing with Calculators   The four basic operations with fractions can be carried out by first changing the fractions to decimals and then performing the given operations. Sometimes a computation with fractions is only approximately equal to one with decimals.

This will occur whenever the decimal representations for the fractions or the computation exceeds the capacity of the calculator. (The symbol $\approx$ means "approximately equal to.")

$$\frac{4}{5} + \frac{1}{3} + \frac{5}{8} \approx .8 + .3333333 + .625 = 1.7583333$$

On some calculators the computations can be stored in memory by pressing $\boxed{M+}$ or a similar key. To evaluate $4/5 + 1/3 + 5/8$, first compute $4 \div 5$ on the calculator and store the result in memory. Then compute $1 \div 3$ and add it to the number in memory by pressing $\boxed{M+}$. Finally, compute $5 \div 8$ and press $\boxed{M+}$ to add it to the number in memory. Pressing the memory recall key, which is sometimes denoted by $\boxed{MR}$, will show the sum of the decimals for these fractions. Each step of this process is shown here.

Steps	Displays
1. $4 \boxed{\div} 5 \boxed{=}$	0.8
2. Press $\boxed{M+}$	0.8
3. $1 \boxed{\div} 3 \boxed{=}$	0.3333333
4. Press $\boxed{M+}$	0.3333333
5. $5 \boxed{\div} 8 \boxed{=}$	0.625
6. Press $\boxed{M+}$	0.625
7. Press $\boxed{MR}$	1.7583333

If the calculator has parentheses, the sum $4/5 + 1/3 + 5/8$ can be computed by entering the fractions and plus signs as they occur from left to right. Each fraction, a/b, is entered as $a \div b$, and the quotients are separated from the sums by parentheses. Here are the steps for computing this sum.

$$(4 \boxed{\div} 5) \boxed{+} (1 \boxed{\div} 3) \boxed{+} (5 \boxed{\div} 8) \boxed{=}$$

On some calculators sums or differences can be computed by merely entering the fractions (as quotients of whole numbers) and the addition or subtraction operations as they occur from left to right. To compute the previous sum, if $4 \boxed{\div} 5 \boxed{+} 1$ is entered onto this type of calculator, 1 is not added to the quotient, $4 \div 5$, but it is divided by 3 when $\boxed{\div} 3$ is entered. Here are the steps for computing the sum on this type of calculator.

$$4 \boxed{\div} 5 \boxed{+} 1 \boxed{\div} 3 \boxed{+} 5 \boxed{\div} 8 \boxed{=}$$

Multiplication of fractions can be carried out on a calculator by simply entering the fractions (as quotients of numbers) and the multiplication operation in the order in which they occur. For example, here are the steps for computing $5/7 \times 2/9$.

$$5 \boxed{\div} 7 \boxed{\times} 2 \boxed{\div} 9 \boxed{=}$$

This sequence of steps produces the correct answer because

$$[(5 \div 7) \times 2] \div 9 = (5 \div 7) \times (2 \div 9)$$

Division cannot be carried out as conveniently on the calculator as can multiplication. To compute $5/7 \div 2/9$, for example, we cannot enter the fractions and the division operation in the same order as we did for multiplication, because

$$[(5 \div 7) \div 2] \div 9 \neq (5 \div 7) \div (2 \div 9)$$

One method of computing $5/7 \div 2/9$ on a calculator is to replace each fraction by a decimal and then divide one decimal by the other. This can be done by first computing $2 \div 9$ and placing the decimal for this quotient in memory storage. Then compute $5 \div 7$ and divide it by the number in storage. Here are the steps and corresponding displays for this computation.

Steps	Displays
1. 2 $\div$ 9 $=$	0.22222222
2. M+	0.22222222
3. 5 $\div$ 7 $=$	0.71428571
4. $\div$	0.71428571
5. MR	0.22222222
6. $=$	3.2142857

On a calculator with parentheses the quotient $5/7 \div 2/9$ can be evaluated by using parentheses around the fractions.

$$(5 \;\boxed{\div}\; 7) \;\boxed{\div}\; (2 \;\boxed{\div}\; 9) \;\boxed{=}$$

Sometimes you will want the sum or difference of two fractions to be written as another fraction. Let's look at an example for addition. First, replace each fraction by a decimal and add the decimals.

$$\frac{3}{8} + \frac{1}{3} \approx .375 + .3333333 = .7083333$$

Next, the smallest common denominator of 3/8 and 1/3 is 24, so we need to solve the following equation for N.

$$.7083333 = \frac{N}{24}$$

To find N multiply $.7083333 \times 24$, and since N is the numerator of a fraction, round off your answer to the nearest whole number. In this example N is equal to 17.

$$.7083333 \times 24 = 16.999992$$

SUPPLEMENT *(Activity Book)*

Activity Set 6.2 Computing with Fraction Bars

Just for Fun: Fraction Games for Operations

Additional Sources

Beck, A., M.N. Bleicher, and D.W. Crowe. *Excursions into Mathematics.* New York: Worth, 1969. pp. 414–34 (Egyptian, Babylonian, and Farey fractions).

Green, G.F. "A model for teaching multiplication of fractional numbers." *The Arithmetic Teacher,* **20** No. 1. (January 1973), 5–9.

National Council of Teachers of Mathematics. *The Rational Numbers.* Reston, Virginia: NCTM, 1972. pp. 55–178 (operations with fractions).

Newman, James R. *The World of Mathematics,* **1.** New York: Simon and Schuster, 1956. "The Rhind Papyrus," pp. 170–78.

Pittman, P.V. "Rapid Mental Squaring of Mixed Numbers." *The Mathematics Teacher,* **70** No. 7 (October 1977), 596–97.

Sherzer, L. "McKay's Theorem." *The Mathematics Teacher,* **66** No. 3 (March 1973), 229–30.

Smith, D.E. *History of Mathematics,* **2.** Lexington, Mass: Ginn, 1925. pp. 223–35.

EXERCISE SET 6.2

Monitoring screens on American Stock Exchange trading floor

1. The photo on the facing page shows several option-monitoring screens on the American Stock Exchange trading floor with the latest computerized stock prices in halves, fourths, eighths, and sixteenths of a dollar. Similar information is contained in this table of American Stock Exchange high and low prices for six companies during one day.

	High Price	Low Price
Canadian Homestead	$8\frac{1}{4}$	$7\frac{5}{16}$
Drew National	$11\frac{1}{8}$	$9\frac{7}{8}$
General Plywood	$3\frac{3}{4}$	
MEM Company	$26\frac{1}{4}$	$25\frac{5}{8}$
Old Town	$7\frac{3}{4}$	$6\frac{3}{8}$
Sea Container		$16\frac{1}{2}$

★ a. General Plywood's low price was 1/8 of a dollar less than its high price for the day. What was its low price?

b. The day's high price for Sea Container was 3/16 of a dollar above its low price. What was the high price?

★ c. What is the difference between the high price and low price for each of these stocks?

Drew National   MEM Company   Old Town

d. How much money would a person have saved if he/she had purchased 1000 shares of MEM Company stock at the low price rather than the high price for the day?

2. Place the edge of a piece of paper on the eighths' line and mark off the length 1 1/8. Then place the beginning of this marked-off length at the 1 1/5 point on the fifths' line and approximate the sum 1 1/5 + 1 1/8. What number on the fifths' line is this sum closest to?

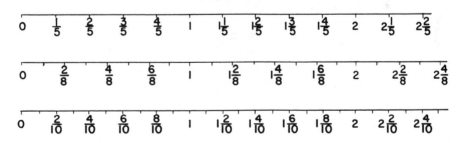

a. Use these number lines to approximate the following sums. Write the number from the given line that the sum is closest to.

$$\frac{4}{5} + 1\frac{3}{8} \text{ (fifths' line)} \qquad \frac{5}{8} + \frac{7}{10} \text{ (eighths' line)} \qquad \frac{3}{10} + 1\frac{4}{5} \text{ (tenths' line)}$$

★ b. Compute these sums and compare them with your approximations.

$$\frac{4}{5} \qquad \frac{5}{8} \qquad 1\frac{4}{5}$$

$$+1\frac{3}{8} \qquad +\frac{7}{10} \qquad +\frac{3}{10}$$

3. Place the edge of a piece of paper on the fifths' line in Exercise 2 and mark off the length 3/5. Then place this marked-off length at the point 1 7/8 on the eighths' line and approximate the difference, 1 7/8 − 3/5. What number on the eighths' line is this difference closest to?

    a. Use the number lines of Exercise 2 to approximate the following differences. Write the number from the given line that the difference is closest to.

$$1\frac{4}{5} - \frac{5}{8} \text{ (fifths' line)} \qquad 1\frac{1}{8} - \frac{7}{10} \text{ (eighths' line)} \qquad \frac{9}{10} - \frac{3}{5} \text{ (tenths' line)}$$

★ b. Compute these differences and compare them with your approximations.

$$1\frac{4}{5} \qquad 1\frac{1}{8} \qquad \frac{9}{10}$$

$$-\frac{5}{8} \qquad -\frac{7}{10} \qquad -\frac{3}{5}$$

4. This picture of a man-eating shark has a length of 45 millimeters. The scale factor from an average size man-eating shark to this picture is 1/120.

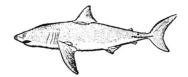

★ a. How long is an average man-eating shark?

    b. A right whale has a length of approximately 12 meters. If the scale factor from this whale to a picture of the whale is 1/300, how long is the picture?

5. A magic square is an array of numbers in which the sum of numbers in each row, column, and diagonal is the same. This sum is called the *magic number*. Write the missing numbers in the following magic squares.

★ a.

		$\frac{2}{3}$
	$\frac{5}{6}$	$1\frac{1}{2}$
1		

b.

$2\frac{1}{4}$	$\frac{3}{8}$	
$2\frac{5}{8}$		$1\frac{1}{8}$
$\frac{3}{4}$		

6. The dark line between the 1/2 bar and the 1/3 bar on this chart represents the difference 1/2 − 1/3. Mark the lines for the differences between the following pairs of fractions: 1/3 and 1/4; 1/4 and 1/5; etc. What seems to be true about these differences?

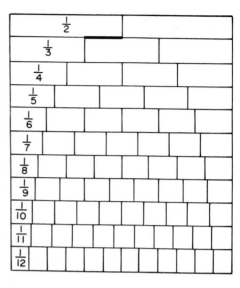

★ a. The differences between the pairs of successive fractions 1/2 − 1/3, 1/3 − 1/4, etc., get smaller and smaller as the denominators increase. Compute these differences and look for a pattern.

$$\frac{1}{2} \qquad \frac{1}{3} \qquad \frac{1}{4} \qquad \frac{1}{5}$$
$$-\frac{1}{3} \qquad -\frac{1}{4} \qquad -\frac{1}{5} \qquad -\frac{1}{6}$$

b. The above chart also shows that the differences between the pairs of successive fractions 2/3 − 1/2, 3/4 − 2/3, 4/5 − 3/4, etc., become smaller and smaller. Compute these differences. Mark these differences on the chart above. How are they related to the differences in part **a**?

$$\frac{2}{3} \qquad \frac{3}{4} \qquad \frac{4}{5} \qquad \frac{5}{6}$$
$$-\frac{1}{2} \qquad -\frac{2}{3} \qquad -\frac{3}{4} \qquad -\frac{4}{5}$$

7. Square I is a magic square. What is its magic number? Multiply each number in Square I by 3 and write these products in the corresponding positions in Square II. Will Square II be a magic square?

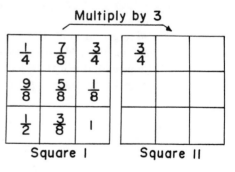

Square I   Square II

8. In computing with fractions there are several types of errors which frequently occur. In this example, 3 2/5 should have been replaced by 2 7/5. Instead, the 1 which was borrowed from the 3 was placed in the numerator to form 2 12/5. Find plausible reasons for the errors in the following examples.

$$3\frac{2}{5} = 2\frac{12}{5}$$
$$-\frac{4}{5} = -\frac{4}{5}$$
$$\overline{}$$
$$2\frac{8}{5} = 3\frac{3}{5}$$

★ a. $\dfrac{1}{4} + \dfrac{5}{6} = \dfrac{6}{10}$ b. $\dfrac{7}{8} - \dfrac{2}{3} = \dfrac{5}{5}$ ★c. $\dfrac{7}{8} \times \dfrac{3}{8} = \dfrac{21}{8}$ d. $\dfrac{3}{4} \div \dfrac{1}{5} = \dfrac{4}{15}$

9. State the number property that is used in each of these equations.

★ a. $\dfrac{3}{7} + \left(\dfrac{2}{9} + \dfrac{1}{3}\right) = \dfrac{3}{7} + \left(\dfrac{1}{3} + \dfrac{2}{9}\right)$

b. $\dfrac{3}{7} + \left(\dfrac{2}{9} \times \dfrac{9}{2}\right) = \dfrac{3}{7} + 1$

★ c. $\dfrac{2}{9} + \left(\dfrac{3}{7} + \dfrac{1}{3}\right) = \left(\dfrac{2}{9} + \dfrac{3}{7}\right) + \dfrac{1}{3}$

d. $\dfrac{5}{6} \times \left(\dfrac{3}{4} + \dfrac{1}{2}\right) = \left(\dfrac{3}{4} + \dfrac{1}{2}\right) \times \dfrac{5}{6}$

★ e. $\dfrac{3}{4} \times \dfrac{5}{6} + \dfrac{1}{2} \times \dfrac{5}{6} = \left(\dfrac{3}{4} + \dfrac{1}{2}\right) \times \dfrac{5}{6}$

10. *Calculator Exercise:* Here are sequences of calculator steps for computing a product, quotient, and sum of fractions. Which of these sequences will not produce the correct answer? Explain why.

a. $80 \times 3/5$
   1. Enter 80
   2. $\boxed{\times}$
   3. $3 \boxed{\div} 5$
   4. $\boxed{=}$

b. $80 \div 3/5$
   1. Enter 80
   2. $\boxed{\div}$
   3. $3 \boxed{\div} 5$
   4. $\boxed{=}$

c. $80 + 3/5$
   1. Enter 80
   2. $\boxed{+}$
   3. $3 \boxed{\div} 5$
   4. $\boxed{=}$

★11. A taxpayer is going to file his federal income tax report by using the Income Averaging method. The eight steps of this method and part of the computation are shown at the top of the next page. Compute the missing amounts.

"You'll be happy to know that nobody in the government is out to get you, nobody's reported you for the finder's fee, nor have we received any anonymous tips. You're here only because we think you've been cheating on your return."

1. 1/3 of base income of $22,000 ...... _____
2. 1/5 of averageable income of $12,000.. _____
3. Line 1 plus line 2 .... _____
4. Tax on line 3 .... $2,318
5. Tax on line 1 .... $1,630
6. Line 4 minus line 5 .... _____
7. Line 6 multiplied by 4 .... _____
8. Line 4 plus line 7 (Total tax) ...... _____

12. A homeowner has a 7-room house and uses 1 room as a home office. On her income tax report this entitles her to subtract 1/7 of the house expenses ($2,840) from the income. How much will she subtract?

House Expenses	
Depreciation	$2,000
Electricity	240
Heat	250
Roof repair	300
Fire insurance ..	50
	$2,840

13. Musical notes which are produced from two strings of equal diameter and tension will vary according to the lengths of the strings. Different fractions of the length of the unit string (see lower string) can be used to produce the notes C, D, E, F, G, A, B, and c (do, re, mi, fa, so, la, ti, do). In particular, if one string is half as long as another, its tone or note will be an octave higher than the longer string.*

```
_____1/2_____ c (one octave higher than C)

_____128/243_____ B

_____16/27_____ A

_____2/3_____ G

_____3/4_____ F

_____64/81_____ E

_____8/9_____ D

_____(unit string)_____1_____ C
```

★ a. Proceeding from the unit string to the top string, all of the strings except two are 8/9 of the length of the previous string. For example, the G string is 8/9 the length of the F string since 8/9 × 3/4 = 2/3. Which strings are 8/9 of the length of the preceding strings?

*See C.F. Linn, *The Golden Mean* (New York: Doubleday, 1974), pp. 9–13, for an elementary explanation of the origin of these fractions.

b. The white piano keys pictured here are the notes of the scale from C to c. All but two of these keys are separated by black keys. How is this observation related to the answer for part **a**?

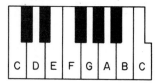

★ c. The eight "c" notes or eight octaves on the piano keyboard are marked in this picture. If the piano wire for each c note were half as long as the wire for the preceding c note, what would be the length of the wire for the highest c as compared to the length of the wire for the lowest c?

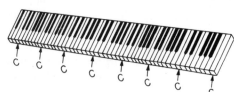

14. The first satisfactory treatment of the operations on fractions is found in the Rhind Papyrus. Most of its 85 problems and solutions require fractions. These problems are preceded by a table containing sums of fractions in the form 1/*m*, which equal fractions of the form 2/*n*. The following sums are from this table. Compute each sum and write the answer as a fraction with a numerator of 2.

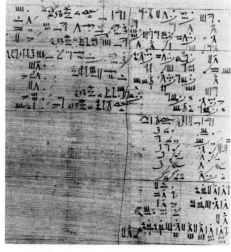

Rhind Papyrus—mathematical text
(ca. 1650 B.C.)

★ a. $\dfrac{1}{4} + \dfrac{1}{28}$

b. $\dfrac{1}{66} + \dfrac{1}{198}$

15. Many of the problems in the Rhind Papyrus, such as Problems 39 and 40 pictured above, concern the rationing of food. Here is an example. "Divide 100 loaves among 5 men in such a way that the shares received shall be an arithmetic sequence and that 1/7 of the sum of the largest three shares shall be equal to the sum of the smallest two." (*Hint:* Find an arithmetic sequence of five numbers whose common difference is 9 1/6.)

16. *The Missing Fraction Puzzle:* In Farmer Brown's will, he bequeathed his 17 horses to his 3 sons in the following manner: 1/2 of his horses to the oldest son, Al; 1/3 of his horses to the middle son, Garry; and 1/9 of his horses to the youngest son, Greg. Being fairly capable with fractions, the boys computed their shares to be 8 1/2 horses, 5 2/3 horses, and 1 8/9 horses. However, each boy was disappointed at the

prospect of getting parts of a horse and they fell to quarrelling about their predicament. At that point Farmer Smith rode up and, after being informed of their dilemma, proposed the following solution. First he donated his horse to make a total of 18 horses. Then he gave 1/2 of the horses to Al, 1/3 of the 18 to Gary, and 1/9 of the 18 to Greg. How many horses did each boy receive? What was the total number of these horses? Seeing their satisfaction with his solution, Farmer Smith jumped on his horse and rode away. Explain why his solution is possible.

★ 17. *Diophantus Puzzle:* Diophantus (ca. 250 B.C.) was a great mathematician who brought fame to Alexandria. The brief record we have of his life is related in the following description: His boyhood lasted 1/6 of his life; his beard grew after 1/12 more; he married after 1/7 more; and his son was born 5 years later; the son lived to half his father's age, and the father died 4 years after his son. How many years did Diophantus live?

6.3 POSITIVE AND NEGATIVE NUMBERS

Historical Development   Whole numbers and fractions were used hundreds of years before negative numbers were developed. As trading became more common, two distinctly different uses of whole numbers were needed—one to indicate credits or gains and one for debits or losses. Conventions were developed to permit the use of whole numbers in both cases. Around 200 B.C. the Chinese were computing credits with red rods and debits with black rods. Similarly, in their writing they used red numerals and black numerals.*

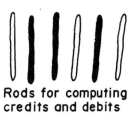

Rods for computing credits and debits

*D.E. Smith, *History of Mathematics,* **2** (Lexington, Mass: Ginn, 1925), pp. 257–58.

The custom today is to reverse the color scheme used by the Chinese. Banks often use red numerals to indicate overdrawn accounts, and we hear the phrase "operating in the red," which refers to business losses. This scoreboard for the 1976 Masters Tournament used red numerals for golf scores below par and green numerals for scores above par.

LEADERS

	HOLE	1	2	3	4	5	6	7	8	9	10	11	12	13	14	15	16	17	18
	PAR	4	5	4	3	4	3	4	5	4	4	4	3	5	4	5	3	4	4
15	FLOYD	15	15	15	14	15	15	15	15	15	15	15	16	16	16	17	17	17	
6	ZIEGLER	6	6	6	6	7	7	7	7	7	7	6	6	7	7	7	7	6	
7	NICKLAUS	7	7	6	6	6	6	6	7	6	6	6	6	7	7	7	6	6	6
5	COODY	5	6	6	6	5	5	5	6	6	5	5	5	5	5	5	5	4	3
4	KITE	4	5	5	5	4	4	5	5	5	4	4	4	4	4	4	3	3	
3	GRAHAM L	3	3	2	2	2	2	2	2	3	3	3	3	2	2	1	1	1	0	1
4	CRENSHAW	3	3	4	4	4	4	5	6	6	6	6	6	8	9	9	9	9	9
2	WEISKOPF	1	2	1	1	1	1	1	1	1	0	0	0	0	0	0	1	0	0
2	CASPER	1	0	0	0	0	1	1	2	2	1	2	2	1	1	1	1		
1	IRWIN	1	0	0	1	0	1	1	0	0	0	0	0	1	2	2	2	3	3

Note: □ indicates that the box has a red numeral.

It is only after hundreds of years of proven necessity that a new type of number can earn its place beside the commonly accepted older numbers. This is especially true of negative numbers. By the seventh century, Hindu mathematicians were using these numbers on a limited basis. They had symbols for negative numbers, such as ⑤ and $\overset{\circ}{5}$ for ⁻5, and rules for computing with them. However, it was another thousand years before the Italian mathematician Jerome Cardan (1501-1576) gave the first significant treatment of negative numbers. Cardan called these new numbers "false" and represented each number by writing "m:" in front of the numeral. For example, he wrote "m:3" for negative three. Other writers of this period called negative numbers "absurd numbers." In 1759, the English mathematician Baron Francis Masères published *Dissertation on the Use of the Negative Sign in Algebra,* in which he expressed the following opinion about negative numbers as solutions to equations:

"... they serve only, as far as I am able to judge, to puzzle the whole doctrine of equations, and to render obscure and mysterious things that are in their own nature exceeding plain and simple ..."

The resistance to negative numbers can be seen as late as 1796 when William Frend, in his text *Principles of Algebra,* argued against their use.

Applications   The concept of positive and negative numbers, also referred to as *signed numbers,* is useful whenever we wish to measure on both sides of a fixed point of reference. The positive numbers indicate one direction, and the negative numbers indicate the opposite direction.

One common example of "opposites" is credits, which are represented by positive numbers, and debits, which are represented by negative numbers. The upper graph shown here is America's *trade balance,* which is the difference between exports and imports. It was positive from 1960 to 1970 and negative in 1971 and 1972. The lower graph

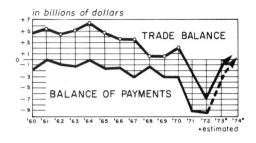

is the *balance of payments,* which is the difference between inflow and outflow of money from this country to others. With the exception of 2 years during the period from 1960 to 1972, America had more money going out than coming in. Its lowest balance was less than ⁻9 billion dollars in 1972.

Another familiar use for positive and negative numbers is in measuring temperatures. The fixed reference point on the Celsius thermometer is 0, the temperature at which water freezes. Temperatures above 0 are positive, and those below 0 are negative.

Scientists often find it convenient to designate a given time as "zero time" and then refer to the time before and after as being negative and positive, respectively. This practice is followed in the launching of rockets. If the time with respect to blast-off is ⁻15 minutes, then it is 15 minutes before the launch. For Apollo 11, the first mission to land men on the moon, countdown began many hours before lift-off, which occurred on July 16, 1969, at 9:32 A.M. Eastern Daylight time. Before this time, minutes and hours were labelled by negative numbers: 9:32 A.M. on July 16 was time 0; and the time following 9:32 A.M. was positive.

Apollo 11 moon launch, July 16, 1969

Sea level is the common reference point for measuring altitudes. Charts and maps which label altitudes below and above sea level use negative and positive numbers. The following chart shows the altitudes in terms of negative numbers for the "floor" of the Atlantic Ocean between South America and Africa.

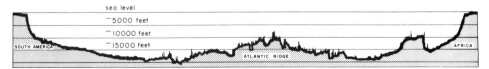

Models for Positive and Negative Numbers

Number Line Model—The number line with a fixed reference point labelled 0 is the common model for positive and negative numbers. For each positive number to the right of 0, there is a negative number to the left, symmetrically opposite. These pairs of numbers, 2 and ⁻2, 3/4 and ⁻3/4, etc., are called *opposites*. More precisely, two numbers are called *opposites* (or *inverses for addition*) if their sum is 0. Sometimes numbers greater than 0, such as 4, 7/8, 9, etc., are labelled as ⁺4, ⁺7/8, and ⁺9, in order to emphasize that they are positive numbers.

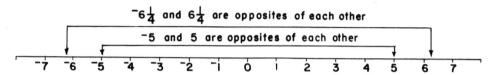

Inequality—For any two numbers (positive or negative) on the number line, the number on the left is *less than* the number on the right. As examples: ⁻8 < ⁻3; ⁻5 1/4 < ⁻3 1/2; and ⁻5 < 0. This is stated more precisely in the following definition for inequality of positive and negative numbers. This definition is similar to the definition for the inequality of whole numbers which was stated on page 88.

Definition: **For any two numbers** m **and** n **(positive or negative),** m **is** *less than* n **($m < n$) if there is a positive number** k, **such that** $m + k = n$.

You may find it helpful, in thinking about inequalities of negative numbers, to recall the applications on the previous pages. Temperatures of ⁻15°C and ⁻6°C are both cold, but ⁻15°C is colder than ⁻6°C: ⁻15 < ⁻6 because ⁻15 + 9 = ⁻6. Similarly, an altitude of ⁻8000 feet is further below sea level than an altitude of ⁻5000 feet: ⁻8000 < ⁻5000 because ⁻8000 + 3000 = ⁻5000.

Integers—The whole numbers together with their opposites are called *integers*.

$$\cdots\ ^-6,\ ^-5,\ ^-4,\ ^-3,\ ^-2,\ ^-1,\ 0,\ 1,\ 2,\ 3,\ 4,\ 5,\ 6,\ \cdots$$

These numbers are also referred to as the positive integers, the negative integers, and 0. The integers are assigned points on the number line by marking off unit lengths to the right and left of 0. This model shows the integers increasing in order from left to right (⁻4 < ⁻3, ⁻3 < ⁻2, ⁻2 < ⁻1, ⁻1 < 0, etc.).

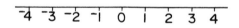

Black and Red Chips Model—The black and red rods which the Chinese used for positive and negative numbers form a concrete model for the integers. In place of rods we will use chips, and the color scheme will be

reversed; that is, black chips will represent positive integers and red chips negative integers. By agreeing that each black chip cancels a red chip, every integer can be represented in an infinite number of ways. Two different sets for 3 are shown at the right on the bottom of the facing page.

On the following pages the elementary definitions and theorems for addition, subtraction, multiplication, and division of integers are illustrated by this model. These same definitions and theorems hold for the basic operations on all positive and negative numbers, including fractions and decimals.

Addition   Addition of whole numbers is illustrated by "putting together" (taking the union of) sets of objects. This same model can also be used to illustrate the addition of integers. To find the sum of two integers we take the

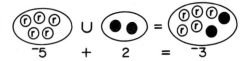

union of their sets, as in the example above on the right. In the union of the sets for ⁻5 and 2, the 2 black chips cancel 2 of the red chips, leaving 3 red chips. This shows that ⁻5 + 2 = ⁻3.

By using black and red chips for computing sums of positive and negative integers, the usual rules for addition can be discovered. For example, to add a positive and a negative integer, the sum will be positive or negative depending on whether there are more black chips or more red chips. Here are the three general cases for addition which involve negative numbers. In each of these theorems n and s are positive numbers.

"Negative plus negative equals negative"	"Positive plus negative equals positive, if $n > s$"	"Positive plus negative equals negative, if $n < s$"
$^-n + {}^-s = {}^-(n + s)$	$n + {}^-s = n - s$	$n + {}^-s = {}^-(s - n)$
$^-3 + {}^-7 = {}^-(3 + 7) = {}^-10$	$13 + {}^-5 = 13 - 5 = 8$	$6 + {}^-11 = {}^-(11 - 6) = {}^-5$

The number line is a more abstract model for illustrating addition of positive and negative numbers. To add two numbers we begin by drawing an arrow from 0 to the point which corresponds to the first number in the summand. Then, if the second number is positive we move to the right on the number line, and if it is negative we move to the left. Here are two examples.

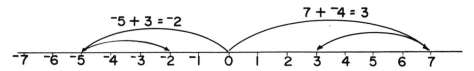

Subtraction   The "take-away" model for subtraction of whole numbers also accommodates the subtraction of integers. The example shown here illustrates ⁻6 take away ⁻2. We begin by representing ⁻6 by 6 red chips and then take away 2 red chips.

$$^-6 - {^-2} = {^-4}$$

© 1957 United Feature Syndicate, Inc.

Traditionally, the "take-away" model for subtraction is used only when one whole number is subtracted from a larger one. However, it is still possible to use this model in cases such as $3 - 5$, where the number being subtracted is the larger one. This can be accomplished by using a suitable representation for 3. For example, instead of representing 3 by black chips as in set A, 3 can be represented by 5 black chips and 2 red chips as in set B. Then, 5 black chips can be taken away, leaving 2 red chips: $3 - 5 = {^-2}$.

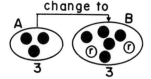

There are two common definitions of subtraction, *adding opposites* and *missing addends,* and both are stated in terms of addition. These definitions can be introduced through the chips model.

Adding Opposites—If instead of removing 5 black chips from set B in the preceding example we put in 5 red chips, the final set will still represent ⁻2. In other words, putting in 5 red chips has the same effect as taking away 5 black chips. This suggests that subtracting 5 is the same as adding its opposite, ⁻5.

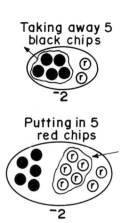

$$3 - 5 = 3 + {^-5} = {^-2}$$

This approach to subtraction is called *adding opposites.* Using this method we can compute differences by replacing them with sums. In general, the difference between any two numbers can be defined in terms of adding opposites.

Definition: For any two numbers, *a* and *b*,

$$a - b = a + {^-b}$$

Missing Addends–The *missing addends* definition which was used to define subtraction of whole numbers can also be used to define subtraction of negative numbers. To use this approach for computing $3 - 5$, we must find the number that can be added to 5 in order to get 3. That is, we must find the missing addend for $\square$ such that $3 = 5 + \square$. Using the chips in the previous example to compute $3 - 5$, we changed from set A to set B so that 5 black chips could be taken away. However, before taking away the 5 black chips, set B shows that $3 = 5 + {}^-2$, and therefore, the missing addend is ${}^-2$. In general, the difference between two numbers can be defined in terms of missing addends.

Definition: **For any two numbers, *a* and *b*,**

$$a - b = \square \quad \text{if and only if} \quad a = b + \square$$

Multiplication

The familiar rules for multiplying with negative numbers, such as, "a negative times a negative is a positive number," are easy enough to remember but difficult to illustrate. There are many different approaches to this topic which attempt to justify the rules for multiplying with negative numbers in an intuitive manner. Three of the more common methods will be explained in the following paragraphs.

Extension of Patterns–This approach to multiplying with negative numbers is based on the assumption that the patterns we observe in the first few equations on the right will continue to hold. In this column of equations we begin with the products of positive integers. For each move down the column, the products on the right side of the equations decrease by 3, until we arrive at $0 \times 3 = 0$. If this pattern is continued into the negative numbers, it suggests that "a negative times a positive is a negative number."

$$
\begin{aligned}
4 \times 3 &= 12 \\
3 \times 3 &= 9 \\
2 \times 3 &= 6 \\
1 \times 3 &= 3 \\
0 \times 3 &= 0 \\
{}^-1 \times 3 &= {}^-3 \\
{}^-2 \times 3 &= {}^-6 \\
{}^-3 \times 3 &= {}^-9
\end{aligned}
$$

In the next column of equations we begin with the fact that "a negative times a positive is a negative number," and extend the pattern to where we are multiplying a negative times a negative number. This time the products on the right side of the equations increase by 4. A continuation of this pattern suggests that "a negative times a negative equals a positive number."

$$
\begin{aligned}
{}^-4 \times 3 &= {}^-12 \\
{}^-4 \times 2 &= {}^-8 \\
{}^-4 \times 1 &= {}^-4 \\
{}^-4 \times 0 &= 0 \\
{}^-4 \times {}^-1 &= 4 \\
{}^-4 \times {}^-2 &= 8 \\
{}^-4 \times {}^-3 &= 12
\end{aligned}
$$

Physical Representations—The "mail carrier model" is perhaps the best known of the physical representations. This model can be illustrated with the black and red chips by thinking of each black chip as a credit of 1 dollar and each red chip as a bill for 1 dollar. Anybody whose financial situation is represented by the chips shown here would have a total worth of 4 dollars.

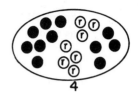

To create a model for multiplication let us agree that in the product $n \times s$, n tells us the number of times we *put in* or *take out* s chips. For example, if n is positive, such as in $3 \times {}^-6$, then we will put in 6 red chips 3 times. If n is negative, such as in ${}^-4 \times {}^-6$, then we will take out 6 red chips 4 times.

We are now ready to use the mail carrier model. Suppose that you have 2 bills, each for 3 dollars. These are represented by the 2 groups of 3 red chips (see figure). If the mail carrier takes away both of these bills, it represents the product ${}^-2 \times {}^-3$. In this case the available funds will increase from 4 dollars to 10 dollars. That is, taking out 6 red chips is equivalent to putting in 6 black chips. This suggests that ${}^-2 \times {}^-3 = 6$.

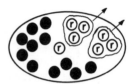

The number represented by this set increases from 4 to 10 by removing 6 red chips.

Deductive Approach—The "patterns approach" and the "mail carrier model" illustrate the reasonableness of the rules for multiplying with negative numbers. These illustrations, however, do not prove that these rules will always hold. To establish these facts we must use deductive reasoning. The following proof that ${}^-2 \times 3 = {}^-6$ is essentially the same as the general proof that a negative number times a positive number is a negative number. One way of proving that ${}^-2 \times 3 = {}^-6$ is to first show that ${}^-2 \times 3 + 6 = 0$.

Proof

$$
\begin{aligned}
{}^-2 \times 3 + 6 &= {}^-2 \times 3 + 2 \times 3 \\
&= ({}^-2 + 2) \times 3 \quad \text{(distributive property)} \\
&= 0 \times 3 \quad \text{(a number plus its opposite equals 0)} \\
&= 0 \quad \text{(0 times any number equals 0)}
\end{aligned}
$$

This sequence of equations shows that ${}^-2 \times 3 + 6 = 0$. Since ${}^-6$ is the only number that can be added to 6 to give 0, ${}^-2 \times 3$ must equal ${}^-6$.

Similar but more general proofs will establish the following theorems for multiplication. In each case n and s are positive numbers.

"Positive times negative equals negative"	"Negative times positive equals negative"	"Negative times negative equals positive"
$n \times {}^-s = {}^-(n \times s)$	${}^-n \times s = {}^-(n \times s)$	${}^-n \times {}^-s = n \times s$
$5 \times {}^-2 = {}^-(5 \times 2) = {}^-10$	${}^-7 \times 3 = {}^-(7 \times 3) = {}^-21$	${}^-4 \times {}^-5 = 4 \times 5 = 20$

Division   Both the partitive and measurement concepts of division will be needed in the following illustrations of division with negative integers.

To illustrate $^-8 \div {}^-2$, we begin with 8 red chips and then measure off or subtract as many groups of 2 red chips as possible. Since there are 4 such groups, $^-8 \div {}^-2 = 4$. This is the *measurement* use of division.

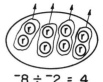

$$^-8 \div {}^-2 = 4$$

To illustrate $^-6 \div 3$, 6 red chips can be divided into 3 equal groups. Since there are 2 red chips in each group, $^-6 \div 3 = {}^-2$. In this illustration the divisor, 3, indicates the number of equal parts into which the set is divided. This is the *partitive* use of division.

$$^-6 \div 3 = {}^-2$$

Examples such as these can be helpful in understanding why "a negative divided by a negative equals a positive number" and "a negative divided by a positive equals a negative number." (*Note:* The black and red chips model is not convenient for illustrating a positive number divided by a negative number.)

The three theorems for division involving negative numbers are stated here. In each case n and s are positive numbers.

"Positive divided by negative equals negative"	"Negative divided by positive equals negative"	"Negative divided by negative equals positive"
$n \div {}^-s = {}^-(n \div s)$	$^-n \div s = {}^-(n \div s)$	$^-n \div {}^-s = n \div s$
$24 \div {}^-6 = {}^-(24 \div 6) = {}^-4$	$^-14 \div 2 = {}^-(14 \div 2) = {}^-7$	$^-30 \div {}^-6 = 30 \div 6 = 5$

Division with negative numbers can be defined in terms of multiplication, just as it was for whole numbers.

Definition: For any two numbers n and s, with $s \neq 0$,

$$n \div s = \square \qquad \textbf{if and only if} \qquad n = s \times \square$$

For example,

$$^-12 \div {}^-4 = \boxed{3} \qquad \text{because} \qquad {}^-12 = {}^-4 \times \boxed{3}$$

It is this inverse relationship between division and multiplication which accounts for the similarity between the rules of signs for division and multiplication with negative numbers.

Properties of Integers   *Inverses for Addition*—Addition of integers has one property which addition does not have for whole numbers. Every integer has an inverse for addition. That is, for any integer k, there is a unique integer ^-k, such that $k + {}^-k = 0$. The integers 8 and $^-8$ are inverses of each other for addition:

$$8 + {}^-8 = 0$$

Commutative Properties—The operations of addition and multiplication are commutative. In particular, these properties hold for negative integers. Here are two examples.

$$^-3 + {}^-5 = {}^-5 + {}^-3 \quad \text{and} \quad {}^-3 \times {}^-5 = {}^-5 \times {}^-3$$

Associative Properties—The operations of addition and multiplication are associative. These properties hold for any combination of three integers. For the integers $^-7$, $^-2$, and $^-6$,

$$(^-7 + {}^-2) + {}^-6 = {}^-7 + ({}^-2 + {}^-6) \quad \text{and} \quad (^-7 \times {}^-2) \times {}^-6 = {}^-7 \times ({}^-2 \times {}^-6)$$

Distributive Property—The distributive property of multiplication over addition holds in the set of integers. Show that this is true in the following example by computing both sides of the equation.

$$^-2 \times (3 + {}^-7) = {}^-2 \times 3 \ + \ {}^-2 \times {}^-7$$

Negative Numbers on Calculators   Most calculators are designed to compute with negative as well as positive numbers. Here are the steps for multiplying 44 times $^-16$.

Steps	Displays
1. Enter 44	44.
2. $\boxed{\times}$	44.
3. $\boxed{-}$ 16	$^-16.$
4. $\boxed{=}$	$^-704.$

On some calculators the minus sign which is used in Step 3 of the previous sequence will cancel (replace) the $\boxed{\times}$ in Step 2. As a result, the calculator will evaluate $44 - 16$ and give an answer of 28. There are several ways around this problem. One is to use the commutative property and multiply $^-16$ times 44. The numbers can now be entered onto the calculator in this order and the correct answer will be obtained. Another approach is to use the fact that a "positive times a negative is negative." Therefore, we can compute 44×16 and then use its negative (additive inverse).

Negative numbers can be added by using the following relationship between addition and subtraction: $a + {}^-b = a - b$. That is, sums of negative numbers can be computed by subtracting positive numbers. The sum $^-118 + {}^-249 + {}^-403$ can be replaced by $^-118 - 249 - 403$. The following steps will work on most calculators.

Steps	Displays
1. Enter $\boxed{-}$ 118	118.
2. $\boxed{-}$	⁻118.
3. Enter 249	249.
4. $\boxed{-}$	⁻367.
5. Enter 403	403.
6. $\boxed{=}$	⁻770.

Some calculators have a button which will *negate* (multiply by negative 1) the number in the display. The $\boxed{CS}$ and $\boxed{+/-}$ are two of these types of buttons. Here are the steps for computing ⁻118 + ⁻249 + ⁻403, using the $\boxed{+/-}$ button.

Steps	Displays
1. 118 $\boxed{+/-}$	⁻118.
2. $\boxed{+}$	⁻118.
3. 249 $\boxed{+/-}$	⁻249.
4. $\boxed{+}$	⁻367.
5. 403 $\boxed{+/-}$	⁻403.
6. $\boxed{=}$	⁻770.

The $\boxed{+/-}$ button is handy if you wish to subtract negative numbers. It negates the display without interfering with the minus operation, as seen in the following steps for computing ⁻15 − ⁻8.

Steps	Displays
1. 15 $\boxed{+/-}$	⁻15.
2. $\boxed{-}$	⁻15.
3. 8 $\boxed{+/-}$	⁻8.
4. $\boxed{=}$	⁻7.

SUPPLEMENT *(Activity Book)*

Activity Set 6.3 Models for Operations with Integers (Illustrations of the four basic operations, with black and red chips)

Just for Fun: Games for Negative Numbers

Additional Sources

Ashlock, R.B. and T.A. West. "Physical representations for signed-number operations." *The Arithmetic Teacher,* **14** No. 7 (November 1967), 549–54.

Bennett, A.B. and G.L. Musser. "A concrete approach to integer addition and subtraction." *The Arithmetic Teacher,* **23** No. 5 (May 1976), 332–36.

Cohen, L.S. "A rationale in working with signed numbers-revisited." *The Arithmetic Teacher,* **13** No. 7 (November 1966), 564–67.

Davis, R.B. *Explorations in Mathematics.* Reading, Mass.: Addison-Wesley, 1967. "Postman Stories," pp. 69–86.

Gardner, M. "The concept of negative numbers and the difficulty of grasping it." *Scientific American,* **236** No. 6 (June 1977), 131–35.

Kline, M. "A Proposal for the High School Mathematics Curriculum." *The Mathematics Teacher,* **59** No. 4 (April 1966), 322–30.

Paterson, J.C. "Fourteen different strategies for multiplication of integers or why $(^-1)(^-1) = 1$." *The Arithmetic Teacher,* **19** No. 5 (May 1972), 396–403.

EXERCISE SET 6.3

1. Antarctica, the only polar continent, is centered near the South Pole and is covered by a huge ice dome reaching a height of almost 4 kilometers (about 13000 ft). One of the hazards of South Polar exploration is the hidden ·crevasses in the ice. This picture shows a crevasse detector operating in the Antarctic.

Ice crevasse detector, Antarctic

★   a. Here are five daytime Celsius temperatures from an Antarctic summer: $^-32°$, $^-27°$, $^-24°$, $^-34°$, and $^-28°$. What is their average?

b. Winter temperatures are usually below $^-75°C$. What is the average of these winter temperatures: $^-84°C$, $^-72°C$, $^-79°C$, $^-81°C$, and $^-78°C$?

c. The coldest place on earth is the Pole of Cold in Antarctica. Its average annual temperature is $^-72°$ Fahrenheit. What is this temperature in degrees Celsius? $C = 5(F - 32)/9$.

2. This table contains the average amount of grain imported and exported by 24 European countries during 1972–75. Each number is in millions of kilograms. The negative numbers represent the amounts imported and the positive numbers the amounts exported.

★ a. How many countries imported more than they exported (include those which had no exports)?

b. Which country imported the most grain during this period?

★ c. A country's balance of trade is the difference between its exports and imports. The grain balance can be found for a given country by adding the negative number from the table to the positive number. Compute the balance for these countries.

Sweden   U.S.S.R.   France   West Germany

	1972–75 average
Austria	−24
Belgium	−1,303 +309
Bulgaria	−27 +295
Czechoslovakia	−812
Denmark	−13 +149
Finland	−14 +88
France	−306 +6,610
Germany, East	−1,645
Germany, West	−2,115 +608
Greece	−63 +4
Hungary	−14 +743
Ireland	−197 +9
Italy	−1,868 +33
Netherlands, The	−1,800 +748
Norway	−334
Poland	−1,540
Portugal	−278
Romania	−18 +660
Spain	−11 +90
Sweden	−15 +532
Switzerland	−371
U.S.S.R.	−8,667 +4,457
United Kingdom	−3,617 +8
Yugoslavia	−384 +1

3. NASA's *Voyager 1* will achieve its closest approach to Jupiter (about 280,000 km) on March 5, 1979. In December, 1978, ⁻80 days before its closest approach, the spacecraft will swivel its narrow-angle television camera to begin its "observatory phase." The white squares shown next are the camera's fields of view at four different approach times.

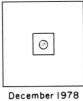

December 1978
⁻80 days

February 1979
⁻29 days

February 1979
⁻18 days

March 1979
⁻7 hours

★ a. How much time has elapsed between the ⁻18-day view and the ⁻7-hour view?

b. As *Voyager 1* moves away from Jupiter, it will examine its moons: Io at ⁺3 hours, Europa at ⁺5 hours, and Ganymede at ⁺14 hours. Thirty-six hours after the ⁻7-hour view it will examine Callisto. How many hours will this be after its closest approach to Jupiter?

4. Illustrate each sum of integers on the corresponding number line.

★ a. 6 + ⁻5 =

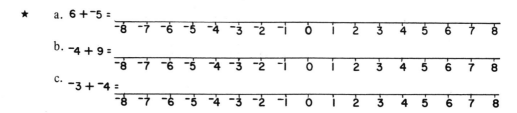

b. ⁻4 + 9 =

c. ⁻3 + ⁻4 =

5. The number line model for subtraction of whole numbers (see page 102) also accommodates several cases of subtraction of integers. As before, the first number is represented by an arrow from 0 to its position on the number line. If the number being subtracted is positive, it is represented by an arrow from right to left. Use this convention to illustrate the following differences.

a. 7 − 5 =

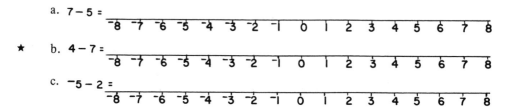

★ b. 4 − 7 =

c. ⁻5 − 2 =

d. A different convention is needed to illustrate subtraction when the number being subtracted is negative. Make up a rule in this case for subtraction on the number line and illustrate the following differences.

3 − ⁻4 =

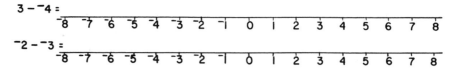

⁻2 − ⁻3 =

6. a. *Addition:* Sketch black and red chips to illustrate this sum. Complete the equation.

3   +   ⁻7   =

b. *Subtraction:* Remove chips from this diagram to compute the difference and then complete the equation.

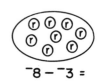

⁻8 − ⁻3 =

c. *Multiplication:* Remove pairs of black chips to compute this product and then complete the equation.

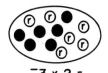

$^-3 \times 2 =$

d. *Division:* Divide these chips into five equivalent sets to compute this quotient. Complete the equation.

$^-15 \div 5 =$

7. Each negative number in this table is the percentage of decrease in the population of a metropolitan area from 1970 to 1975, and each positive number is the percentage of increase. (The 1975 populations are estimated.)

★ a. What fraction of these 60 areas had a decrease in population from 1970 to 1975?

b. How much greater was the percentage decrease in population for Cleveland than the percentage decrease for Buffalo?

★ c. How much greater was the percentage decrease for New York City than the percentage decrease for the Los Angeles-Long Beach area?

	Population		Percent change 1970–75
Name	1970 census	1975 estimate	
New York-Newark-Jersey City SCSA	17,033,367	16,848,000	⁻1.1
New York City	9,973,716	9,567,000	⁻4.1
Nassau-Suffolk	2,555,868	2,750,800	7.6
Newark	2,057,468	2,061,300	0.2
Bridgeport	792,814	791,900	⁻0.1
Jersey City	607,839	601,400	⁻1.1
New Brunswick-Perth Amboy	583,813	600,600	2.9
Long Branch-Asbury Park	461,849	473,000	2.4
Los Angeles-Long Beach-Anaheim SCSA	9,983,017	10,316,600	3.3
Los Angeles-Long Beach	7,041,980	6,944,900	⁻1.4
Anaheim-Santa Ana-Garden Grove	1,421,233	1,710,200	20.3
Riverside-San Bernardino-Ontario	1,141,307	1,223,400	7.2
Oxnard-Simi Valley-Ventura	378,497	438,100	15.7
Chicago-Gary SCSA	7,610,978	7,623,300	0.2
Chicago	6,977,611	6,982,900	0.1
Gary-Hammond-East Chicago	633,367	640,400	1.1
Philadelphia-Wilmington-Trenton SCSA	5,627,719	5,764,500	2.4
Philadelphia	4,824,110	4,933,400	2.3
Wilmington	499,493	515,300	3.2
Trenton	304,116	315,800	3.8
Detroit-Ann Arbor SCSA	4,669,154	4,701,100	0.7
Detroit	4,435,051	4,444,700	0.2
Ann Arbor	234,103	256,400	9.5
San Francisco-Oakland-San Jose SCSA	4,423,797	4,579,800	3.5
San Francisco-Oakland	3,107,355	3,128,800	0.7
San Jose	1,065,313	1,173,400	10.1
Vallejo-Fairfield-Napa	251,129	277,600	10.5
Boston-Lawrence-Lowell SCSA	3,848,593	3,914,600	1.7
Washington, D.C.	2,909,355	3,029,600	4.1
Cleveland-Akron-Lorain SCSA	2,999,811	2,912,300	⁻2.9
Cleveland	2,063,729	1,975,400	⁻4.3
Akron	679,239	668,200	⁻1.6
Lorain-Elyria	256,843	268,700	4.6
Dallas-Fort Worth	2,378,353	2,535,500	0.7
Houston-Galveston SCSA	2,169,128	2,438,100	12.4
Houston	1,999,316	2,256,300	12.8
Galveston-Texas City	169,812	181,800	7.1
St. Louis	2,410,492	2,392,500	⁻0.7
Pittsburgh	2,401,362	2,315,900	⁻3.6
Miami-Fort Lauderdale SCSA	1,887,892	2,301,100	21.9
Miami	1,267,792	1,438,600	13.5
Fort Lauderdale-Hollywood	620,100	862,500	39.1
Baltimore	2,071,016	2,136,900	3.2
Minneapolis-St. Paul	1,965,391	2,027,500	3.2
Seattle-Tacoma SCSA	1,836,949	1,821,500	⁻0.8
Seattle-Everett	1,424,605	1,411,700	⁻0.9
Tacoma	412,344	409,800	⁻0.6
Atlanta	1,595,517	1,793,800	12.4
Cincinnati-Hamilton SCSA	1,613,414	1,628,600	0.9
Cincinnati	1,387,207	1,384,500	⁻0.2
Hamilton-Middletown	226,207	244,100	7.9
Milwaukee-Racine SCSA	1,574,722	1,602,300	1.8
Milwaukee	1,403,884	1,426,400	1.6
Racine	170,838	175,900	2.9
San Diego	1,357,854	1,587,500	16.9
Denver-Boulder	1,239,477	1,404,300	13.3
Tampa-St. Petersburg	1,088,549	1,365,400	25.4
Buffalo	1,349,211	1,327,200	⁻1.6
Kansas City	1,273,926	1,295,000	1.6
Indianapolis	1,111,352	1,147,400	3.2

★ 8. If n is a positive integer less than 10 and s is a negative integer greater than $^-5$, what is the greatest possible value for $n \times s$?

9. If m is an integer which is greater than or equal to $^-8$ and less than or equal to 3 ($^-8 \leqslant m \leqslant 3$), and n is an integer from $^-21$ to $^-11$ ($^-21 \leqslant n \leqslant ^-11$), what are the least and greatest values that $m + n$ can have?

10. The following condensed narrative is from "Why I don't have any examples of negative numbers," by James E. Schultz, and appeared in the May 1973 *Arithmetic Teacher*. Find the examples which could be described by negative numbers.

... I intended to start looking for examples two days ago when I had plans to go to the library. But as you remember, that was the day when the temperature was 5 degrees below zero and my car wouldn't start. Last night I finally made it to the library, but everything went wrong. I started at the main desk on the second floor but was directed to take the elevator down three floors to the reference room in the basement. ... After I got home I decided to read the paper. When I read in an article that the Penn Central Railroad went 300 million in the red last year, I quickly turned to the stock market column, where I found that my stock in the company had gone down two points a share.

11. Find the missing number for each equation.

★   a. $4 + \square = {}^-10$ b. ${}^-3 + \square = {}^-11$

★   c. $6 - \square = 10$ d. ${}^-4 - \square = 7$

12. Compute each side of these equations. Which number properties of the integers are exemplified?

★   a. ${}^-4 \times (16 + {}^-9)/({}^-6 + 13) = {}^-4 \times (16 + {}^-9)/(13 + {}^-6)$

      b. ${}^-4 \times ({}^-3 + 3) + {}^-17 = ({}^-4 \times {}^-3) + ({}^-4 \times 3) + {}^-17$

★   c. $({}^-8 + 7) + 2 \times ({}^-6 \times {}^-5) = ({}^-8 + 7) + (2 \times {}^-6) \times {}^-5$

      d. ${}^-3 \times (16 \times {}^-5)/({}^-14 + 2) = {}^-3 \times ({}^-5 \times 16)/({}^-14 + 2)$

13. Extend the patterns in each of these columns of equations by writing the next three equations. What multiplication rule for negative numbers is suggested by the last few equations that you have written in each column?

     a. $5 \times 3 = 15$ b. $3 \times 6 = 18$ c. ${}^-3 \times 3 = {}^-9$

        $5 \times 2 = 10$ $2 \times 6 = 12$ ${}^-3 \times 2 = {}^-6$

        $5 \times 1 = 5$ $1 \times 6 = 6$ ${}^-3 \times 1 = {}^-3$

        $5 \times 0 = 0$ $0 \times 6 = 0$ ${}^-3 \times 0 = 0$

★ 14. For each of the numbers in the table on the right, write its negative (inverse for addition) and its reciprocal (inverse for multiplication).

Number	$\frac{7}{8}$	${}^-4$	$\frac{{}^-1}{2}$	10
Negative				
Reciprocal				

15. Compute the following products and quotients.

★   a. ${}^-6 \times 2 =$ b. ${}^-8 \times {}^-3 =$ ★ c. $1/3 \times {}^-12 =$

      d. $24 \div {}^-6 =$ ★ e. ${}^-20 \div 4 =$ f. ${}^-6 \div 1/3 =$

16. Find the missing number for each equation.

★   a. $^-6 \times \square = ^-12$ b. $^-1/5 \times \square = 1$

★   c. $^-15 \div \square = ^-3$ d. $24 \div \square = ^-8$

17. A number k is defined to be less than m $(k < m)$ if there is a positive number $\square$ such that $k + \square = m$. Find the replacement for $\square$ to show that each of the following inequalities is true.

★   a. $^-3 < ^-2$ b. $^-14 < 3$ c. $^-7 < 1$
         $^-3 + \square = ^-2$ $^-14 + \square = 3$ $^-7 + \square = 1$

18. For each number written above the number line, locate its reciprocal (inverse for multiplication).

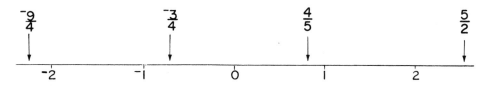

★   a. What general statement can be made about the reciprocals of nonzero numbers in the interval from $^-1$ to 1?

      b. What general statement can be made about the reciprocals of numbers outside of the interval from $^-1$ to 1?

      c. Does 0 have a reciprocal?

19. Do the computing on each side of the equality signs in parts **a** through **g**. Use these examples to determine which two of the given properties do not hold for the integers.

★   a. Addition is commutative: $^-17 + ^-6 \overset{?}{=} ^-6 + ^-17$

      b. Addition is associative: $^-8 + (^-10 + ^-4) \overset{?}{=} (^-8 + ^-10) + ^-4$

★   c. Subtraction is associative: $^-8 - (6 - ^-4) \overset{?}{=} (^-8 - 6) - ^-4$

      d. Multiplication is commutative: $^-4 \times 6 \overset{?}{=} 6 \times ^-4$

★   e. Division is commutative: $^-8 \div 2 \overset{?}{=} 2 \div ^-8$

      f. Multiplication is associative: $^-5 \times (^-6 \times ^-2) \overset{?}{=} (^-5 \times ^-6) \times ^-2$

      g. Multiplication distributes over addition: $^-5 \times (3 + ^-4) \overset{?}{=} (^-5 \times 3) + (^-5 \times ^-4)$

20. *Magic Squares:* Place the following integers in the missing boxes to form a 3 by 3 magic square. The magic number will be $^-6$.

        $^-10, ^-8, ^-6, ^-4, 0, 2, 4, 6$

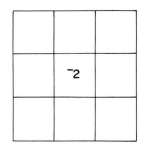

21. *Calculator Exercise:* Calculators which add or multiply by constants will generate arithmetic and geometric sequences for negative numbers.

★   a. Enter ⁻3 $+$ and then repeatedly press $=$ to get an arithmetic sequence. List the first eight numbers.

      b. Enter ⁻3 $\times$ and repeatedly press $=$ to get a geometric sequence. List the first eight numbers. Why is every other number positive in this sequence?

22. *Calculator Exercise:* These two sequences of steps use memory storage, $M+$, and memory recall, MR , for subtracting negative numbers. Which one of these sequences computes ⁻15 − ⁻6 = ⁻9? What number will show in the other display?

a.	Steps	Displays	b.	Steps	Displays
1.	$-$ 15 $=$	⁻15.	1.	$-$ 6 $=$	⁻6.
2.	$M+$	⁻15.	2.	$M+$	⁻6.
3.	C	0.	3.	C	0.
4.	$-$ 6 $=$	⁻6.	4.	$-$ 15 $=$	⁻15.
5.	$-$	⁻6.	5.	$-$	⁻15.
6.	MR	⁻15.	6.	MR	⁻6.
7.	$=$	?	7.	$=$	?

CHAPTER SEVEN

DECIMALS: RATIONAL AND IRRATIONAL NUMBERS

He is unworthy of the name of man who is ignorant of the fact that the diagonal of a square is incommensurable with its side.

Plato

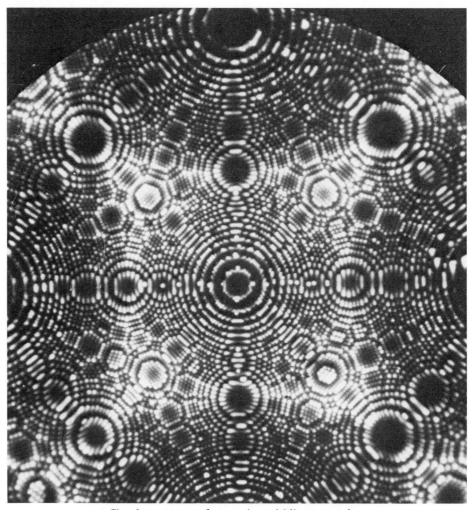

Circular patterns of atoms in an iridium crystal,
magnified more than a million times by a field ion microscope

7.1 DECIMALS AND RATIONAL NUMBERS

Each dot in this remarkable picture is an atom in an iridium crystal. These circular patterns show the order and symmetry governing atomic structures. The diameters of atoms, and even the diameters of electrons contained in atoms, can be measured by decimals. Each atom in this picture has a diameter of .000000027 centimeter, and the diameter of an electron is .00000000000056354 centimeter.

The use of decimals is not restricted to that of describing small objects. The gross national product (GNP) and the national income (NI) for three 5-year periods are expressed to an accuracy of tenths of a billion dollars in the following table.

	1965	1970	1975
GNP (billions)	$688.1	$982.4	$1516.3
NI (billions)	$566.0	$798.4	$1207.6

Historical Development   The person most responsible for our use of decimals is Simon Stevin, a Dutchman. In 1585 Stevin wrote *La Disme,* the first book on the use of decimals. He not only stated the rules for computing with decimals but also pointed out their practical applications. Stevin showed that business calculations can be performed as easily as if they involve only whole numbers. He recommended that the government adopt the decimal system and enforce its use.

As decimals gained acceptance in the sixteenth and seventeenth centuries, there were a variety of notations. Many writers used a vertical bar in place of a decimal point. Here are some examples of how 27.847 was written during this period.

27 | 847   27(847)   27 |847   27 847

27847 . . . ③   27,8^{i}4^{ii}7^{iii} 27847   27⓪8①4②7③

Today, there are still variations in the use of decimal notation. The English place their decimal point higher above the line than it is written in the United States. In other European countries a comma is used in place of a decimal point. A comma and raised numeral denote a decimal in Scandinavian countries.

United States	England	Europe	Scandinavian countries
82.17	82·17	82,17	82,^{17}

Decimal Terminology and Notation   The word "decimal" comes from the Latin "decem," meaning ten. Technically, any number written in base ten positional numeration is a decimal. However, "decimal" is more often used to refer only to numbers such as 17.38 or 104.5, which are expressed with decimal points. The number of digits to the right of the decimal point is called the *number of decimal places.* There are 2 decimal places in 17.08 and 1 decimal place in 104.5.

The positions of the digits to the *left* of the decimal point represent increasing powers of 10 (1, 10, 10^2, 10^3, . . .). The positions to the *right* of the decimal point represent decreasing powers of 10 (10^{-1}, 10^{-2}, 10^{-3}, . . .) or reciprocals of powers of 10 (1/10, 1/10^2, 1/10^3, . . .). In the decimal 5473.286 the "2" represents 2/10, the "8" denotes 8/100, and the "6" stands for 6/1000.

Just as in the case of whole numbers, decimals can be written in expanded form in several ways.

5473.286

$$5 \times 10^3 + 4 \times 10^2 + 7 \times 10 + 3 + 2 \times \frac{1}{10} + 8 \times \frac{1}{10^2} + 6 \times \frac{1}{10^3}$$

5473.286

$$5000 + 400 + 70 + 3 + .2 + .08 + .006$$

5473.286

5 thousands + 4 hundreds + 7 tens + 3 units + 2 tenths + 8 hundredths + 6 thousandths

Reading and Writing Decimals—The digits to the left of the decimal point are read as a whole number (see page 60) and the decimal point is read as "and." The digits to the right of the point are also read as a whole number, and then followed by the name of the place value of the last digit. For example, 1208.0925 is read, "one thousand, two hundred eight, and nine hundred twenty-five ten-thousandths."

1 2 0 8 . 0 9 2 5

One thousand, two hundred eight

and

nine hundred twenty-five ten-thousandths

When writing an amount of money, the decimal part of a dollar is in hundredths. Notice that on this bank check it is unnecessary to write "dollars" or "cents." The amount is in terms of dollars, and this unit is printed at the end of the line upon which the amount of money is written. Some people write the decimal part of a dollar as a fraction. For example, the amount on this check might have been written as "one hundred seventy-seven and 24/100."

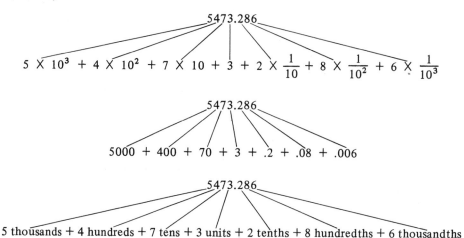

Rational Numbers   Decimals, percents, fractions, and ratios are different types of notation for the same numbers. The following four numerals represent the same rational number.

.25	25%	1/4	1:4
Decimal	Percent	Fraction	Ratio

These examples show that a rational number can be represented by several different types of numerals. Regardless of the notation, if a number can be written as the quotient of two integers, it is called a rational number.

Definition: **A rational number is a number that can be written in the form *a/b*, where *a* and *b* are integers and *b* ≠ 0.**

Whenever the denominator of *a/b* is 1, the rational number equals an integer: $0/1 = 0$, $6/1 = 6$, $^-4/1 = ^-4$, etc. Therefore, the rational numbers contain the integers. Here are some more examples of rational numbers. Each is equal to a quotient of integers, *a/b*.

$$.45 = \frac{9}{20} \qquad ^-9 = \frac{^-9}{1} \qquad .333 \cdots = \frac{1}{3} \qquad 62\% = \frac{62}{100} \qquad \frac{1.6}{5} = \frac{8}{25}$$

Every rational number can be written as a terminating decimal or as an infinite repeating decimal. A *terminating decimal* is a decimal with a finite number of decimal places. Here are some rational numbers and their terminating decimals.

$$\frac{3}{4} = .75 \qquad \frac{1}{8} = .125 \qquad \frac{17}{80} = .2125 \qquad \frac{46}{25} = 1.84$$

A rational number whose decimal does not terminate after a finite number of decimal places is called a *nonterminating* or *infinite repeating decimal.* In such decimals, there is a pattern of digits that repeats. This repeating pattern is indicated by three dots or by placing a bar over the numerals that repeat.

$$\frac{7}{6} = 1.16666 \cdots \qquad \frac{1}{11} = .090909 \cdots \qquad \frac{3744}{9900} = .37818181 \cdots$$

$$= 1.1\overline{6} \qquad\qquad = .\overline{09} \qquad\qquad = .37\overline{81}$$

Models for Decimals

Abacus—Decimals can be illustrated on an abacus by using columns for powers of 10 and for reciprocals of powers of 10. From left to right the columns shown on this abacus represent thousands, hundreds, tens, units, tenths, hundredths, thousandths, and ten-thousandths. The 2 markers on the 1/10 column represent 2/10; the 8 markers on the $1/10^2$ column represent 8/100; and the 6 markers on the $1/10^3$ column represent 6/1000.

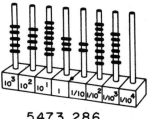

5473.286

A marker from any column can be replaced by putting 10 markers on the column to the right. Conversely, 10 markers from any column can be replaced by 1 marker on the column to the left. This shifting of markers from column to column is called *regrouping*. For example, 10 markers on the $1/10^3$ column can be regrouped or replaced by putting 1 marker on the $1/10^2$ column.

$$10 \times \frac{1}{10^3} = \frac{10}{10^3} = \frac{1}{10^2}$$

Number Line—The number line is a common model for illustrating decimals. We can locate .372 by subdividing the unit interval into tenths, hundredths, and thousandths. First, the interval from 0 to 1 is divided into 10 equal parts. These are tenths of a unit, and .372 is between .3 and .4. Second, the interval from .3 to .4 is divided into 10 equal parts. These are hundredths of a unit, and .372 is between .37 and .38. Finally, if the interval from .37 to .38 is divided into 10 equal parts, each will be one-thousandth of a unit (see enlarged segment below number line). The second one-thousandth mark in this interval corresponds to .372.

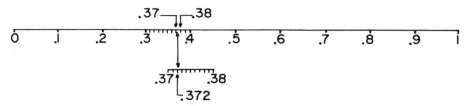

This process of repeatedly subdividing intervals into 10 parts can be used to locate any terminating decimal in a finite number of steps. Infinite repeating decimals cannot be located in this manner. For example, subdividing intervals into 10 parts to locate .333 ... could go on indefinitely without producing the point on the number line which corresponds to this decimal. There is, however, another way of locating infinite repeating decimals on the number line. Every infinite repeating decimal can be written as a rational number of the form r/s (see Examples 1 and 2 on pages 320 and 321). The point for r/s can be located by dividing the unit interval into s equal parts and counting off r of these parts. The decimal .3333 ... is equal to 1/3, and this can be located by dividing the unit interval into three equal parts.

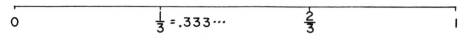

Decimals are used for negative as well as for positive numbers. This graph contains increasing and decreasing changes in wholesale prices in tenths of a percent from May 1975 to May 1976. The change was negative two-tenths in January and negative four-tenths in February.

For each positive decimal, whether terminating or infinite repeating, there is a corresponding negative decimal, such that the sum of the two decimals is zero. When two numbers have a sum of zero they are called *opposites* (or *inverses for addition*) of each other. The January and March price changes in this graph are opposites. Several decimals and their opposites are shown on the following number line.

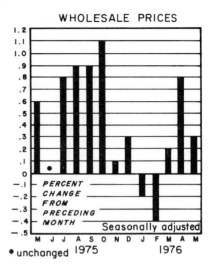

WHOLESALE PRICES

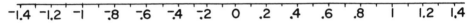

Inequality of Decimals   The definition for inequalities of decimals is the same as the definition that was stated for the inequality of positive and negative numbers on page 298. For any two decimals, if there is a positive number which can be added to the first decimal to get the second, then the first decimal is less than the second. Consider .372 and .6. Since .372 + .228 = .6, .372 is less than .6. This definition, however, is not a practical test for determining inequality. It is much easier to see that .372 is less than .6 by comparing the tenths digits in these decimals: .3 is less than .6 (3/10 < 6/10). The "72" in ".372" represents 72/1000, which is less than .1 and not great enough to affect the inequality. Remember how we located .372 on the number line on the previous page, by subdividing smaller and smaller intervals? The "72" in ".372" shows us that .372 lies between .3 and .4 on the number line, and therefore .372 lies to the left of .6.

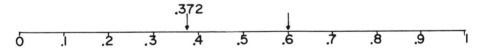

In general, for positive decimals that are both less than 1, the decimal with the greater digit in the tenths place will be the greater number. If these digits are equal, this test is applied to the hundredths digits, etc.

Converting between Fractions and Decimals   Every number of the form a/b can be written as either a terminating or an infinite repeating decimal, and conversely, every terminating or infinite repeating decimal can be written in the form a/b, where a and b are integers and $b \neq 0$.

From Fractions to Decimals—Every rational number of the form a/b can be changed to a decimal by dividing the numerator by the denominator. For example, $4/7 = 4 \div 7 = .\overline{571428}$, as shown here by the long division algorithm. In this case, the decimal is infinite repeating. This can be seen by the two circled 40s, which show that after six steps in the division process, all of the numbers will be repeated. On the other hand, when the numerator of 3/8 is divided by its denominator, the division algorithm shows that the decimal terminates after three digits.

$$
\begin{array}{r}
.5714285 \\
7\overline{)4.0000000} \\
35 \\
\hline
50 \\
49 \\
\hline
10 \\
7 \\
\hline
30 \\
28 \\
\hline
20 \\
14 \\
\hline
60 \\
56 \\
\hline
40 \\
35 \\
\end{array}
$$

Division Algorithm

It is fairly easy to see why every rational number r/s can be represented either by an infinite repeating decimal or by a terminating decimal. When r is divided by s, the remainders are always less than s. If a remainder of zero occurs in the division process, as it does when dividing 3 by 8, then the decimal terminates. If there is no zero remainder then eventually a remainder will be repeated, in which case the digits in the quotient will also start repeating.

$$
\begin{array}{r}
.375 \\
8\overline{)3.000} \\
2\,4 \\
\hline
60 \\
56 \\
\hline
40 \\
40 \\
\hline
\end{array}
$$

Division Algorithm

From Decimals to Fractions—Terminating decimals can be written as fractions whose denominators are powers of 10 ($.92 = 92/10^2$ and $.4718 = 4718/10^4$). The power of 10 will be the number of decimal places in the decimal. The expanded form of .378 shows why this decimal equals 378/1000. The common denominator of the fractions in the top equation is 1000, and their sum is 378/1000.

$$.378 = \frac{3}{10} + \frac{7}{100} + \frac{8}{1000}$$

$$= \frac{300}{1000} + \frac{70}{1000} + \frac{8}{1000}$$

$$= \frac{378}{1000}$$

Infinite repeating decimals can be converted to fractions by following a sequence of steps similar to those in the following examples.

EXAMPLE 1

1. Represent the decimal by x.

    $x = .656565 \cdots$

2. Multiply both sides of the equation by 10^2. (It is necessary to move the decimal point past the first repeating pattern.)

    $100x = 65.6565 \cdots$

3. Subtract Equation 1 from Equation 2. (By doing this the infinite repeating part of the decimal is subtracted off.) $99x = 65$

4. Solve for x. $x = \dfrac{65}{99}$

EXAMPLE 2

1. Represent the decimal by x. $x = .2222 \cdots$
2. Multiply both sides of the equation by 10. $10x = 2.222 \cdots$
3. Subtract Equation 1 from Equation 2. $9x = 2$
4. Solve for x. $x = \dfrac{2}{9}$

Both of these examples can be checked by dividing the numerators by their denominators. In each case you will obtain the original decimal.

Since a rational number can always be represented as the quotient of two integers, a/b, or as a decimal, we have two different types of numerals that represent the same numbers. In the following paragraphs we will refer to both types of numerals, fractions and decimals, and you may think of them as interchangeable.

Density of Rational Numbers   Between any two terminating or infinite repeating decimals on the number line, there is always another such decimal. It follows from this, that between any two decimals there are an infinite number of decimals. This crowding together of decimals is referred to by saying that the rational numbers are *dense*.

One method of locating a decimal that is between two given decimals is to add the two decimals and divide by 2. If we begin with 1.4 and 1.82, their sum divided by 2 is 1.61. Next, 1.4 + 1.61, divided by 2, is a decimal between 1.4 and 1.61. If this process is continued, the resulting sequence of decimals will get closer and closer (but never equal) to 1.4.

In spite of the abundance of rational numbers, there are still an infinite number of points left on the number line which do not correspond to these numbers (see Section 7.4, Irrational and Real Numbers).

Rounding Off   To round off a decimal at a given place value or to a given number of decimal places, locate the given place value and check the digit to its right. If the digit to the right is 5 or greater, then all digits to the right are dropped off and the place value is increased by 1. Otherwise, all digits to the right of the given place value are merely dropped off. Here are some examples.

1. Rounding to 2 decimal places (rounding to hundredths)

$$\underset{\downarrow}{\text{Hundredths}}$$
$$1.6825 \longrightarrow 1.68$$

2. Rounding to 1 decimal place (rounding to tenths)

$$\underset{\downarrow}{\text{Tenths}}$$
$$1.6825 \longrightarrow 1.7$$

3. Rounding to 3 decimal places (rounding to thousandths)

$$\underset{\downarrow}{\text{Thousandths}}$$
$$1.6825 \longrightarrow 1.683$$

Decimals on Calculators   To enter a decimal onto a calculator the keys are pressed for the digits from the highest to lowest place values with the decimal point being entered between the units and tenths digits. The digits usually appear in the right side of the display and are pushed to the left as each new digit is entered.

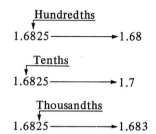

Some calculators have a *fixed decimal point display*, which means that all decimals are automatically rounded off. The Hewlett-Packard HP-21 rounds off to 2 decimal places, and the digits enter the left side of the display. If the keys for 123.4289 are pressed and then the enter key, Enter , the number 123.43 will appear in the display. Similarly, all computations are rounded off to 2 decimal places. If $2 \div 3$ is computed on this calculator, the decimal 0.67 will appear in the display. Even though the display shows only a 2-place decimal, internally the calculator maintains the value of the number to 10 decimal places. In the example of $2 \div 3$, the calculator works internally with the decimal .6666666667. The fixed decimal point display has the advantage of showing all answers to the same number of decimal places.

Some calculators have a key for rounding off to different numbers of decimal places. On the HP-21 this key is DSP . Pressing DSP · 4 will round off decimals on this calculator to 4 decimal places. Pressing DSP · 0 will round off all decimals to the nearest whole number.

Some calculators automatically round off decimals which exceed the full display of the calculator. On these calculators if 2 is divided by 3, the decimal 0.6666666667 will show in the display. Almost all calculators which round off at a digit before 5 will increase this digit, as described previously. For example, 55/99 is equal to the infinite repeating decimal .5555 . . . . If 55 is divided by 99 on a calculator which rounds off, 0.5555555556 will show in the display.

There are calculators that use two or three more digits than shown in their display. If you compute $1 \div 17$ on a calculator with 8 places for digits and the calculator does not round off, you will get 0.0588235. The next three digits, 2, 9, and 4, are kept in the calculator for computing but are not displayed. To find the first "hidden digit," multiply by 10. The display will then show 0.5882352. To find the next hidden digit, multiply by 10 and subtract 5. The display will then show 0.8823529. To find the last hidden digit multiply by 10 and subtract 8. The display will show 0.8235294.

Sometimes a decimal may appear to terminate or repeat at a certain point when actually it doesn't. If you compute 1/17 on a calculator, the first 7 decimal places are .0588235 but this last "5" is not the beginning of a repeating pattern.

Let's see how a calculator that displays eight digits can be used to find more digits in the decimal representation of 1/17. The long division algorithm shows the remainder is 9 after obtaining 6 decimal places in the quotient. This remainder can be found by multiplying .058823 by 17 and subtracting the result from 1.

$$.058823 \times 17 = 0.999991$$

This product, 0.999991, differs from 1 by .000009, which shows that the next step in the division algorithm is to divide 9 by 17. (*Note:* To avoid problems due to hidden digits or rounding off, 17 was multiplied by .058823 rather than by .0588235. Similarly, the last digit in each display will not be used in the following steps.)

```
              .058823
        17) 1.000000
              85
              150
              136
              140
              136
               40
               34
               60
               51
Remainder       9
```

Dividing 9 by 17 on the calculator gives 0.5294117, the next seven digits in the decimal representation for 1/17. Since there is no repeating pattern of decimals, we will multiply .529411 by 17 to get the next remainder.

$$17 \times .529411 = 8.999987$$

This product differs from 9 by .000013, so the next remainder is 13. Dividing once more by 17, $13 \div 17$ will show where the pattern begins to repeat.

$$13 \div 17 = 0.7647058$$

The pattern of digits for the infinite repeating decimal for 1/17 is .0588235294117647.

SUPPLEMENT *(Activity Book)*

Activity Set 7.1 Decimals on the Abacus and Calculator

<u>Additional Sources</u>

Anderson, J.T. "Periodic Decimals." *The Mathematics Teacher,* **67** No. 6 (October 1974), 504–9.

Beck, A., M.N. Bleicher and D.W. Crowe. *Excursions into Mathematics.* New York: Worth, 1969. "Decimal Fractions," pp. 444–70.

Hilferty, M. "Some Convenient Fractions for Work with Repeating Decimals." *The Mathematics Teacher,* **65** No. 3 (March 1972), 240–41.

Hobbs, B.F. and C.H. Burris. "Minicalculators and repeating decimals." *The Arithmetic Teacher,* **25** No. 7 (April 1978), 18–20.

Hutching, M.R. "Investigating the Nature of Period Decimals." *The Mathematics Teacher,* **65** No. 4 (April 1972), 325–27.

Manchester, M. "Decimal Expansions of Rational Numbers." *The Mathematics Teacher,* **65** No. 8 (December 1972), 698–702.

Sgroi, J.T. "Patterns of Repeating Decimals: A Subject Worth Repeating." *The Mathematics Teacher,* **70** No. 7 (October 1977), 604–5.

EXERCISE SET 7.1

1. Intervals of time can be measured by this cesium-beam atomic clock to an accuracy of .0000000000001 of a second. (This is equivalent to an accuracy of within 1 second every 300,000 years.) This strange clock, which is 6 m long, is operated at Boulder, Colorado, by the National Bureau of Standards. The frequency of cesium waves is 9,192,631,770 cycles per second. If the clock is adjusted to emit a pulse every 9,192,632 cycles (9,192,631,770 rounded off to the nearest thousand and divided by one thousand) it provides time intervals of .001 (one-thousandth) of a second,

Cesium-beam atomic clock

★ a. What is 9,192,631,770 rounded off to the nearest million? The answer divided by one million is the number of cycles the clock is adjusted to in order to provide intervals of .000001 (one-millionth) of a second.

b. What number must 9,192,631,770 be rounded off to, and divided by, in order to provide intervals of .000000001 (one-billionth) of a second?

2. The first book written about decimals was by Simon Stevin in 1585. He used small circled numerals between the digits. Among his examples he defined 27 + 847/1000 to be 27⓪8①4②7③. Explain how the decimal point could have evolved from this notation.

    a. In England this decimal is written as 27·847 and in the United States as 27.847. Which of these notations is closer to that used by Simon Stevin?

    b. What advantage is there in the English location of the decimal point as compared to the decimal point location which is used in the United States?

3. The sketch that follows is from a study showing the influence of surface temperature on air currents.

★   a. What is the highest surface temperature on this graph?

    b. What is the difference between the highest and lowest surface temperatures?

★   c. What is the average surface temperature (sum of the six temperatures divided by 6)?

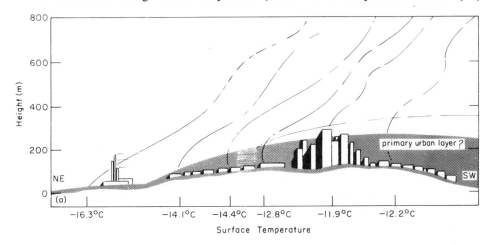

4. Draw an arrow from the numbers to their corresponding points on the number line. For each point that is labelled with a letter, write the number which corresponds to the point.

    .07   .72   1.40   1.68

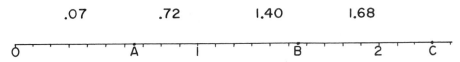

5. Write the names of these numbers.

★   a. 360.2876   b. .99999   c. 7.060   d. 49.00158

★   e. $347.96 _____ Dollars

    f. $23.50 _____ Dollars

6. Regroup the markers on this abacus so that there are fewer than ten on each column. Then write the number that is represented.

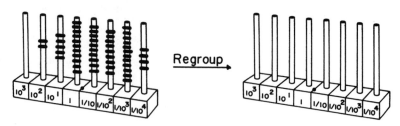

Regroup

★ 7. Find the decimal representations for each of the following fractions.

$$\frac{7}{12} \qquad \frac{1}{8} \qquad \frac{5}{6} \qquad \frac{3}{7} \qquad \frac{11}{20} \qquad \frac{6}{25}$$

★   a. The denominators of three of these fractions only have factors of 2 or of 5. Do these fractions have finite or infinite repeating decimals?

b. Make a conjecture about the type of fractions that will be equal to terminating decimals.

8. Express the following decimals as fractions.

★   a. .837   b. $.\overline{5}$ ★ c. $.\overline{64}$ d. $4.\overline{1}$

9. Round off these decimals to the given number of decimal places.

★   a. 2 places   b. 3 places   ★ c. 2 places   d. 4 places
    .4372   .7272 $\cdots$ 3.715   .06999

10. The following sums are expanded forms of decimals. Explain how these sums can be easily computed. Check your answers by entering these numbers into a calculator.

★   a. .8 + .02 + .003 + .0006

b. 4 + .1 + .09 + .003

c. 80 + 6 + .4 + .07

11. *Calculator Exercise:* The fraction 1/19 is represented by an infinite repeating decimal that has 18 digits before the pattern repeats. Use a calculator and the method described in the text to find this repeating pattern of digits.

12. *Calculator Exercise:* The rational numbers are dense, which means that between any two rational numbers there is always another one.

★   a. Use a calculator to find the number that is halfway between .4172 and .436 by adding these decimals and dividing by 2.

b. Add your answer in part **a** to .436 and divide by 2.

★ c. Add your answer in part **b** to .436 and divide by 2.

d. Add your answer in part **c** to .436 and divide by 2.

★ e. If you continue obtaining numbers in this manner, they will get closer and closer to what number?

13. *Calculator Exercise:* Every decimal except 0 has a *reciprocal.* The product of the decimal times its reciprocal is 1. The reciprocal of 2.318 is 1/2.318, which in decimal form to 7 places is .4314064. Use a calculator to compute the reciprocals of the following decimals.

★ a. 2.4   b. .48   c. .0086

14. *Calculator Exercise:* Use a calculator that adds a constant number by repeatedly pressing $\boxed{=}$.

★ a. Enter .01 $\boxed{+}$ on the calculator. What number will show on the display after $\boxed{=}$ is pressed: 10 times? 65 times? 100 times? 124 times?

b. Enter .1 $\boxed{+}$ on the calculator. What number will show on the display after $\boxed{=}$ is pressed: 10 times? 43 times? 132 times?

15. Write the decimals for these fractions by continuing the decimal patterns. Divide the numerators by the denominators to check your answers.

$$\frac{1}{9} = .111 \cdots, \quad \frac{2}{9} = .222 \cdots, \quad \frac{3}{9} = .333 \cdots, \quad \frac{4}{9} =$$

$$\frac{5}{9} = \qquad \frac{6}{9} = \qquad \frac{7}{9} = \qquad \frac{8}{9} =$$

★ a. Write .999 . . . as a fraction with a denominator of 9.

★ b. Is .999 . . . equal to 1?

★ 16. The 1978 edition of the *Guinness Book of World Records* contains an evolution of sports records for the twentieth century. Two categories of competition are shown in the table. For each distance, replace the feet and inches by feet to 2 decimal places and the minutes and seconds by minutes to 2 decimal places. (*Hint:* Divide the inches by 12 and the seconds by 60.)

	Start of the Century	Middle of the Century	Recent Records
High Jump	M. Sweeney 6' 5 5/8" U.S. 1895	Lester Steers 6' 11" U.S. 1941	V. Yashchenko 7' 7 3/4" U.S.S.R. 1977
Fastest Mile	W. G. George 4m 12.8s U.K. 1886	Gunder Hägg 4m 1.3s Sweden 1945	John Walker 3m 49.4s N.Z. 1975

17. The decimals for 1/11, 2/11, 3/11, etc., each repeat after 2 places. By computing a few more of these decimals you will see a pattern that will help you to complete the table.

$\dfrac{1}{11} = .\overline{09}$	$\dfrac{2}{11} = .\overline{18}$	$\dfrac{3}{11} =$	$\dfrac{4}{11} =$	$\dfrac{5}{11} =$	$\dfrac{6}{11} =$	$\dfrac{7}{11} =$	$\dfrac{8}{11} =$	$\dfrac{9}{11} =$	$\dfrac{10}{11} =$

A phototimer finish of a one-mile race
between Marty Liquori and Jim Ryun

7.2 OPERATIONS WITH DECIMALS

Decimals to hundredths and thousandths of a second are used in electronic timers for athletic competition. The above picture shows a phototimer finish of a one-mile race in which Marty Liquori of Villanova beat Jim Ryun of the Oregon Track Club at the 1971 International Freedom Games at Philadelphia's Franklin Field. Both runners were officially clocked in 3:54.6 seconds by officials using hand-operated stopwatches. However, the phototimer clocked Liquori in 3:54.54 seconds and Ryun in 3:54.75 seconds, a difference of .21 second. What you see is not a simultaneous photograph of the two runners but a continuous photograph of the finish line showing each runner as he crossed it.

Addition   In addition of whole numbers the digits are aligned so that units are placed over units, tens over tens, etc. This same idea is extended to the addition of decimals. By writing the numerals so that their decimal points are lined up, tenths will be over tenths, hundredths over hundredths, etc. The addition of digits by columns can then be carried out just as in the case of whole numbers.

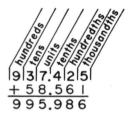

$$
\begin{array}{r}
937.425 \\
+\ 58.561 \\
\hline
995.986
\end{array}
$$

This is essentially the method of adding decimals which Simon Stevin used. The example shown here is taken from his book. The ⓪ marks the units digits, ① the tenths digits, ② the hundredths digits, and ③ the thousandths digits. In our notation this sum is 941.304.

$$
\begin{array}{r}
⓪①②③ \\
27847 \\
37675 \\
875782 \\
\hline
941304
\end{array}
$$

Addition can be illustrated on the abacus by representing the numbers to be added on the columns. In this example, 430.353 is represented on the upper portion of the columns and 145.472 beneath. The sum is computed by counting the total number of markers in each column.

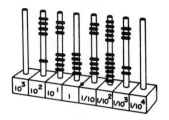

$$430.353 + 145.472$$

When the sum of the digits in any column is greater than 9, regrouping or "carrying" is necessary. In this example the sum of the digits in the hundredths column is 12, so a 1 is regrouped and added to $3 + 4$ in the tenths column. On the abacus this regrouping is accomplished by removing 10 markers from the $1/10^2$ column and carrying 1 marker to the $1/10$ column. This "1" which is regrouped represents $10/100$, or $1/10$. The following equations show the sum in the hundredths column of this example and the regrouping.

$$
\begin{array}{r}
1 \\
430.353 \\
+ \ 145.472 \\
\hline
575.825
\end{array}
$$

$$\frac{5}{100} + \frac{7}{100} = \frac{12}{100} = \frac{10}{100} + \frac{2}{100} = \left(\frac{1}{10}\right) + \left(\frac{2}{100}\right)$$

"1" carried to tenths column

"2" recorded in hundredths column

Subtraction   To subtract two decimals the digits are lined up with tenths under tenths, hundredths under hundredths, etc., just as for adding decimals. Subtraction then takes place from right to left. In this example, thousandths are subtracted from thousandths, hundredths from hundredths, and so on, to the tens column.

$$
\begin{array}{r}
47.658 \\
- \ 16.421 \\
\hline
31.237
\end{array}
$$

To illustrate subtraction on the abacus, only one number is represented on the columns. Then we "take away" the appropriate number of markers from each column to compute the difference. For example, to compute $77.6528 - 65.9386$, we begin by removing 6 markers from the $1/10^4$ column.

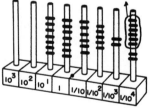

$$77.6528 - 65.9386$$

When regrouping or "borrowing" is necessary, it is done just as it is for subtracting whole numbers. In this example the "2" in the thousandths column is changed to a "12" by borrowing 1 from 5. On the abacus this regrouping is done by replacing 1 marker from the $1/10^2$ column by 10 markers on the $1/10^3$ column. This can be done because $1/100 = 10/1000$. The following equations show the change in the hundredths column and the regrouping.

$$
\begin{array}{c}
4 \\
77.6\!\!\not{8}28 \\
-\,65.9386 \\
\hline
42
\end{array}
$$

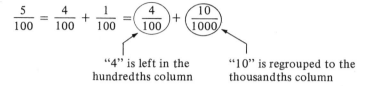

"4" is left in the hundredths column

"10" is regrouped to the thousandths column

To continue this example, regrouping is also needed to subtract 9 from 6 in the tenths column. In this case, 1 is regrouped from the 7 in the units column and used as 10 tenths in the tenths column. These equations show the change in the units column and the regrouping.

$$
\begin{array}{c}
6\ \ \ 4 \\
7\!\!\not{7}.6\!\!\not{8}28 \\
-\,65.9386 \\
\hline
11.7142
\end{array}
$$

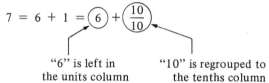

"6" is left in the units column

"10" is regrouped to the tenths column

Multiplication

The product of a whole number times a decimal can be illustrated on the abacus by multiplying the numbers of markers on each column separately. To multiply by 2 the number of markers on each column is doubled. The regrouping can be done as you multiply or after the multiplication has been carried out. In this example the markers for 352.46 have been doubled. After regrouping, these markers will represent 704.92. Compare the steps for computing this product on the abacus with those used in the standard multiplication algorithm.

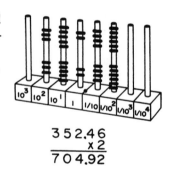

$$
\begin{array}{r}
3\,5\,2.4\,6 \\
\times\,2 \\
\hline
7\,0\,4.9\,2
\end{array}
$$

Multiplying by Decimals—One of the notations for locating decimal points in the sixteenth century was that of writing circled indices to the right of the numerals. For example, 27.487 was represented as 27487 ... ③, and 9.21 was 921 ... ②. This notation is especially convenient when computing the product of two decimals. The whole numbers

$$
\begin{array}{r}
27487 \ldots ③ \\
\times\ 921 \ldots ② \\
\hline
27487 \\
54974 \\
247383 \\
\hline
25315527 \ldots ⑤
\end{array}
$$

are multiplied, and then the numbers in circles are added to determine the location of the decimal point. This particular notation points out the close relationship between computing products of whole numbers and computing products of decimals. Using our present notation the decimal point should be placed between the "3" and "1" in this product.

Our present-day algorithm for multiplying decimals is essentially the same as the sixteenth century example. Two numbers are multiplied as though they were whole numbers and then the decimal point is located in the product. The digits *do not* have to be positioned so that units are above units, tenths above tenths, etc., as they are for addition and subtraction of decimals. The number of decimal places in the answer is the total number of decimal places in the original two numbers.

$$
\begin{array}{r}
27.487 \\
\times\ 9.21 \\
\hline
27487 \\
54974 \\
247383 \\
\hline
253.15527
\end{array}
$$

To understand this algorithm it will help to examine the rules for multiplying and dividing by powers of 10. In the following example, 43.286 is expressed in expanded form and multiplied by 10^2.

$$
10^2 \times 43.286 = 10^2 \times \left(4 \times 10 + 3 + 2 \times \frac{1}{10} + 8 \times \frac{1}{10^2} + 6 \times \frac{1}{10^3} \right)
$$

$$
= (4 \times 10^2 \times 10) + (3 \times 10^2) + \left(2 \times 10^2 \times \frac{1}{10} \right)
$$

$$
+ \left(8 \times 10^2 \times \frac{1}{10^2} \right) + \left(6 \times 10^2 \times \frac{1}{10^3} \right)
$$

$$
= 4 \times 10^3 + 3 \times 10^2 + 2 \times 10 + 8 + 6 \times \frac{1}{10}
$$

$$
= 4328.6
$$

Multiplying by 10^2 increases the powers of 10 by 2 for each digit in 43.286 and has the effect of moving the decimal point 2 places to the right.

Similarly, dividing by a power of 10 decreases the powers of 10 for each digit in a numeral and has the effect of moving the decimal point to the left. In the following equations, $743.28 \div 10^2$ is computed by using the expanded form for 743.28.

$$\frac{743.28}{10^2} = \frac{1}{10^2} \times \left(7 \times 10^2 + 4 \times 10 + 3 + \frac{2}{10} + \frac{8}{10^2}\right)$$

$$= \left(7 \times \frac{1}{10^2} \times 10^2\right) + \left(4 \times \frac{1}{10^2} \times 10\right) + \left(3 \times \frac{1}{10^2}\right)$$

$$+ \left(\frac{2}{10} \times \frac{1}{10^2}\right) + \left(\frac{8}{10^2} \times \frac{1}{10^2}\right)$$

$$= 7 + \frac{4}{10} + \frac{3}{10^2} + \frac{2}{10^3} + \frac{8}{10^4}$$

$$= 7.4328$$

In general, for positive values of n, we can prove: (1) to multiply a decimal by 10^n, move the decimal point n places to the right; and (2) to divide a decimal by 10^n, move the decimal point n places to the left.

These rules for computing with powers of 10 are used in the following equations to show why we count off 5 places in locating the decimal point in the product of 27.487×9.21. Furthermore, the product of the two numerators, 27487×921, shows why we multiply decimals as though they were whole numbers.

$$27.487 \times 9.21 = \frac{27,487}{10^3} \times \frac{921}{10^2} = \frac{25,315,527}{10^5} = 253.15527$$

Division   The algorithm for division will be examined in two cases: (1) dividing a decimal by a whole number; and (2) dividing a decimal by a decimal.

Dividing by Whole Numbers—When a decimal on the abacus is divided by a single-digit number, the division can be performed separately on the numbers of markers in each column. In this example, 97.56 is divided by 3, beginning with the 9 markers on the tens column.

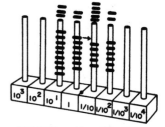

$97.56 \div 3 = 32.52$

1. 9 tens $\div$ 3 = 3 tens, so 3 markers are placed above the tens column to represent 30.

2. $7 \div 3 = 2$, with 1 remaining, so 2 markers are placed above the units column. The remaining marker is regrouped as 10 markers on the tenths column.

3. There are now 15 tenths: 15 tenths $\div$ 3 = 5 tenths, so 5 markers are placed above the tenths column.

4. 6 hundredths $\div$ 3 = 2 hundredths, so 2 markers are placed above the hundredths column.

Notice that in this example on the abacus we did not make any special use of the location of the decimal point. That is, the division process and regrouping are the same as though we were dividing a whole number by 3.

Four similar steps are carried out when computing
$97.56 \div 3$ by the division algorithm. Division is performed without regard to the decimal point, just as
though we were computing $9756 \div 3$. Then the decimal point is placed in the quotient directly above
the decimal point in 97.56. The following equations
show why this quotient can be computed by dividing
a whole number by a whole number. In the first
equation the decimal point is removed; in the second
equation, 9756 is divided by 3; and in the last equation the decimal point is replaced.

$$\frac{97.56}{3} = \frac{9756}{3} \times \frac{1}{10^2}$$

$$= 3252 \times \frac{1}{10^2}$$

$$= 32.52$$

```
    3 2.5 2
3/9 7.5 6
  9
  ─
  7
  6
  ─
  1 5
  1 5
    6
    6
    ─
```

Dividing by Decimals—In the long division algorithm
for dividing with decimals we actually never do divide
by a decimal. Before dividing, a slight adjustment is
made so that the divisor is always a whole number.
For example, the algorithm shown here indicates
that we are to divide 1.504 by .32. Before dividing,
however, the decimal points in .32 and 1.504 are
moved 2 places to the right. This has the effect of
changing the divisor and the dividend so that we
are dividing 150.4 by 32, as shown in the lower
algorithm.

```
.32 /1.504
```
changed to

```
        4.7
32. /150.4
    128
    22 4
    22 4
```

The rule for dividing by a decimal is to count the number of decimal places in the
divisor and then move the decimal points in the divisor and the dividend this many places
to the right. In the previous example the decimal points in .32 and 1.504 were moved 2
places to the right because .32 has 2 decimal places. The justification for this process of
shifting decimal points is illustrated in the following equations.

$$\frac{1.504}{.32} = \frac{1.504 \times 10^2}{.32 \times 10^2} = \frac{150.4}{32.}$$

These equations show that the answer to $1.504 \div .32$ is the same as that for $150.4 \div 32$.
No further adjustment is needed as long as we shift the decimal points in both the divisor
and the dividend by the same amount. In this manner, dividing a decimal by a decimal can
always be carried out by dividing a decimal by a whole number.

Approximate Computations   There are times when approximate computations are as useful as exact computations. To make a decision regarding a purchase, for example, all we may need is a "rough idea" of the cost. Suppose you are interested in the total cost of a stereo system with these components and prices: tape deck, $219.50; turntable with cartridge, $179; two speakers, $284; and receiver $335.89. An approximate computation with these costs will save time and can be done in your head. Here are the exact sum and the approximate sum for numbers rounded off to the hundreds place.

Exact Sum	Approximate Sum
$ 219.50	$ 200
$ 179.00	$ 200
$ 284.00	$ 300
$ 335.89	$ 300
$1018.39	$1000

Whenever exact answers are required, approximate computations can serve as a guide for detecting large errors. One common source of error is that of misplacing a decimal point in the answer. This can usually be discovered by rounding off each number to its highest nonzero place value. For example,

$$342.8 \times .046 = 15.7688$$

can be checked by rounding off 342.8 to 300 and .046 to .05. The product of 300 and .05 is 15, which is a good approximation to the original product of 15.7688. If the decimal point in 15.7688 had been misplaced, such as in 1.57688 or in 157.688, the relatively large difference between these numbers and 15 would indicate an error.

Another source of error is that of misplacing the decimal point when a number is entered into a calculator. Suppose the previous product is computed on a calculator and "342.8" is mistakenly entered as "34.28." In this case, the product

$$34.28 \times .046 = 1.57688$$

is sufficiently different from the approximate answer of 15 to indicate an error.

Rounding off will help to detect large errors but not small ones. Here are three more examples of approximate computations from rounding off.

Exact	Rounded off	Exact	Rounded off
430.48	400	194.15	200
92.81 →	90	− 76.62 →	− 80
+ 160.36	+ 200	117.53	120
683.65	690		

Exact	Rounded off

```
        1 6 2.                 1 5 0.
.3 8/6 1.5 6      →       .4/6 0.0
   3 8                      4
   2 3 5                    2 0
   2 2 8                    2 0
     7 6
     7 6
```

Decimal Operations on Calculators   The steps for carrying out the four basic operations with decimals on a calculator are the same as for whole numbers. A pair of numbers and their operations are entered in the order in which they appear (from left to right) when written horizontally. However, when addition and subtraction are combined with multiplication and division, care must be taken regarding the order of the operations. As in the case of whole numbers, multiplication and division are performed before addition and subtraction. Consider the following example of computing income tax. According to Schedule Y in the 1977 Internal Revenue Service forms, the tax on earnings greater than $11,200 and less than $15,200 is $1,380 plus 22% of the amount over $11,200. Therefore, the tax on $12,900 is

$$\$1380 + .22 \times \$1700$$

On some calculators this cannot be computed by entering the numbers and operations into the calculator as they appear from left to right, because the .22 would be added to the 1380. One method of determining this tax is to compute .22 × 1700 and then add 1380. Another method is to place 1380 in storage, compute .22 × 1700, and then perform the addition. Here are the steps for finding the answer by this method.

	Steps	Displays
1.	Enter 1380	1380.
2.	M+	1380.
3.	.22 × 1700 =	374.
4.	+ MR	1380.
5.	=	1754.

On some calculators arithmetic and geometric sequences of decimals can be generated by repeatedly using the equality key, as was demonstrated in Chapter 3 with whole numbers. Geometric sequences of whole numbers increase so rapidly that the numbers very quickly exceed the capacity of calculators not having scientific notation. With decimals, however, if a number close to 1 is chosen as a constant multiplier to generate a geometric sequence, the resulting numbers increase very slowly. The sequence obtained from the calculator steps in the following example begins with 14.6, and each new number is obtained by multiplying the previous number by the constant 1.06. There are 10 numbers in this sequence before the numbers become greater than 26. Notice that the constant multiplier 1.06 is entered first. On some calculators the second number which is entered into the calculator will be the constant multiplier.

	Steps	Displays
1.	Enter 1.06	1.06
2.	×	1.06
3.	Enter 14.6	14.6
4.	=	15.476
5.	=	16.40456
6.	=	17.388833
7.	=	18.432162

The sequence obtained by repeatedly multiplying 14.6 by 1.06 can be used as an example of rising costs due to an annual 6% rate of inflation. If an item costs $14.60, a 6% rate of inflation will increase its cost to 1.06 × $14.60, or $15.48, as shown in Step 4 above. The numbers in the calculator displays for Steps 5, 6, and 7 give the year-by-year increases in cost due to this rate of inflation. After 4 years this item will cost $18.43.

Here is another example of how a calculator that will multiply several different numbers by the same number can be applied to solving practical problems. Suppose an item costs $16.75 to rent per hour and you wish to find the costs for these times: 14.5 hours, 8.2 hours, 5.4 hours, 12 hours, and 13.1 hours. The 16.75 × which is entered in the first step of the next sequence remains in effect throughout Steps 2, 3, 4, and 5. For example, to multiply 8.2 by 16.75 in Step 2, it is necessary only to enter 8.2 = . If you wish the total of these costs, they can be added together in the memory storage by pressing M+ after each = in the following steps.

	Steps	Displays
1.	16.75 × 14.5 =	242.875
2.	8.2 =	137.35
3.	5.4 =	90.45
4.	12 =	201.
5.	13.1 =	219.425

SUPPLEMENT *(Activity Book)*

Activity Set 7.2 Decimal Operations on the Abacus

Just for Fun: Decimal Tricks with Calculators

Additional Sources

Henry, B. "Do we need separate rules to compute in decimal notation." *The Arithmetric Teacher,* **18** No. 1 (January 1971), 40–42.

National Council of Teachers of Mathematics. *The Rational Numbers.* Reston, Virginia: NCTM, 1972. pp. 179–238 (operations with decimals).

Prielipp, R.W. "Decimals." *The Arithmetic Teacher,* **23** No. 4 (April 1976), 285–88.

School Mathematics Study Group. *Studies in Mathematics, 6.* New Haven: Yale University, 1961. pp. 138–51.

Smith, D.E. *History of Mathematics, 2.* Lexington, Mass: Ginn, 1925. "Ratio Proportion and the Rule of Three," pp. 477–91 and "Percentage and the Use of Percent," pp. 247–50.

EXERCISE SET 7.2

1. A 1976 survey by *U.S. News and World Report* based on 6448 returned questionnaires produced the ratings shown in this list.

   ★ a. What is the difference between the ratings for the following pairs of businesses?

Airlines	Retail food chains
vs.	vs.
Railroads	Service stations
Banks	Life-insurance companies
vs.	vs.
Oil and Gas Companies	Automobile dealers

   b. Add all 20 ratings and divide by 20 to get an average rating for these businesses.

HOW BUSINESSES RATE WITH CUSTOMERS

Survey participants were asked to rate the performance of 20 major industries. In the scoring system, 1 is the equivalent of a poor job, and 7 the equivalent of an excellent job. The ranking:

	Average Rating
1. Airlines	5.17
2. Banks	5.05
3. Trucking companies	4.61
4. Large department stores	4.59
5. Tire manufacturers	4.57
6. Retail food chains	4.53
7. Appliance manufacturers	4.44
8. Steel manufacturers	4.37
9. Life-insurance companies	4.31
10. Drug manufacturers	4.30
11. Food manufacturers	4.23
12. Electric utilities	4.16
13. Construction companies	4.13
14. Gas utilities	4.13
15. Service stations	3.98
16. Appliance-repair services	3.54
17. Automobile manufacturers	3.53
18. Oil and gas companies	3.53
19. Automobile dealers	3.53
20. Railroads	3.16

2. The gelosia method of multiplying whole numbers (see page 107) can also be used for multiplying decimals. The two numbers to be multiplied, 16.34 and 5.87, are written at the top and right side as in the case of whole numbers. The partial products are then written in the squares, and the product is the sum of the numbers in the diagonals. The arrows indicate the method of locating the decimal point.

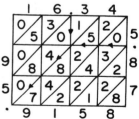

Use this method to compute the following products.

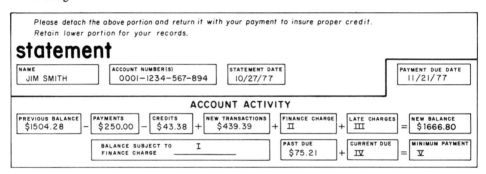

★ a.
```
  4  7 . 8
                3
                2
                .
                6
```

b.
```
. O  O  8
             3
             .
             4
```

c.
```
. 4  7
          6
          .
          2
```

3. *Credit Cards:* Fill in the missing amounts on this credit card by completing parts **a** through **e**.

Please detach the above portion and return it with your payment to insure proper credit.
Retain lower portion for your records.

statement

NAME	ACCOUNT NUMBER(S)	STATEMENT DATE		PAYMENT DUE DATE
JIM SMITH	0001–1234–567–894	10/27/77		11/21/77

ACCOUNT ACTIVITY

PREVIOUS BALANCE		PAYMENTS		CREDITS		NEW TRANSACTIONS		FINANCE CHARGE		LATE CHARGES		NEW BALANCE
$1504.28	−	$250.00	−	$43.38	+	$439.39	+	II	+	III	=	$1666.80

	BALANCE SUBJECT TO FINANCE CHARGE	I		PAST DUE		CURRENT DUE		MINIMUM PAYMENT
				$75.21	+	IV	=	V

★ a. The balance subject to the finance charge (box I) is the difference between the previous balance and the sum of the payments and credits. Compute $250 + $43.38, and subtract this from the balance of $1504.28. Write this amount in box I.

b. The monthly finance charge on the amount in box I is determined by the following rule: .0125 times the first $500; .0095 times the next $500; and .0083 times the amount over $1000, which in this example is .0083 times $210.90. Compute these three products and, if necessary, round off to the nearest penny. Add these products and record the sum in box II.

★ c. The late charges (box III) are .05 times the amount past due, $75.21. Compute this product and round your answer to the nearest penny. Enter this amount in box III.

d. If you add the previous balance, $1504.28; the new transactions, $439.39; the finance charge in box II; and the late charges in box III; and subtract the payments, $250.00, and the credits, $43.38, then you should get the new balance of $1666.80. The current amount due is .05 times $1666.80. Compute this product and enter it into box IV.

★ e. The minimum payment (box V), which is due within 25 days after the date of this statement, is the past amount due, $75.21, plus the amount currently due from box IV. Compute this sum and enter it into box V.

4. *Payroll Deductions:* A job that pays $5.50 an hour amounts to a gross income of $220 for a 40-hour week. Compute the deductions in parts **a** through **c** and enter them into the table.

★ a. The federal income tax on $220 for a single person is .175 times $220.

   b. The state income tax is .029 times $220.

★ c. The FICA or social security tax is .0605 times $220.

   d. Compute the sum of the three deductions in parts **a**, **b**, and **c**. How many working hours, at $5.50 per hour, are needed to pay for these deductions?

★ e. Subtract the sum of the deductions in parts **a**, **b**, and **c** from the gross earnings in order to find the net or "take-home" pay.

Total gross earnings $220.00

Federal income tax _____

State income tax _____

FICA tax _____

Net pay _____

5. *Electricity Costs:* The basic unit for measuring electricity is the *kilowatt-hour* (kWh). This is the amount of electrical energy required to operate a 1000-watt appliance for 1 hour. For example, it takes 1 kilowatt-hour of electricity to light ten 100-watt bulbs for 1 hour. This table contains the average number of kilowatt-hours for operating the appliances for 1 month.

★ a. What is the sum of the kilowatt-hours for operating these eight appliances for 1 month? Record this at the bottom of the table.

   b. At a cost of $.04 for each kilowatt-hour, what is the monthly cost for operating each of these appliances? Round off your answers to the nearest penny and record the amounts in the table.

Appliance	Kilowatt-hours per month	Cost per month
Microwave oven	15.8	_____
Range with oven	97.6	_____
Refrigerator	94.7	_____
Frostless refrigerator	152.4	_____
Water heater	400.0	_____
Radio	7.5	_____
Television (black-white)	29.6	_____
Television (color)	55.0	_____
Total	Sum	

★ c. Compute the sum of the monthly costs in part **b** for the eight appliances.

   d. Compute the difference in the monthly costs of electricity for black and white television and color television.

   e. How much less does it cost to operate a refrigerator for 1 year than a frostless refrigerator?

6. In the example shown here, a "1" is regrouped from the hundredths column to the tenths column. This can be explained by the equations

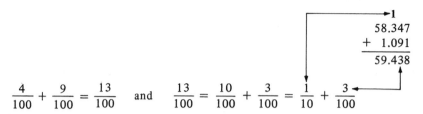

$$\frac{4}{100} + \frac{9}{100} = \frac{13}{100} \quad \text{and} \quad \frac{13}{100} = \frac{10}{100} + \frac{3}{100} = \frac{1}{10} + \frac{3}{100}$$

In each of the following addition exercises, there is one sum of digits which requires regrouping to the next column. Write the number which is to be regrouped and use equations to explain the regrouping.

★ a. 4.821
      + 61.73

   b. .367
       .015
       + .509

   c. 7.00048
       + .1738

7. In each of the following exercises, there is one column for which regrouping is needed before subtracting. Mark this column and use equations to explain how the regrouping takes place.

★ a. 66.43
      − 41.72

   b. .046
       − .018

   c. 5.63
       − .17

8. The normal body temperature is 37°C. When you are sick it may go as high as 40°C. Even when you are healthy your temperature varies during the day. Frank took his temperature seven times one day and got these results.

7 A.M.	9 A.M.	11 A.M.	1 P.M.	3 P.M.	5 P.M.	7 P.M.
36.8°C	37.6°C	37.3°C	38.2°C	37.7°C	37.2°C	37°C

a. What is the difference between the highest and lowest temperatures?

★ b. What was his average temperature for the 12-hour period? (Add the temperatures and divide by 7.)

★ 9. The markers on the top half and bottom half of this abacus represent 35.654 and 125.347 respectively. The sum of these numbers can be obtained by regrouping the markers so that there are fewer than ten on each column. Show this regrouping on the other abacus. What is this sum?

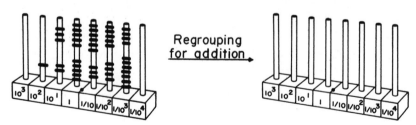

Regrouping for addition

10. The abacus on the left represents 84.0273. In order to compute 84.0273 − 6.5409, regrouping is necessary several times. Show this regrouping by drawing markers on the abacus on the right. Then compute 84.0273 − 6.5409 by removing markers from this abacus. What is this difference?

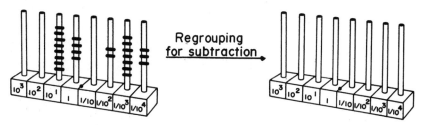

Regrouping for subtraction ▸

11. *Approximate Computation:* Round off each number to its highest place value and carry out the approximate computation. Use these results to predict which exercise is done incorrectly. Check this exercise by exact computation.

★   a. 258.35 + 32.70 + 706.42 = 997.47

      b. 481.302 − 168.625 = 312.677

★   c. 634.5 × 3.62 = 2296.89

      d. 4.3542 ÷ .82 = 53.1

12. *Casting Out Nines:* This method was used on page 152 for checking addition, subtraction, and multiplication of whole numbers. By disregarding the decimal points and treating each decimal as a whole number, we can check these operations on decimals by casting out nines. In this example the sum of the excesses is 8,

	Excess
3.46	4
17.08	7
+ 9.42	+6
29.96	8

(4 + 7 + 6 = 17 and 1 + 7 = 8), and the nines excess of the sum is 8 (2 + 9 + 9 + 6 = 26 and 2 + 6 = 8). Use casting out nines to determine which of the following computations have errors. (*Note:* Casting out nines will not indicate misplaced decimal points.)

a.   34.276
     46.302
   + 89.477
    170.055

b.   3218.46
   − 913.97
     2305.49

c.   .00148
   × 36.47
   .0539756

13. *Calculator Exercise:* In this geometric sequence each number is obtained by multiplying .85 times the previous number.

$$2, \quad 1.7, \quad 1.445, \quad 1.22825, \quad 1.0440125$$

★   a. What is the next number in this sequence?

      b. How many numbers will there be in this sequence before reaching a number that is less than .1?

★ c. Eventually, if the preceding sequence is continued, 0 will be displayed on your calculator. How many numbers will there be in this sequence before reaching a number that is less than .01?

d. Will there ever be a number in this sequence that is equal to 0?

14. *Calculator Exercise:* If these steps are carried out on some calculators, each number from Step 2 on will be multiplied by 1.02. Fill in the displays for Steps 4 and 5.

Steps	Displays
1. Enter 1.02	1.02
2. ☒	1.02
3. ═	1.0404
4. ═	
5. ═	

★ a. What will happen to the numbers in the display if these steps are continued indefinitely?

b. Will your answer in part **a** be true for any number greater than 1 which is entered in Step 1?

★ c. If a number less than 1 is entered in Step 1 and the remaining steps are continued indefinitely, what will eventually happen to the numbers in the display?

15. *Calculator Exercise:* Costs due to an annual inflation rate of 4% can be determined by repeatedly multiplying the cost of an item by 1.04. Use this rate of inflation to answer the following questions.

a. How much will the cost of a $545 washing machine increase after five years, due to inflation?

★ b. In how many years will the cost of the washing machine in part **a** be doubled?

c. If a $16,500 annual salary is increased 4% each year to keep pace with inflation, what will this salary be after 3 years?

16. In the first 6 months of 1976, Canada had drilled 2557 new oil wells totalling 8.2 million feet.

★ a. What is the average depth of each of these wells? Compute your answer to 2 decimal places.

b. Canada's expectations for the last half of 1976 were to have 2688 more wells and 8.6 million more feet of drilling. What would be the average depth of these wells?

★ c. What is Canada's expected total of new wells and number of feet drilled for 1976?

17. The greatest record spree ever occurred in the 1976 Olympic swimming competition when world records were set in 22 out of 26 events. In one of these events, an East German, Petra Thümer, set a world record in the 400-meter freestyle, winning in 4:09.89 (4 minutes and 9.89 seconds).

★   a. Thümer's time was 1.87 seconds faster than the old record. What was the time for the old record?

      b. In the 200-meter freestyle a world record was set by another East German, Kornelia Ender, in the time of 1:59.26. If this rate of speed could be maintained, how long would it take to swim the 400-meter event? Compare this with Thümer's time for the 400-meter event.

      c. In the 1964 Olympics in Tokyo, Don Schollander, an American, had a winning time of 4:12.2 in the 400-meter freestyle. How many seconds faster was Thümer's time for the 400-meter event?

18. The 4 × 400 relay gold medals were won by the United States team of Herm Frazier (45.2 seconds), Benny Brown (44.7), Fred Newhouse (43.9), and Maxie Parks (45.0).

★   a. What was the difference between Frazier's time and Newhouse's?

      b. What was the total time in minutes and seconds for the 4 × 400 relay?

      c. What was the average of these four times?

19. This "pie diagram" shows the amounts of money in billions of dollars spent on health in the U.S. in 1975.

★   a. What was the total amount of money spent for the year?

      b. How many times greater was the amount spent on hospital bills as compared to the dentists' fees?

★   c. Which amount is greater: the money spent on hospital bills or the total amount paid for dentists, physicians, drugs, and nursing homes?

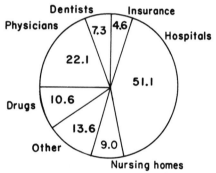

Note: The numbers in the pie diagram are not percents.

Constar communications satellite—
handles long-distance telephone calls.

7.3 RATIO, PERCENT, AND SCIENTIFIC NOTATION

Engineers in this picture are checking out the second Constar communications satellite before launching. It handles long-distance phone calls for American Telephone and Telegraph and for General Telephone and Electronics. About 20 million long-distance phone calls are made in the United States every day.

The table that follows shows the countries that had over 100,000 telephones in service in 1975. There are several countries with more than a 100% increase in the number of telephones since 1965. A 100% increase is twice as many telephones, and a 200% increase is three times as many. Which countries had more than three times as many phones in 1975 as in 1965?

The numbers in the columns showing "telephones per 100 population" are ratios. In Japan there were 37.88, or approximately 38, telephones for each 100 people. In Egypt, there were between 1 and 2 (1.37) telephones per 100 people.

Country	Number of tele-phones	Percent-age increase over 1965	Tele-phones per 100 popula-tion	Country	Number of tele-phones	Percent-age increase over 1965	Tele-phones per 100 popula-tion
Algeria	229,673	64.5	1.38	Luxembourg	141,686	77.7	39.69
Argentina	2,373,665	61.2	9.41	Malaysia	259,405	97.0	2.21
Australia*	4,999,982	87.3	37.49	Mexico	2,546,186	251.2	4.37
Austria	1,986,733	112.2	26.37	Morocco	189,000	33.2	1.13
Belgium	2,666,701	81.6	27.32	Netherlands, The	4,678,945	114.6	34.41
Brazil	2,651,728	107.6	2.50	New Zealand	1,494,587	55.3	48.12
Bulgaria	718,325	188.6	8.18	Nigeria	111,478	62.2	0.16
Canada	12,454,331	77.4	54.96	Norway	1,355,142	56.0	33.90
Chile	447,014	69.8	4.26	Pakistan*	195,325	42.2	0.29
Colombia	1,186,205	189.6	4.74	Panama	139,241	194.9	8.61
Cuba†	274,949	19.1	3.16	Peru	333,346	143.5	2.14
Czechoslovakia	2,480,801	77.4	16.83	Philippines	446,262	170.7	1.09
Denmark	2,183,847	88.2	42.48	Poland	2,399,249	101.0	7.09
Ecuador	167,505	285.1	2.58	Portugal	1,011,177	93.7	11.67
Egypt	503,200	67.0	1.37	Puerto Rico	466,465	129.7	15.25
Finland	1,678,873	115.8	35.78	Rhodesia	171,881	69.8	2.77
France	12,405,000	117.5	23.52	Romania	1,076,566	152.4	5.10
Germany, East	2,451,011	54.5	15.04	Singapore	280,280	222.6	12.53
Germany, West	18,767,033	129.8	30.25	South Africa	1,935,831	70.8	7.77
Greece	1,862,050	331.7	20.71	Spain	7,042,968	178.7	19.96
Hong Kong	988,545	350.0	22.75	Sweden	5,178,082	116.9	63.32
Hungary	1,013,731	88.2	9.65	Switzerland	3,790,351	77.8	59.46
India	1,689,528	122.3	0.29	Syria*	143,320	83.7	2.08
Indonesia	284,831	65.5	0.23	Taiwan	900,605	509.2	5.68
Iran	805,560	344.7	2.40	Thailand*	270,840	248.4	0.66
Iraq	152,932	135.3	1.42	Tunisia	114,250	126.8	2.03
Ireland	393,879	81.4	12.78	Turkey	899,923	192.1	2.30
Israel	735,156	241.9	21.57	U.S.S.R.	15,782,000	122.3	6.23
Italy	13,695,006	147.7	24.62	United Kingdom	20,342,457	104.2	36.26
Japan	41,904,960	242.1	37.88	United States	143,971,718	62.1	66.02
Kenya	113,688	112.1	0.88	Uruguay	247,923	33.8	8.97
Korea, South*	1,014,016	269.6	3.09	Venezuela	554,197	113.0	4.65
Kuwait	108,587	317.0	11.00	Yugoslavia	1,142,883	209.0	5.38
Lebanon‡	227,000	116.2	7.67				

*1974.
†1972.
‡1973.

Ratios and Rates   Ratio is one of the most useful ideas in everyday mathematics. A *ratio* is a pair of positive numbers that is used to compare two sets. For example, in 1976 the ratio of numbers of U.S. Air Force personnel to Army personnel was 3 to 4. This tells us that for every three people in the Air Force there were four people in the Army. This ratio gives the relative sizes of these sets but not the actual number of personnel.

There are two common notations for ratios. The ratio of Air Force personnel to Army personnel can be written as 3:4 or 3/4 and is read as "3 to 4."

Definition: For any two positive numbers, *a* and *b*, the ratio of *a* to *b* is *a/b*. This is sometimes written as *a:b*.

Notice that *a* and *b* do not have to be whole numbers. You will see examples of ratios of decimals in the paragraphs on proportion on page 348.

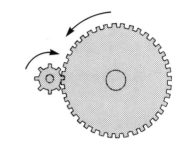

Comparing the relative size of large sets by the use of small numbers is the primary use of ratios. There are 8 teeth on the small gear shown here and 40 teeth on the large gear. This is a ratio of 1 to 5 from the number of teeth on the small gear to the number of teeth on the large gear, because 8/40 = 1/5. Every time the small gear revolves 5 times, the large gear revolves once. This ratio of turns from the small gear to the large gear is 5 to 1 or 5/1.

Gears such as these which step-down the number of revolutions per minute were used by farmers as they operated windmills for pumping water. Between 1880 and 1930 more than 6.5 million of these windmills were built in the United States.

Sometimes the two numbers of a ratio are divided to obtain one number. As an example, there is a federal law that the ratio of the length to the width of an official United States flag must be 1.9. This means that no matter what the size of the flag, if the length is divided by the width, the result should be 1.9.

*Rates—*A rate is a special kind of ratio in which the two sets being compared have different units. Kilometers per hour, cost per hour, and births per day are all examples of rates. This graph shows a rate of 3 cents for each 2 kilowatt-hours. For 2 kilowatt-hours it costs 3 cents; for 4 kilowatt-hours it costs 6 cents; etc.

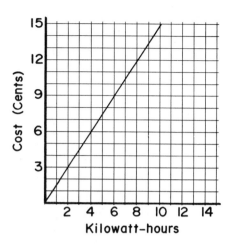

Proportions

Each ratio gives rise to infinitely many pairs of numbers all of which are *equal ratios*. For example, in 1973 the ratio of juveniles to adults who were prosecuted for burglary was 11 to 9. This means that for every 11 juveniles there were 9 adults, for every 22 there were 18, etc.

Juvenile Burglars	Adult Burglars
11	9
22	18
33	27
44	36
.	.
.	.
.	.

An equality of two ratios is called a *proportion.* In the previous example, 9/11 and 18/22 are equal ratios, and 9/11 = 18/22 is an example of a proportion.

Definition: For any two equal ratios, *a/b* and *c/d, a/b* = *c/d* is called a *proportion*.

We can use the test for equality of fractions to determine whether we have a proportion. That is,

$$\frac{a}{b} = \frac{c}{d} \qquad \text{if and only if} \qquad ad = bc$$

For example, 9/11 = 18/22 because 9 × 22 = 11 × 18.

Proportions are useful in problem solving. Typically, three of the four numbers in a proportion are given and the fourth is to be found. For this reason the rule for solving proportions has historically been called the "Rule of Three." This rule was so highly prized by merchants of the past that it was referred to as the *golden rule.* Here is a typical example. If 2.3 kilograms of flour cost $1.20, how much will you save buying 9 kilograms of flour at $4.45? Let's compute the cost of 9 kilograms of flour based on the rate of 2.3 kilograms for $1.20. Using ratios of "dollars to kilograms," and x for the unknown cost, the price for 9 kilograms of flour can be found by solving the following proportion.

$$\frac{1.20}{2.3} = \frac{x}{9}$$

This proportion can be changed to $2.3x = 10.80$ by using the preceding test for equality of ratios. This equation is satisfied by $x \approx \$4.70$. Therefore, based on the cost of 2.3 kilograms of flour, 9 kilograms will cost $4.70. The $4.45 price represents a savings of 25 cents.

A proportion may be set up in many different ways and still produce the correct answer. The previous problem may be solved by any one of the following proportions. The important thing is that we remain consistent on both sides of the equation. The same value of $x \approx \$4.70$ satisfies each of these equations.

Dollars to kilograms	Dollars to kilograms	Kilograms to dollars	Kilograms to dollars	Kilograms to kilograms	Dollars to dollars
$\dfrac{1.20}{2.3} = \dfrac{x}{9}$		$\dfrac{2.3}{1.20} = \dfrac{9}{x}$		$\dfrac{2.3}{9} = \dfrac{1.20}{x}$	

Percent

Percent has evolved as an outgrowth of the use of fractions with denominators of 100. The Roman method of taxation involved hundredths, or fractions which could easily be changed to hundredths. As examples, there was a sales tax of 1/100 on merchandise and a tax of 1/25 on slaves.

The word "percent" comes from the Latin "per centum," meaning "out of a hundred," and the symbol % has evolved from the seventeenth century use of "per $\frac{o}{o}$." Percent was first used in the fifteenth through seventeenth centuries for computing interest, profits, and losses. Currently, it has much broader applications as illustrated by the accompanying news clippings.

A percent can be changed to a decimal in two easy steps. First write the percent as a fraction with a deonominator of 100, as shown in the following examples. (Notice the similarity between the percent symbol "%" and the numeral "100.") Second, a fraction with a denominator of 100 can be written as a decimal by moving the decimal point in the numerator 2 places to the left.

$$42\% = \frac{42}{100} = .42 \qquad 5.2\% = \frac{5.2}{100} = .052 \qquad 132\% = \frac{132}{100} = 1.32$$

Conversely, to write a decimal as a percent, first write it as a fraction with a denominator of 100 and then as a percent.

$$.647 = \frac{64.7}{100} = 64.7\% \qquad .1735 = \frac{17.35}{100} = 17.35\%$$

Most computations with percents fall into two categories. In one case we are given a percent in order to determine a part of the whole. Consider the percentages of injuries to different parts of the body by National Football League players in 1974. In games and practices, 1169 players were injured. The number of players with a particular type of injury can be found by converting the percent to a decimal and multiplying by 1169. The following computation shows the number of players with shoulder injuries.

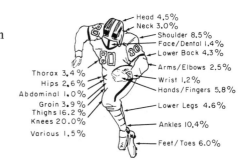

National Football League 1974 Injuries

$$8.5\% = \frac{8.5}{100} = .085 \qquad \text{and} \qquad .085 \times 1169 = 99.365$$

Rounded off to the nearest whole number, there were 99 players with shoulder injuries. In the second case we determine a percent by comparing a part to the whole. In the preceding football study, 189 players suffered injuries to the thigh. Thus, the fractional part of players with thigh injuries was 189/1169. This fraction can be converted to a decimal and then to a percent.

$$189 \div 1169 \approx .161676 \qquad \text{or approximately} \qquad 16.2\%$$

Scientific Notation   It has been remarked that the human race appears to be about "half-way in size" between many of the smallest and largest things we can measure. Through microscopes we look at objects that are many times smaller than people, and through telescopes we see stars and galaxies that are many times bigger. For example, the diameter of hydrogen atoms is .0000000000106 meters, and the diameter of the largest stars is 2,770,000,000,000 meters. Such very small and very large numbers can be expressed by using powers of 10. The diameter of this star in meters can be written as

$$2.77 \times 1,000,000,000,000 = 2.77 \times 10^{12}$$

The diameter of the atom in meters can be written as

$$\frac{1.06}{100,000,000,000} = \frac{1.06}{10^{11}} = 1.06 \times 10^{-11}$$

because $10^{-11} = 1/10^{11}$. In general, for any numbers x and n, with $x \neq 0$,

$$x^{-n} = \frac{1}{x^n}$$

Any positive number can be written as the product of a number from 1 to 10 and a power of 10. This method of writing numbers is called *scientific notation* or *scientific form.* The number from 1 to 10 is called the *mantissa* and the exponent of 10 is called the *characteristic.* The product 2.77×10^{12} is a number in scientific notation. The mantissa is 2.77 and the characteristic is 12. The following table contains five examples of numbers written in positional numeration and scientific notation.

	Positional Numeration	Scientific Notation
Years since age of dinosaurs	150,000,000	1.5×10^8
Seconds of half-life of U-238	142,000,000,000,000,000	1.42×10^{17}
Wave length of gamma ray (m)	.0000000000003048	3.048×10^{-13}
Size of viruses (cm)	.000000914	9.14×10^{-7}
Orbital velocity of earth (kph)	41290	4.129×10^4

Computing with Scientific Notation   This graph shows the world's rapidly increasing population. It wasn't until 1825 that the population reached one billion (1×10^9), and by 1975 it was four billion (4×10^9). Since there are about 3.64×10^3 square meters of cultivated land (land with crops) per person, the total amount of this land in square meters is

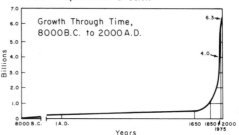

World Population Growth

$$(4 \times 10^9) \times (3.64 \times 10^3)$$

Rearranging these numbers and using the rule for adding exponents, this product can be written as

$$(4 \times 3.64) \times (10^9 \times 10^3) = 4 \times 3.64 \times 10^{12}$$

Finally, the product of the two mantissas (4 × 3.64) is computed, and the answer is written in scientific notation.

$$(4 \times 3.64) \times 10^{12} = 14.56 \times 10^{12} = 1.456 \times 10^{13}$$

This example shows that the products of numbers in scientific notation can be found by multiplying numbers from 1 to 10 (the mantissas) and adding exponents of 10 (the characteristics). This method of computing products holds whether the exponents are positive or negative. Let's consider an example with a negative exponent. There is a type of bacteria, called *Escherichia coli,* which is sometimes detected in swimming pools. These bacteria multiply rapidly, and under ideal conditions one bacteria would be replaced by a population of 4.8×10^8 after 30 hours. Since each bacteria weighs 2×10^{-12} grams, the total weight of the population in grams would be

$$(4.8 \times 10^8) \times (2 \times 10^{-12}) = (4.8 \times 2) \times 10^{8 + {}^{-}12} = 9.6 \times 10^{-4}$$

This is approximately 1/1000 of a gram or 1 milligram.

The convenience of scientific notation for computing is due primarily to the rules for adding and subtracting exponents. These rules, which were stated for whole numbers on page 128, can be extended to fractions and decimals. For any rational number b, and any integers n and m,

$$b^n \times b^m = b^{n+m} \quad \text{and} \quad b^n \div b^m = b^{n-m}, \text{ for } b \neq 0$$

Percent and Scientific Notation on Calculators   One method of determining the percentage of a given amount with a calculator, whether or not the calculator has a percent key, is to convert the percent to a decimal and multiply times the given number. To illustrate this, 6% of the 1169 National Football League injuries in 1974 were foot injuries. Here are the steps for determining 6% of 1169.

Steps		Displays
1.	Enter .06	0.06
2.	×	0.06
3.	Enter 1169	1169.
4.	=	70.14

The most common occurrences of percent in day-to-day affairs are in discounts and sales taxes. In the case of discounts we want to subtract a certain percent of the original cost. What is the cost, for example, of an object which is listed for $36.50 with a 15% discount? Many calculators are designed to compute this cost using the steps that follow. When the 15% is entered in Step 3, the display shows the amount that is to be subtracted from $36.50. Pressing $\boxed{=}$ in Step 4 shows the discounted price.

	Steps		Displays

	Steps		Displays
1.	Enter 36.50		36.50
2.	$-$		36.50
3.	Enter 15 $\boxed{\%}$		⁻5.475
4.	$\boxed{=}$		31.025

The cost of an object plus the sales tax is computed in a similar way with the minus sign in Step 2 replaced by a plus sign. The cost of the previous item ($31.03) plus a 6% sales tax is found by the following steps.

	Steps		Displays
1.	Enter 31.03		31.03
2.	$+$		31.03
3.	Enter 6 $\boxed{\%}$		1.8618
4.	$\boxed{=}$		32.8918

Calculators which add a given percentage, as in the previous example, make it easy to compute bank interests on saving accounts. Suppose you deposit $500 in a bank that pays 6% interest compounded annually. This means that at the end of each year you earn 6% on the money which was in the savings for the past year. After the first year (or beginning of the second year) you will have $500 plus 6% of $500, which according to Step 2 in the following calculator sequence is $530. Step 3 shows the amount of your savings at the beginning of the third year, and each succeeding step shows the annual amount in the savings account.

	Steps		Displays	
1.	Enter 500		500.	
2.	$+$ 6 $\boxed{\%}$ $\boxed{=}$		530.	(Beginning of 2nd year)
3.	$+$ 6 $\boxed{\%}$ $\boxed{=}$		561.8	(Beginning of 3rd year)
4.	$+$ 6 $\boxed{\%}$ $\boxed{=}$		595.508	(Beginning of 4th year)

***Scientific Notation*—**Some calculators have a special button for displaying numbers in scientific notation. These may be labelled $\boxed{\text{EE}}$ or $\boxed{\text{EEX}}$ or something similar, where E stands for exponent. If 749,300,000 is entered on such a calculator and the button for

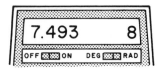

exponential or scientific notation is pressed, the mantissa, 7.493, and the characteristic or exponent, 8, will appear in the display to denote the number for 7.493×10^8. The base 10 will not appear in the display. On some calculators the mantissa, 7.493, and the characteristic, 8, must be entered separately. The EE button is used to enter 10^8, and then this number is multiplied by 7.493.

Calculators with scientific notation will automatically display numbers in this form whenever the number is too large or too small for the display. Suppose, for example, you computed 473,200 times 639,000, which is 302,374,800,000. If the calculator does not represent numbers in scientific notation, the display will be exceeded and flashing lights or some other type of signal will appear. A calculator with scientific notation will represent this product, which equals 3.023748×10^{11}, by the display shown here.

| 3.023748 | 11 |

Similarly, if the number is too small for the standard display, it will be represented by a mantissa and a negative power of 10. Consider the product .0004 times .000006, which is .0000000024, or 2.4×10^{-9} in scientific notation. If you compute this on a calculator whose display has only 8 places for digits and no scientific notation, it will show a product of 0. On a calculator with scientific notation, a mantissa of 2.4 and a characteristic of $^-9$ will appear in the display, as shown here.

| 2.4 | $^-9$ |

SUPPLEMENT (*Activity Book*)

Activity Set 7.3 Computing with the Calculator

Just for Fun: Number Search

Additional Sources

Davis, P.J. *The Lore of Large Numbers.* New York: Random House, 1961. pp. 28–50 (exponents and scientific notation).

Herron, J.D. and G.H. Wheatley. "A Unit Factor Method for Solving Proportion Problems." *The Mathematics Teacher,* **71** No. 1 (January 1978), 18–21.

Schmalz, R. "The Teaching of Percent." *The Mathematics Teacher,* **70** No. 4 (April 1977), 340–43.

School Mathematics Study Group. *Studies in Mathematics,* **6**. New Haven: Yale University, 1961. pp. 129–38, 151–65 (ratio, percent, and decimals).

1. This picture shows the space shuttle *Enterprise* riding piggyback on top of a 747 jumbo jet during test flights in 1977.

   ★ a. The 747 has a cruising speed of 595 mph. At this rate how long will it take to travel from the east coast to the west coast of the United States, which is approximately 3300 miles?

   b. The *Enterprise* has an orbiting speed of approximately 17,600 mph. At this rate, how long will it take to travel from the east coast to the west coast?

Space shuttle orbiter atop 747 jumbo jet in test flight at Edwards Air Force Base, Calif

2. This public service ad in the *Eugene Register-Guard* points out the need for car pooling to reduce traffic.

   a. According to this ad, what fraction of the car seats are empty during the morning's rush hour?

   b. In a city the size of Los Angeles, there would be 9,000,000 empty seats during the rush hour. How many seats are filled?

In this morning's rush hour, empty seats outnumbered full seats 4 to 1.

In a city the size of Los Angeles, that's 9,000,000 empty seats in cars jammed up on the freeways.

Think about that while you're sitting in traffic.

Share the ride with a friend. It sure beats driving alone.

Ad Council

Presented as a public service by

Eugene Register-Guard
Daily and Sunday

★ 3. Fill in the missing numbers in this table.

Distance from Sun in Miles

Planet	Scientific Notation	Positional Numeration
Mercury	3.6002×10^7	
Venus		67,273,000
Earth	9.3003×10^7	
Mars		141,709,000
Jupiter	4.83881×10^8	
Saturn		887,151,000
Uranus	1.784838×10^9	
Neptune		2,796,693,000
Pluto	3.669699×10^9	

4. Write the following numbers in scientific notation.

★ a. Size of minute insects in centimeters

.032

b. Length of day in seconds

86,400

c. Number of years since earth's formation

3,250,000,000

5. Write the following numbers in positional numeration.

★ a. Wavelength of X rays in centimeters

3.048×10^{-9}

b. Average length of solar year in seconds

3.15569×10^7

★ c. Total number of possible bridge hands

6.35×10^{11}

6. Write the answers to parts **a**, **b**, and **c** in scientific notation.

★ a. The velocity of a jet plane is 1.76×10^3 kph, and the escape velocity of a rocket from earth is 22.7 times faster. Find the rocket's velocity by computing $1.76 \times 10^3 \times 22.7$.

b. The earth travels 10^9 km each year about the sun, in approximately 9×10^3 hr. Compute $10^9 \div (9 \times 10^3)$ to determine its speed in kilometers per hour.

★ c. A *light-year,* the distance that light travels in 1 year, is 9.4488×10^{13} km. The sun is 2.7×10^4 light-years from the center of our galaxy. Find this distance in kilometers by computing $9.4488 \times 10^{13} \times 2.7 \times 10^4$.

d. At one point in *Voyager 1*'s journey to Jupiter, its radio waves travelled 7.44×10^8 km to reach the earth. These waves travel at a speed of 3.1×10^5 kilometers per second. Compute $7.44 \times 10^8 \div 3.1 \times 10^5$ to determine the number of seconds for these signals to reach the earth. How many minutes is this?

★ 7. *Discounts:* The cost of a $9.85 item which is being discounted at 12% can be determined by subtracting 12% of $9.85 from $9.85. The following equations show that the cost of this item can also be found by taking 88% of $9.85. What number property is used in the first of these two equations?

$$9.85 - (.12 \times 9.85) = (1 - .12) \times 9.85 = .88 \times 9.85$$

Use one of these two methods for computing the cost after discount of these items.

★   a. Portable typewriter $209.50 (15% off)

   b. Backpacker sleeping bag $153.95 (20% off)

   c. Snowshoes $86 (28% off)

8. *Sales Taxes:* The total cost of a $15.70 item plus a 6% sales tax can be determined by adding 6% of $15.70 to $15.70. The total cost can also be found by multiplying 1.06 times $15.70, as shown by these equations. What number property is used in the first of these two equations?

$$15.70 + (.06 \times 15.70) = (1 + .06) \times 15.70 = 1.06 \times 15.70$$

Use one of these two methods for computing the cost plus the sales tax of these items.

★   a. Fishing tackle outfit $48.60 (4% tax)

   b. Ten-speed bike $189 (5% tax)

   c. Cassette tape recorder $69.95 (6% tax)

9. Car sales for the first quarter of 1976 were far beyond expectations. The Upturn in Sales chart shows that during this 3-month period, 10.3 million (10,300,000) cars and passenger vans were sold. According to the What's Selling Best chart, the intermediate-size cars were the most popular with 27.4% of the market. For each type of car listed, compute the number of cars that were sold during the first three months of 1976.

★   a. Intermediate   b. Compact   ★ c. Imports   d. Passenger Vans

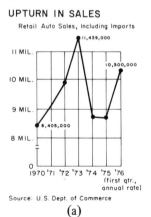

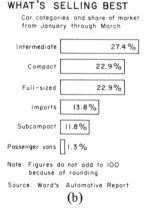

10. Here are the results of two polls, a 1970 and a 1975 Harris Survey, which asked a nationwide cross section of 1519 men and women the following question:

"All in all, do you favor or oppose most of the efforts to strengthen and change women's status in society?"

★ a. How many of the people surveyed in 1970:

   1. Favored strengthening?
   2. Opposed strengthening?

b. How many of the people surveyed in 1975:

   1. Favored strengthening?
   2. Opposed strengthening?

c. From 1970 to 1975 was there a greater increase in the percent of men or percent of women who favored a strengthening of women's status in society?

EFFORTS TO STRENGTHEN AND CHANGE WOMEN'S STATUS IN SOCIETY

	Favor	Oppose	Not Sure
1975			
Total Public	63%	25%	12%
Women	65%	25%	10%
Men	59%	25%	16%
1970			
Total Public	42%	41%	17%
Women	40%	42%	18%
Men	44%	39%	17%

11. The controversy over unit pricing has existed for many years. Without a price per unit, such as cost per gram or cost per ounce, consumers are often confused about the better buy. For example, would you pay more per ounce for the 63-cent package mix pictured here or for the 46-cent package? Since 63 divided by 48 is 1.3125, each ounce of the larger package costs approximately 1.3 cents. How much does each ounce of the smaller package cost? Which package is the better buy?

For each of the following pairs of packages, which is the better buy?

48 ounces for 63¢   32 ounces for 46¢

★ a. Betty Crocker Complete Buttermilk

   Small Size   40 ounces at 69¢

   Large Size   56 ounces at 85¢

b. Bisquick Variety Baking Mix

   Small Size   20 ounces at 31¢

   Large Size   32 ounces at 55¢

c. If the large box of Hungry Jack has enough mix for 115 4-inch pancakes, how many 4-inch pancakes can be made from the small box?

12. This National Association of Home Builders chart shows how the burden of purchasing a home has been reduced over the years. In 1890 the median (a type of average; see Section 10.3) price of a home was $4422 and the average yearly family income was $455. This ratio of house cost to income is 9.8. This means it took almost 10 years of annual salary to purchase a home. By 1940 it took only 5 years of an annual salary to purchase a home. Compute the missing numbers in this table for the following years.

★ a. 1960 b. 1965 ★c. 1975

Ratio of median sales price to annual family income
total U.S. housing starts 1890–1975

Year	Median Sales Price	Annual Family Income	Ratio Median Income
1890	$ 4,422.	$ 455.	$ 9.8
1895	4,500.	462.	9.9
1900	4,881.	490.	9.9
1905	4,311.	554.	7.7
1910	5,377.	630.	8.5
1915	5,159.	687.	7.5
1920	6,296.	1,489.	4.2
1925	7,809.	1,434.	5.4
1930	7,146.	1,360.	5.2
1935	6,296.	1,137.	5.5
1940	6,558.	1,300.	5.0
1945	7,476.	2,189.	3.4
1950	9,446.	3,319.	2.8
1955	13,386.	4,418.	3.0
1960	16,652.	☐	2.9
1965	☐	6,957.	2.8
1970	23,400.	9,867.	2.3
1975	39,300.	13,991.	☐

Source: Available upon request; NAHB Economics Department

13. This 100-foot-high windmill was built by NASA near Sandusky, Ohio. It is designed to generate 100 kW, enough electricity to power 25 homes.

★ a. The rotor is designed to rotate at a constant 40 revolutions per minute. It drives a generator which turns 1800 rpm. This increase in speed is accomplished by step-up gearing. Write the ratio of 40 to 1800 using the smallest positive whole numbers.

b. *Tip-speed ratio* is a comparison of the speed of the blade tips to the speed of the wind. Ratios of 5 to 1 and 8 to 1 are not uncommon for high-speed windmills. The blade-tip speed of NASA's rotor is 180 mph in a 20 mph wind. What is this tip-speed ratio?

Windmill near Sandusky, Ohio, for generating electricity

14. This graph shows the world's increasing population for developed and underdeveloped countries. By 1975, there were 4 billion (4×10^9) people on earth.

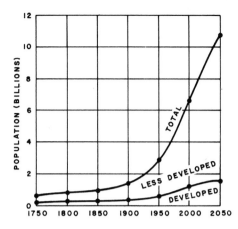

a. If the world's gross product for 1975 averaged 1000 (10^3) dollars for each person on earth, how much was the gross product? (Write your answer in scientific notation.)

b. Approximately 3 billion (3×10^9) people live in underdeveloped countries where the average number of dollars per person for 1975 was only 200 (2×10^2). How much was the gross world product for the underdeveloped countries in 1975?

15. The following diagram shows each step in the nuclear fuel cycle, from mining to the reactor.

★ a. It takes 125,900 tons of uranium ore to produce 239 tons of uranium oxide (yellow cake). The weight of the yellow cake is what percent of the weight of the uranium ore?

b. It takes 300 tons of uranium hexafluoride gas to produce 42.3 tons of enriched uranium. The weight of the enriched uranium is what percent of the weight of the hexafluoride gas?

★ c. The enriched uranium supplies 28.3 tons of power-plant fuel when it is converted back to a solid. This supply loses 4% of its weight in running the nuclear reactor for 1 year. What is the weight of the uranium which is left over for reprocessing?

Mining	Conversion	Fuel assembly
It takes 125,900 tons of uranium ore to produce one year's fuel for a 1,000-megawatt reactor	The yellowcake, combined with fluorine, produces 300 tons of uranium hexafluoride gas	Converted back to a solid the enriched uranium supplies 28.3 tons of power-plant fuel

Milling	Enrichment	Reactor
The ore yields 239 tons of uranium oxide or yellowcake	The gas yields 42.3 tons of enriched uranium leaving 257.5 tons of depleted uranium	Running a reactor for a year the 28.3 tons of fuel loses 4% of its weight

16. *Calculator Exercise:* One thousand dollars is placed in a savings account which pays 6% interest compounded annually.

★   a. How much money is in the account at the end of 1 year?

     b. If the original deposit and the money earned from interest are left in the bank for 5 years, what will be the total savings?

★   c. In how many years will the $1000 deposit be doubled?

     d. If $500 had been placed in this savings account rather than $1000, in how many years would it be doubled?

★ 17. *Calculator Exercise:* This table contains the numbers of elementary school students and teachers in 1975 for several states. The student-teacher ratio for each state is the number of students divided by the number of teachers. Compute these ratios to the nearest tenth.

     a. Which state has the best student-teacher ratio?

     b. Which has the poorest?

★   c. Which two states have the same student-teacher ratio?

State	Number of Students	Number of Teachers	Student-Teacher Ratio
Alabama	383,386	17,878	
Florida	881,325	37,003	
Hawaii	92,100	4,404	
Iowa	314,947	16,069	
Maine	172,000	7,000	
Missouri	645,886	24,666	
Oregon	278,452	11,680	
Wyoming	44,140	2,573	

18. *Calculator Exercise:* Assume that you have a calculator with scientific notation and that by repeatedly pressing $=$, as shown in the following steps, each new number is obtained by multiplying the previous one by 1000 or 10^3. Fill in the displays for Steps 6 and 7. If the calculator represents numbers up to 10^{100}, what is the number of the step for which the calculator display will be exceeded?

Steps		Displays
1. Enter 1000		1000.
2. Press $\times$		1000.
3. $=$		1000000.
4. $=$		1. 09
5. $=$		1. 12
6. $=$		
7. $=$		

19. *Calculator Exercise:* Population density is the ratio of the number of people to each unit of land area. The population density of California in 1970 was 127.6 people per square mile. This is an increase of 27.2 people per square mile from 1960 to 1970.

California Population in 1970	Square Miles in California	Population Density
19,953,134 ÷	156,361 =	127.6

Compute the population densities for the 1970 populations of the states in the following table.

State	1970 Population	Land Area in Square Miles	1960 Population Density	1970 Population Density
California	19,953,134	156,361	100.4	127.6
New Jersey	7,168,164	7,521	805.5	
Texas	11,196,730	262,134	36.4	
Alaska	302,173	566,432	.4	
Montana	694,409	145,587	4.6	
Rhode Island	949,723	1,049	819.3	

★ a. Which of these six states had the greatest increase in the number of people per square mile from 1960 to 1970?

b. In 1970 New Jersey became the most densely populated state, overtaking Rhode Island which had held the title since the first census in 1790. How many more people did New Jersey have per square mile than Rhode Island in 1970? Round your answer to the nearest whole number.

c. The U.S. population in 1970 was 203,184,772. If this number of people were distributed equally over the 50 states with its land area of 3,536,855 square miles, what would be the U.S. population density for 1970?

20. The total production of all the world's oil companies in 1975 was 19,229,705,990 barrels. For some of the larger companies their percent of the total output is shown in this diagram.

★   a. How many barrels were produced by the 25 largest oil producers?

     b. The largest oil company produced 8.5% of the total. How many barrels was this?

★   c. How many barrels of oil did Union Oil Company produce if their share of the pie is 2.5%?

     d. Several oil companies each produced .3% of the total. How many barrels did each of these companies produce?

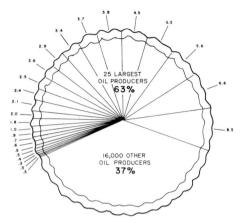

21. A study of 7057 felony arrests in Washington, D.C., in 1974 showed the following:
4517 cases were dismissed or dropped;
1835 cases were solved by pleas of guilty;
 705 cases were brought to trial.

★   a. Determine the percentage for each of the three categories. Round your answer to the nearest whole number percent.

     b. Of the cases that were brought to trial, 210 were acquitted. What is the total number of cases that were dismissed, dropped, or acquitted? This total is what percent of the 7057 cases?

"WE HAVE REASON TO BELIEVE BINGLEMAN IS AN IRRATIONAL NUMBER HIM SELF."

7.4 IRRATIONAL AND REAL NUMBERS

Irrational Numbers   The following number line shows the locations of a few rational numbers. Each rational number corresponds to a point on the number line, and between any two such numbers, no matter how close, there is always another rational number. Crowded together as they are, it would seem that there is no room left for any new types of numbers.

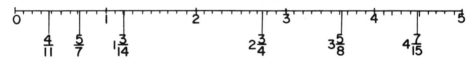

There are, however, points on the number line which correspond to numbers that are not rational. The $\sqrt{2}$ is such an example. This is the number which when multiplied by itself equals 2. The following equations show that $\sqrt{2}$ is just a little bigger than 1.4.

$$(1.4)^2 = 1.96$$
$$(1.41)^2 = 1.9881$$
$$(1.414)^2 = 1.999396$$
$$(1.4142)^2 = 1.99996164$$

The beginning of the decimal representation for $\sqrt{2}$ is 1.4142135, but no matter how many places are computed there will be no repeating pattern as in the case of rational numbers.* Such infinite nonrepeating decimal numbers are called *irrational numbers*. It is easy to exhibit examples of this type of decimal. In the following numeral, each "7"

*For a proof that $\sqrt{2}$ is irrational, see R. Courant and H. Robbins, *What Is Mathematics* (New York: Oxford University Press, 1941), pp. 59–60.

is preceded by one more zero than the previous "7": .07007000700007 . . . . While there is a pattern here, there is no repeating pattern of digits, as in the case of a rational number. Therefore this is an irrational number.

Numbers which are not rational were first recognized by the Pythagoreans of the fifth century B.C. Previously, they had thought that the length of any line segment was a rational number, and most of their proofs depended on this assumption. Of even greater importance to them was the belief that all practical and theoretical affairs could be explained by whole numbers and ratios of whole numbers. The discovery of line segments whose lengths were not expressible by rational numbers caused a logical scandal which threatened to destroy the Pythagorean philosophy. According to one legend, they attempted to keep the matter a secret by taking its discoverer, Hippacus, on a sea voyage from which he never returned.

The Greeks used the word "expressible" if a line segment had a rational length and "inexpressible" if it did not. Our words "rational" and "irrational" are derived from these words. It wasn't until the nineteenth century, more than 2000 years after the discovery of inexpressible lengths, that our number system was extended to include irrational numbers. This was accomplished by the German mathematician Richard Dedekind (1831–1916) who carefully defined irrational numbers and removed the ambiguities surrounding them.

Pythagorean Theorem   It is assumed by some scholars that the discovery of irrational numbers arose in connection with the Pythagorean theorem. This theorem states that for any right triangle, the sum of the squares of the legs is equal to the square of the hypotenuse. The *legs* of a right triangle are the perpendicular sides, and the *hypotenuse* is the side opposite the 90-degree

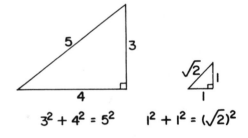

$$3^2 + 4^2 = 5^2 \qquad 1^2 + 1^2 = (\sqrt{2})^2$$

angle. For a right triangle with legs of lengths 3 and 4, the hypotenuse has length 5, because $3^2 + 4^2 = 5^2$. If the legs are both 1, the hypotenuse is the irrational number $\sqrt{2}$ since $1^2 + 1^2 = (\sqrt{2})^2$.

The first proof of the Pythagorean theorem is thought to have been given by Pythagoras (ca. 540 B.C.). According to legend, when Pythagoras discovered this theorem he was so overjoyed that he offered a sacrifice of oxen. The theorem was used, however, many years before by the Babylonians and Egyptians. It is illustrated on this 4000-year-old Babylonian tablet, which contains a square and its diagonals. The numbers on this tablet show that the Babylonians had computed $\sqrt{2}$ to be 1.414213, which is the correct value to 6 decimal places.

There are many proofs of the Pythagorean theorem. *The Pythagorean Proposition*, a book by E.S. Loomis, contains 370 such proofs. The proof suggested by

Babylonian stone tablet with approximation of $\sqrt{2}$

the following squares was known by the Greeks and may have been the one given by Pythagoras. The triangles with sides *a, b,* and *c* are all right triangles. The relationship, $a^2 + b^2 = c^2$, can be proven by the following observations. Square I has an area of $a^2 + b^2 + 4T$, where T is the area of each triangle. Square II has an area of $c^2 + 4T$. Both squares have sides of length $a + b$, and therefore their areas are equal. Setting these areas equal to each other, $a^2 + b^2 + 4T = c^2 + 4T$, shows that $a^2 + b^2 = c^2$.

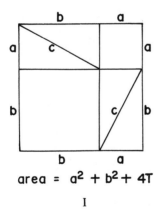

area = $a^2 + b^2 + 4T$

I

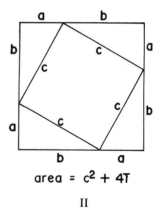

area = $c^2 + 4T$

II

The converse of the Pythagorean theorem also holds. If the sum of the squares of two sides of a triangle equals the square of the third side, then the triangle must be a right triangle. This means that if you used a rope with 30 knotted intervals of equal length and formed a triangle of sides 5, 12, and 13, it would be a right triangle. This fact was undoubtedly known by the ancient Egyptians and used by them to form right triangles. The top part of the next photo

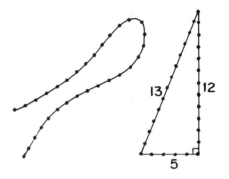

shows Egyptian surveyors, called "rope stretchers," with a piece of rope. Mark off a piece of string with 30 equal intervals and try forming a triangle whose sides have lengths of 5, 12, and 13.

An Egyptian painting (about fifteenth century B.C.) depicting the needs of an advanced society. The upper part shows surveyors with rope.

Square Roots and Other Irrational Numbers

The square root of a number n is defined as the number which when multiplied by itself equals n. The square root of 5.29 is 2.3, since $2.3 \times 2.3 = 5.29$. The symbol for the square root of a number n is $\sqrt{n}$. The term "square root" comes from the word "root." It was first represented by "r" then by $\sqrt{}$ and finally by $\sqrt{}$.

5.29 cm² | $\sqrt{5.29}$ cm

$\sqrt{5.29} = 2.3$

The square roots of the square numbers 1, 4, 9, 16, 25, ... are always whole numbers. The square roots of all other whole numbers greater than 0 are irrational: $\sqrt{2}, \sqrt{3}, \sqrt{5}, \sqrt{6}, \sqrt{7}, \sqrt{8}, \ldots$. These numbers all have infinite nonrepeating decimals.

Even though we cannot write the complete decimal for an irrational number, these numbers should not be thought of as mysterious or illusive. They are the lengths of line segments, as illustrated by the following triangles. The legs of the first triangle each have a length of 1, and so, by the Pythagorean theorem, the hypotenuse is $\sqrt{2}$. The legs of the second triangle are 1 and $\sqrt{2}$, and the hypotenuse is $\sqrt{3}$. Next, we can use legs of length $\sqrt{2}$ and $\sqrt{3}$ to obtain a hypotenuse of $\sqrt{5}$. The lengths of these hypotenuses have been transferred to the number line below the triangles. Line segments of lengths $\sqrt{6}, \sqrt{7}, \sqrt{8}$, etc., can be constructed in a similar manner.

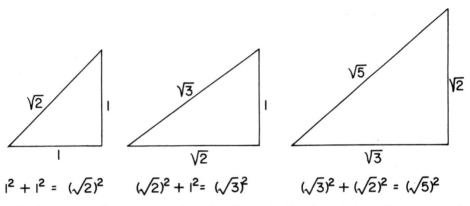

$$1^2 + 1^2 = (\sqrt{2})^2 \qquad (\sqrt{2})^2 + 1^2 = (\sqrt{3})^2 \qquad (\sqrt{3})^2 + (\sqrt{2})^2 = (\sqrt{5})^2$$

In this manner, line segments whose lengths are irrational numbers can be constructed and used to locate these numbers on the number line.

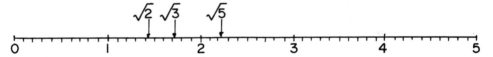

The cube root of a number n is denoted by $\sqrt[3]{n}$. This is the number s such that $s \times s \times s$ equals n. If a cube has a volume of 15 cubic centimeters then each side of the cube has length $\sqrt[3]{15}$. The cube roots of cube numbers 1, 8, 27, 64, 125, etc., are whole numbers. The cube roots of all other whole numbers are irrational numbers. For example, the cube roots of 4, 10, and 35 are all infinite nonrepeating decimals. Try cubing the following decimals to see how close you get to 4, 10, and 35.

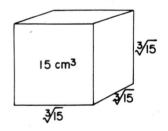

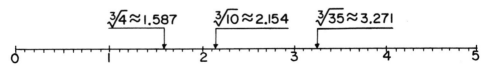

The most celebrated of all irrational numbers is pi, which is denoted by π. This number is the ratio of the circumference of a circle to its diameter. To 4 decimal places it is 3.1416. Pi has had a long and interesting history. In the ancient Orient, π was frequently taken to be 3, and this value also occurs in the Bible (I Kings, Chapter 7, and II Chronicles, Chapter 2). There have been many attempts to compute π. Archimedes computed π to 2 decimal places, and in 1841, Zacharias Dase computed π to 200 places. In 1873 William Shanks of England computed π to 707 places. In 1946 D.F. Ferguson of England discovered errors starting with the 528th place in Shank's value for π, and a year later he gave a corrected value of π to 710 places. In recent years, electronic computers have calculated π to more than one million places. Among the curiosities connected with π are the word devices for remembering the first few decimal places. In the following sentence the number of letters in each word is a digit in π.

$$3. \quad 1 \quad 4 \quad 1 \quad 5 \quad \quad 9 \quad \quad 2 \quad \quad 6$$

May I have a large container of coffee?

Real Numbers The irrational numbers together with the rational numbers form the set of real numbers. The accompanying diagram shows the relationships between the familiar sets of numbers. The set of rational numbers and the set of irrational numbers are disjoint and their union is the set of real numbers, R. The rational numbers, Q, contain the whole numbers, $W = \{0, 1, 2, 3, \ldots\}$, and the integers, $Z = \{0, \pm1, \pm2, \pm3, \ldots\}$. Viewed in another way, the sets W, Z, Q, and R form an increasing sequence of subsets: The set of whole numbers is contained in the set of integers, $W \subset Z$; the set of integers is contained in the set of rational numbers, $Z \subset Q$; and the set of rational numbers is contained in the set of real numbers, $Q \subset R$. The next diagram shows another way to picture the various relationships between these types of numbers.

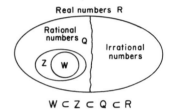

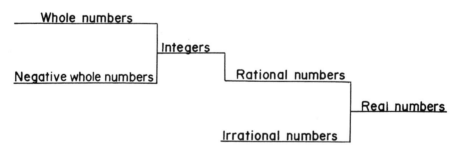

*This mnemonic and others are given by H.W. Eves, *An Introduction to the History of Mathematics,* 3rd ed. (New York: Holt, Rinehart and Winston, 1969) p. 94.

Properties of Real Numbers   The commutative, associative, and distributive properties which were stated in Chapter 3 and 6 hold for the whole numbers, integers, rational numbers, and real numbers. There are other properties which do not hold for all of these number systems. The integers have inverses for addition (because they have negative numbers) but the whole numbers do not. The rational numbers have inverses for multiplication (because they have reciprocals) but

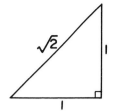

the integers do not. The real numbers have the property called *completeness,* but the rational numbers do not have this property. Intuitively, completeness means that for every line segment there is a real number which equals its length. If we limit ourselves to the rational numbers, this is not true. For example, we have seen that there is no rational number corresponding to the length of the hypotenuse of a right triangle whose legs have lengths of 1 unit.

Expressed in a slightly different way, completeness of the real numbers means that there is a one-to-one correspondence between the real numbers and the points on a line. Because of this relationship, a line called the *real number line* is used as a model for the real numbers. Once a zero point is labelled and a unit is selected, each real number can be assigned to a point on the line. Each positive real number is assigned to a point to the right of zero such that the real number is the distance from this point to the zero point. The negative of this number corresponds to a point symmetrically to the left of zero. Here are a few examples.

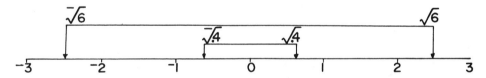

Here are 12 properties which hold for the real number system.

Closure for Addition—For any two real numbers, their sum is another real number. This is stated by saying that *addition is closed* on the set of real numbers. Closure can be illustrated best by a set of numbers on which addition is not closed. Consider the set of odd numbers, $\{1, 3, 5, 7, \ldots\}$. The sum of two odd numbers is not another odd number, so addition is not closed on this set. Addition is closed on the sets of whole numbers, integers, and rational numbers.

Closure for Multiplication—For any two real numbers, their product is another real number. That is, *multiplication is closed* on the set of real numbers. Multiplication may or may not be closed, depending on the given set. The product of any two negative numbers is not another negative number, so multiplication is not closed on the set of negative numbers. Multiplication is closed on the sets of whole numbers, integers, and rational numbers.

Addition Is Commutative—For any real numbers r and s, $r + s = s + r$.

Multiplication Is Commutative—For any real numbers r and s, $r \times s = s \times r$.

Addition Is Associative—For any real numbers r, s, and t, $(r + s) + t = r + (s + t)$.

Multiplication Is Associative—For any real numbers r, s, and t, $(r \times s) \times t = r \times (s \times t)$.

Identity for Addition—For any real number r, $0 + r = r$. Zero is called the *identity for addition* because when it is added to any number, there is "no change." That is, the identity of the number is left unchanged.

Identity for Multiplication—For any real number r, $1 \times r = r$. One is called the *identity for multiplication* because multiplying by 1 leaves the identity of each number unchanged.

Inverses for Addition—For any real number r, there is a unique real number ^-r, called its *negative* or *inverse for addition*, such that $r + {}^-r = 0$.

Inverses for Multiplication—For any real number r, not equal to 0, there is a unique real number $1/r$, called its *reciprocal* or *inverse for multiplication*, such that $r \times 1/r = 1$.

Multiplication Is Distributive over Addition—For any real numbers r, s, and t, $r \times (s + t) = r \times s + r \times t$.

Completeness Property—For any line segment there is a real number which equals its length.

Operations with Irrational Numbers   At first it is difficult to imagine how to perform arithmetical operations with numbers which cannot be expressed exactly in decimal notation. One solution is to replace irrational numbers by rational approximations. Thus, for example, we see both 3.14 and 3.1416 being used for π. For a circle whose diameter is 2.8 centimeters, its circumference will be approximately 8.8 centimeters. This is sufficient accuracy for most purposes.

Another solution is to write products and sums without doing the computing. For example, the sides of the square on this centimeter dot grid each have a length of $\sqrt{2}$ centimeters. The perimeter is denoted by $4\sqrt{2}$, which means 4 times $\sqrt{2}$. It is fairly easy to prove that $4\sqrt{2}$ is an irrational number. The argument is as follows. We know by closure for multiplication of real numbers that $4\sqrt{2}$ is a real number, so it is either rational or irrational. Let's suppose it is a rational number and denote it by r. That is, $r = 4\sqrt{2}$. Multiplying both sides of this equation by $1/4$, we get

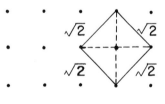

$\sqrt{2} = r \times 1/4$. Now by closure for multiplication of rational numbers, $r \times 1/4$ is a rational number. However, this can't be true because $\sqrt{2}$ is an irrational number. Since the assumption that $4\sqrt{2}$ is rational leads to a contradiction, $4\sqrt{2}$ must be irrational. A similar argument can be used to prove that the product of any nonzero rational number with an irrational number is an irrational number.

As another example, the triangle whose sides have lengths $1, 2$, and $\sqrt{5}$ has $3 + \sqrt{5}$ as a perimeter. This is also an irrational number, as can be proven by an argument similar to the previous one. In general, the sum of a rational number and an irrational number is an irrational number.

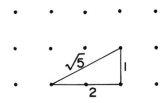

The perimeter of the square, $4\sqrt{2}$, and the triangle, $3 + \sqrt{5}$, can be approximated by substituting a rational approximation for $\sqrt{2}$ and $\sqrt{5}$, respectively. There are cases, however, in which we can compute with irrational numbers without replacing them by decimal approximations. The area of the square whose sides have length $\sqrt{2}$ (see previous page) is $\sqrt{2} \times \sqrt{2}$. By the definition of square root, $\sqrt{2} \times \sqrt{2} = 2$. We can also see that the area of this square is 2, by dividing it into 4 smaller half-squares.

Let's consider another example of a product of two irrational numbers. The rectangle in this figure has a length of $\sqrt{18}$ and a width of $\sqrt{2}$. Its area, using the formula for the area of a rectangle, is $\sqrt{18} \times \sqrt{2}$. Using a second method, this area can be seen to be 6 square centimeters by dividing the rectangle into 12 small half-squares. These two methods indicate that $\sqrt{18} \times \sqrt{2} = 6$. Since $6 = \sqrt{36} = \sqrt{18 \times 2}$, we see that $\sqrt{18} \times \sqrt{2} = \sqrt{18 \times 2}$. In this example the product of the square roots of two numbers is equal to the square root of the product of the two numbers. This result is stated in the following theorem.

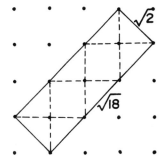

Theorem: For any positive numbers a and b, $\sqrt{a} \times \sqrt{b} = \sqrt{a \times b}$.

Often, as in the preceding examples, $\sqrt{a}$ and $\sqrt{b}$ will be irrational numbers but their product will be a rational number. In addition to computing the products of square roots, this theorem is useful for simplifying square roots. For example, $\sqrt{18} = \sqrt{9 \times 2} = \sqrt{9} \times \sqrt{2} = 3\sqrt{2}$. A square root can always be simplified if the number under the root sign has a factor that is a square number.

Irrational Numbers and Calculators   Some calculators have buttons for irrational numbers. One such button is for π. However, since the decimal representation for an irrational number has an infinite number of digits, the calculator computes with an approximate value for π. Pressing $\boxed{\pi}$ on a calculator with 10 places for digits will give the number shown in the display on this calculator.

Many calculators have a square root key, $\boxed{\sqrt{x}}$. If the square root of a number is irrational, the decimal which appears in the display will be an approximation. Here are the steps for obtaining $\sqrt{2}$.

Steps	Displays
1. Enter 2	2.
2. Press $\boxed{\sqrt{x}}$	2.
3. $\boxed{=}$	1.414213562

By definition, $\sqrt{2}$ is that number which when multiplied by itself produces 2. If the decimal in the previous display is multiplied by itself, $(1.414213562)^2$, the product is a decimal with 19 digits, whose first 12 digits are 1.99999999894. On a calculator which has 10 positions for digits and which automatically rounds off, 1.999999999 will appear in the display. If, however, the calculator has 9 or fewer positions for digits and automatically rounds off, the product will be rounded off to the number 2.

Some calculators have a button for finding *any* root (cube root, fourth root, etc.) of a positive number. One common button for this is $\boxed{\sqrt[x]{y}}$. Here are the steps for finding the cube root of 12 by using this button. The number in Step 4 is only an approximation because $\sqrt[3]{12}$ is an irrational number.

Steps	Displays
1. Enter 12	12.
2. Press $\boxed{\sqrt[x]{y}}$	12.
3. Enter 3	3.
4. $\boxed{=}$	2.289428485

If this decimal is cubed, the result may or may not be equal to 12. This will depend on the number places for digits in the display and whether or not the calculator rounds off.

SUPPLEMENT *(Activity Book)*

Activity Set 7.4 Irrational Numbers on the Geoboard
Just for Fun: Golden Rectangles

Additional Sources

Dantzig, T. *Number, The Language of Science.* 4th ed. New York: Macmillan, 1954. "The Act of Becoming," pp. 139–63.

Eves, H.W. "Irrationality of $\sqrt{2}$." *The Mathematics Teacher,* **38** No. 7 (November 1945), 317–18.

Jones, P.S. "Irrationals or Incommensurables I: Their Discovery and a Logical Scandal." *The Mathematics Teacher,* **49** No. 2 (February 1956), 123–27.

Kline, M. *Mathematical Thought from Ancient to Modern Time.* New York:

Oxford University Press, 1972. pp. 8, 32–33, 80–81, 191–92, 251–52, 982–87 (irrational numbers).

Niven, I. *Numbers: Rational and Irrational.* New York: Random House, 1961. pp. 21–51.

Rothbart, A. and B. Paulsell. "Pythagorean Triples: A New Easy-to-Derive Formula with Some Geometric Applications." *The Mathematics Teacher,* **67** No. 3 (March 1974), pp. 215–18.

There once was a lady named Lou,
Who computed the square root of 2.
    When no pattern repeated
    She gave up defeated,
Two million digits is all she would do.

$$\sqrt{2} = 1.414213562 \cdots$$

EXERCISE SET 7.4

★ 1. For each number in the left-hand column, determine what type of number it is and put checks in the appropriate columns. For example, $^{-}3$ is an integer, a rational number, and a real number.

	Whole numbers	Integers	Rational numbers	Real numbers
$^{-}3$		✓	✓	✓
$\frac{1}{8}$				
$\sqrt{3}$				
π				
14				
$\frac{16}{4}$				
$.\overline{82}$				

2. Assuming that the decimal patterns continue, which of the following numbers are rational and which are irrational?

★   a. .103410341034 · · ·   ★ c. .01001000100001 · · ·

      b. .12123123412345 · · ·   d. 4.001001001001 · · ·

3. We know that $\sqrt{a} \times \sqrt{b} = \sqrt{ab}$ for all positive numbers a and b. A similar condition for the sums of square roots would be: $\sqrt{a} + \sqrt{b} = \sqrt{a+b}$. Find two numbers for a and b to show that this equation does not hold. Use decimals to evaluate both sides of the equation.

4. Simplify the following square roots so that the smallest possible whole number is left under the square root symbol.

★   a. $\sqrt{45}$ b. $\sqrt{48}$ ★ c. $\sqrt{60}$

5. Compute each of the following products. Write your answers as whole numbers.

★   a. $\sqrt{20} \times \sqrt{5}$ b. $\sqrt{6} \times \sqrt{24}$ ★ c. $\sqrt{7} \times \sqrt{28}$ d. $\sqrt{3} \times \sqrt{12}$

6. Compute the following square roots and cube roots to 1 decimal place.

★   a. $\sqrt{20}$ b. $\sqrt{58}$ ★ c. $\sqrt[3]{12}$ d. $\sqrt[3]{20}$

7. Use the Pythagorean theorem to find the missing length for each of the right triangles. Compute each answer to 1 decimal place.

★   a.   b.   ★ c.   d.

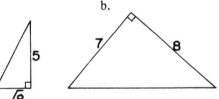 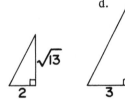

8. Any three whole numbers a, b, and c, such that $a^2 + b^2 = c^2$, are called *Pythagorean triples*. The following equations can be used to find such numbers, by using whole numbers for u and v.

$$a = 2uv, \quad b = u^2 - v^2, \quad c = u^2 + v^2$$

For example, if $u = 5$ and $v = 3$, then a, b, and c have the values shown in the table. Use the values of u and v in the table to find the remaining values for a, b, and c. Check your answers by showing that $a^2 + b^2 = c^2$.

u	v	a	b	c
2	1			
3	2			
4	3			
4	2			
5	3	30	16	34

9. The numbers $\sqrt{2}, \sqrt{3}, \sqrt{5}, \sqrt{6}$, and $\sqrt{7}$ can be located on a number line by constructing triangles whose hypotenuses have these lengths. A right triangle having legs of length 1 and a hypotenuse of length $\sqrt{2}$ has been constructed on the top line. The arrow indicates where this length should be marked off on the line. Next, we can use the length $\sqrt{2}$ on the second line to construct a triangle whose legs have the lengths $\sqrt{2}$ and 1. The hypotenuse of this triangle is $\sqrt{3}$.

★   a. Continue this process of constructing triangles to locate $\sqrt{5}, \sqrt{6}$, and $\sqrt{7}$ on the next three lines.

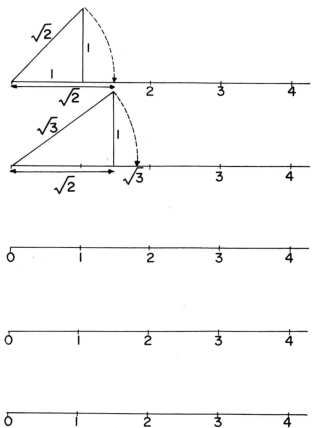

b. Locate $\sqrt{2}, \sqrt{3}, \sqrt{5}, \sqrt{6}$, and $\sqrt{7}$ on the following number line by using the lengths which are marked off on the previous lines.

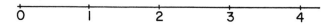

10. *Rationalizing the Denominator:* Quotients of real numbers, such as $2 \div \sqrt{3}$, are often written as fractions, $2/\sqrt{3}$. The process of replacing an irrational denominator by a rational denominator is called *rationalizing the denominator.*

$$\frac{2}{\sqrt{3}} = \frac{2 \times \sqrt{3}}{\sqrt{3} \times \sqrt{3}} = \frac{2\sqrt{3}}{3}$$

Rationalize the following denominators of these fractions.

★   a. $\dfrac{3}{\sqrt{7}}$ b. $\dfrac{3}{2\sqrt{6}}$ ★c. $\dfrac{5}{\sqrt{5}}$ d. $\dfrac{^-1}{\sqrt{2}}$

★11. Compute the square roots to 1 decimal place of the numbers between 0 and 1 on the following number line. Locate their square roots on the line. What general statement can be made about the square roots of numbers between 0 and 1?

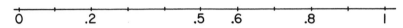

12. *Calculator Exercise:* Use a calculator with a square root key to carry out Steps 1 through 4. Write in the numbers for the display in Step 5.

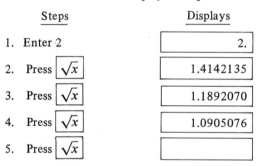

Steps	Displays
1. Enter 2	2.
2. Press $\sqrt{x}$	1.4142135
3. Press $\sqrt{x}$	1.1892070
4. Press $\sqrt{x}$	1.0905076
5. Press $\sqrt{x}$	

★   a. If you continue to press the square root key in this example, all the numbers in the display will eventually be equal. What is this number?

★   b. What number will eventually show in the display of a calculator if you enter a number less than 1 and repeatedly press the square root key?

13. Which of the following operations are closed for the given sets? If an operation is not closed, provide an example to show this.

★   a. Subtraction on the set of whole numbers.

    b. Division by nonzero numbers on the set of rational numbers.

    c. Multiplication on the set of irrational numbers.

★   d. Addition on the set of integers.

14. State the property of the real numbers which is used in each equation.

★   a. $3(cb)y + 3c(2bz) = 3(cb)y + 3(c2)bz$

     b. $3c(by) + 3c(2bz) = 3c(by + 2bz)$

★   c. $4[2k + 5(k + y)] = 4[5(k + y) + 2k]$

     d. $(2k + 5) + (6a + 7)/bc = 1(2k + 5) + (6a + 7)/bc$

     e. $4 + [2k + 5(k + y)] = (4 + 2k) + 5(k + y)$

15. Which of the 12 properties that are listed in the text for the real numbers do not hold for the following sets of numbers?

★   a. Whole numbers   b. Integers   ★ c. Rational numbers

16. Another way to consider the differences between the various sets of numbers is to consider the types of equations which can be solved. Solve each equation for x, and name the set or sets of numbers $(W, Z, Q,$ and $R)$ which contain the solution(s).

★   a. $15 - x = 22$

     b. $3x + 2 = 9$

     c. $17 + x = 12$

★   d. $x^2 + 14 = 17$

     e. $2x + 14 = 36$

17. Use the fact that the sum of an irrational number and a rational number is an irrational number, to construct an infinite sequence of irrational numbers. Describe a one-to-one correspondence between this set of irrational numbers and the set of whole numbers. (*Hint:* Select an irrational number and add whole numbers to it.)

18. This spiral of right triangles, somewhat resembling a cross section of the seashell of the chambered nautilus (see page 413), shows the square roots of consecutive whole numbers. The first triangle has two legs of unit length and a hypotenuse of $\sqrt{2}$. This hypotenuse is the leg of the next triangle which has a hypotenuse of $\sqrt{3}$. Each triangle uses the hypotenuse of the preceding triangle as a leg.

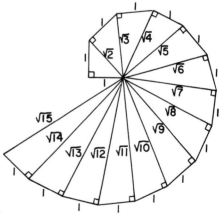

     a. For each number n on the horizontal axis on the following page, plot $\sqrt{n}$ on the vertical axis. The lengths of $\sqrt{n}$ can be measured from the spiral of triangles.

★   b. Will this graph continue to rise as *n* increases?

★   c. Connect the points of the graph with a curve. Use the graph to approximate $\sqrt{7.5}$. Could this graph be used to approximate the square root of any non-negative number less than 12?

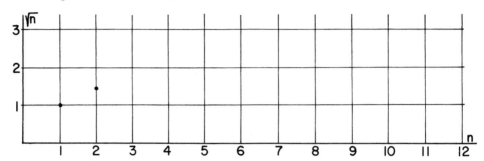

19. *Calculator Exercise:* Here is a brief chronology of some early approximations for π. Divide the numerators of these fractions by their denominators and compare the decimals with the following value of π: To 15 decimal places, $\pi = 3.141592653589793$. Which one of these fractions is closest to the value of π? Which two of these fractions equals the same decimal?

★   a. Archimedes (240 B.C.) $\dfrac{223}{71}$

b. Claudius Ptolemy (A.D. 150) $\dfrac{377}{120}$

★   c. Tsu Ch'ung-chih (A.D. 480) $\dfrac{355}{113}$

d. Aryabhata (A.D. 530) $\dfrac{62832}{20000}$

★   e. Bhāskara (A.D. 1150) $\dfrac{3927}{1250}$

20. *Infinity of Squares:* The inner square shown here was obtained by connecting the midpoints of the sides of the outer square. The outer square is 2 by 2 and has an area of 4. The inner square is $\sqrt{2}$ by $\sqrt{2}$ and has an area of 2.

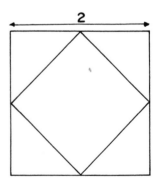

a. Continue forming smaller squares by connecting the midpoints of sides of squares. Find the lengths of the sides of the next four squares and their areas. Write your answers in the table.

★   b. Use the pattern from the table to determine the area and dimensions of the 20th square in this sequence of squares.

	Square 1	Square 2	Square 3	Square 4	Square 5	Square 6
Length of side	2 by 2	$\sqrt{2}$ by $\sqrt{2}$				
Area	4	2				

CHAPTER EIGHT

GEOMETRY WITH COORDINATES

The advancement and perfection of mathematics are intimately connected with the prosperity of the State.

Napoleon

Nine-man star with five people making approaches for slots, over California

8.1 FUNCTIONS AND COORDINATES

There are two ideas that underlie every branch of mathematics: one is that of a set, and the other is called a function. Roughly speaking, a *function* is a relationship between the elements of two sets. The distance a sky diver falls, for example, is related to the time that has elapsed during the jump. Since the distance fallen depends on time, the distance is said to be a function of the time. At the end of the first second the sky diver has fallen 16 feet, and after 2 seconds the distance has increased to 62 feet. (Speed increases during the first part of the jump.) If the sky diver wants to free-fall for 30 seconds and then open the parachute at a safe altitude of 2500 feet, the jump must be taken from a height of 7115 feet. (4615 + 2500 = 7115)

Distance Fallen in Free-Fall
Stable Spread Position

Seconds	Distance	Seconds	Distance
1	16	16	2179
2	62	17	2353
3	138	18	2527
4	242	19	2701
5	366	20	2875
6	504	21	3049
7	652	22	3223
8	808	23	3397
9	971	24	3571
10	1138	25	3745
11	1309	26	3919
12	1483	27	4093
13	1657	28	4267
14	1831	29	4441
15	2005	30	4615

The idea of a function, as illustrated in the sky diving example, is the same as that which you may hear being used in daily affairs. The gas mileage of a truck is a function of its speed (gas mileage is poorer at higher speeds); the growth rate of a baby is a function of his/her age (growth rates are greatest during the first few months); and the class a prize fighter is placed in is a function of weight (there are eight classes from flyweight, not over 112 pounds, to heavyweight, over 175 pounds).

Functions   The table of distances and times for sky diving matches each time from 1 to 30 seconds with a unique (one and only one) distance. For example, 3 seconds corresponds to just one distance and not two or more. These conditions of matching and uniqueness are the essential features in the definition of a function.

Time	Distance
1 ⟶	16
2 ⟶	62
3 ⟶	138
.	∘
.	∘
.	∘
30 ⟶	4615

Definition: A *function* is a matching of each element in one set to a unique element of a second set.

The first set of numbers is called the *domain,* and the second set is called the *range.* In the previous example, the domain is the set of numbers from 1 to 30 and the range is the set of 30 distances. Since the times and distances are continually changing, they are called *variables.* The elements of the domain are called *independent variables,* and those in the range are the *dependent variables.* In the preceding and the following tables the independent variables are time and the dependent variables are distance.

The distance that a sky diver falls in any given second can be computed by subtracting the successive distances in the previous table. For example, the diver falls $242 - 138$, or 104 feet, from the third to the fourth second. These differences show that for the first 11 seconds, the sky diver falls a little further during each second. By the 12th second the rate of fall stabilizes at 174 feet per second (120 mph). The domain of this function is the set of numbers from 1 to 30, and the range is the set of 12 different distances. In this function the time periods from 12 to 30 seconds are each paired with the same distance of 174 feet. This type of function is called a *many-to-one function* because there are two or more numbers in the domain which correspond to the same number in the range. This type of matching is still a function because each number in the domain corresponds to only one number in the range.

Seconds	Distance Fallen Each Second
1	16
2	46
3	76
4	104
5	124
6	138
7	148
8	156
9	163
10	167
11	171
12	174
13	174
14	174
15	174
.	.
.	.
.	.
30	174

At this point you may find it helpful to examine the following correspondences. The first correspondence is one-to-one and is a function from four people to their weight. The second correspondence is many-to-one and is a function from four children to their mother. The third correspondence is not a function because there is an element in the first set, namely, Pete, which is matched with two elements in the second set, his phone numbers.

Person	Weight		Person	Mother		Person	Telephone Number
Mary →	50		Kris →	Roxanne		John ⟶	868-2930
Harry →	56		Marcy →	Roxanne		Pete ⟹	344-2284
Tom →	54		Andrea →	Roxanne			344-9502
Jane →	52		Nicky →	Roxanne		Helen ⟶	431-2803

Functions and Equations   For every function there are two sets and some means of matching up the elements of one set with those of the other. In the two sky diving examples the matching of domain and range numbers is given by tables. Another method of defining or specifying a matching between sets of numbers is by equations. The independent variable in an equation is usually represented by x and the dependent variable by y, although this is not necessary. The equation $y = 3x + 2$, for example, matches each x with $3x + 2$, and, in particular, matches the first eight whole numbers with the numbers 2, 5, 8, etc., as shown in the accompanying table.

Domain x	Range y
0 ⟶	2
1 ⟶	5
2 ⟶	8
3 ⟶	11
4 ⟶	14
5 ⟶	17
6 ⟶	20
7 ⟶	23

$$y = 3x + 2$$

Here are three more examples of equations showing a few replacements for the variable x and the matching numbers for y.

EXAMPLE 1	**EXAMPLE 2**	**EXAMPLE 3**
$y = 2x - 1$	$y = x^3 + 2x - 6$	$y = x^2 + 6$

x	y		x	y		x	y
0 ⟶	⁻1		0 ⟶	⁻6		0 ⟶	6
1 ⟶	1		1 ⟶	⁻3		1 ⟶	7
2 ⟶	3		2 ⟶	6		2 ⟶	10
3 ⟶	5		3 ⟶	27		3 ⟶	15

Functions and equations are represented geometrically by plotting pairs of dependent and independent variables (x,y) on coordinate systems. The rectangular and polar coordinate systems are explained in the following paragraphs.

Rectangular Coordinates   There are many types of coordinate systems, each with its own method of locating points with respect to some frame of reference. Points on the earth's surface are located by a global coordinate system whose frame of reference is the equator and the zero meridian (see page 189). Without a frame of reference the location of a point gives little or no information, as the fellow in this cartoon is about to discover.

Points on a plane, such as those on this sheet of paper, can be located by specifying their distance from two perpendicular lines. These reference lines are called the *x-axis* and the *y-axis*. Every point in the plane, including those on the axes, can be located by an ordered pair of numbers. The first number of the pair (x,y) is called the *x-coordinate* and is the distance measured horizontally. Positive numbers are used for the distances to the right of the *y*-axis and negative numbers for distances to the left. The second number of the pair (x,y) gives the distance measured vertically and is called the *y-coordinate*. Positive numbers are used above the *x*-axis and negative numbers below. The intersection of the *x* and *y* axes is called the *origin* and has coordinates (0,0).

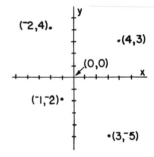

This method of locating points is called the *rectangular coordinate system* or *Cartesian coordinate system*. The name "Cartesian" is in honor of René Descartes (1596–1650), the French mathematician and philosopher who first described the use of coordinates for representing geometric figures. The link which Descartes provided between algebra and geometry is one of the greatest mathematical achievements of all times.

Polar Coordinates   At Massachusetts Institute of Technology's Artificial Intelligence Laboratory children draw geometric figures by giving instructions to a mechanical robot called a "turtle." As the turtle moves across large sheets of paper, a pen at its center traces the path. Using a desk computer the children command the turtle to move from one point to another by giving it an angle to turn through and a distance to move. This method of moving from one point to another is the basic idea of polar coordinates. With rectangular coordinates a point is located by horizontal and vertical distances, but with polar coordinates a point's position is given by an angle and by a distance.

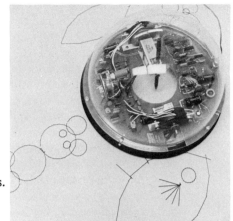

A robot at M.I.T.'s artificial intelligence lab

The circular coordinate system shown here is called a *polar grid*. The point B has coordinates $(3, 60°)$ because it is 3 units from the center or origin of the grid, and $60°$ in a counterclockwise direction from the *zero ray* $\overrightarrow{OP}$. Point A has polar coordinates $(4, 340°)$. In general, the coordinates of each point are (r, θ): where r is the distance from the origin to the point, and θ is the angle whose sides are the ray $\overrightarrow{OP}$ and the ray from the origin through the given point.

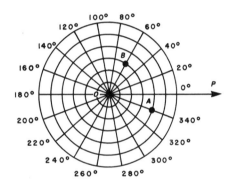

Angles measured in a counterclockwise direction from $\overrightarrow{OP}$ are positive, as shown in the above figure, and those measured clockwise are negative. Thus, an angle of $^-20°$ denotes the same ray from the origin as does an angle of $340°$. Unlike the rectangular coordinate system, where each point has a unique pair of coordinates, there are many coordinates for each point in a polar coordinate system. Point A with coordinates $(4, 340°)$ also has coordinates $(4, ^-20°)$. Furthermore, it is possible to use angles greater than $360°$. An angle of $420°$, for example, is one complete turn around the circle plus $60°$ more. Therefore, $(3, 420°)$ designates point B as well as $(3, 60°)$.

Slopes of Lines   Highway engineers measure the slope or steepness of a road by comparing each 100 feet of horizontal distance with the corresponding vertical rise. The Federal Highway Administration recommends a maximum of 12 feet vertically for each 100 feet of horizontal distance. Many secondary roads and streets are much steeper. Filbert Street in San Francisco has a vertical rise of approximately 1 foot for each 3 feet of horizontal distance. By comparison, the east and west walls of the Daytona International Speedway have a vertical rise of 3 feet for each 5 feet of horizontal distance. These banks enable a car to turn at the ends of the speedway while maintaining speeds of 180 to 200 miles per hour.

East wall of the Daytona International Speedway with a slope of 31°

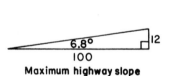

Maximum highway slope

Filbert Street slope

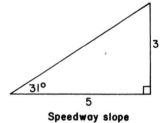

Speedway slope

The slope of a line on a rectangular coordinate system is also measured by horizontal and vertical distances. The horizontal distance, which is the difference between the *x* coordinates of two points on a line, is called the *run*. The vertical distance is the difference between the *y* coordinates of the two points and is called the *rise*. The ratio of these distances, rise/run, is the *slope* of the line. In this example, the points ($^-$2,2) and (1,4) were used to find the slope of 2/3. For any two points on the same line the slopes will be equal.

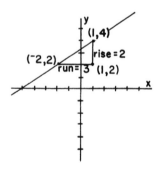

To distinguish between lines such as *L* and *K*, which are inclined in opposite directions, lines running from lower left to upper right have a *positive slope* and those running from upper left to lower right have a *negative slope*.

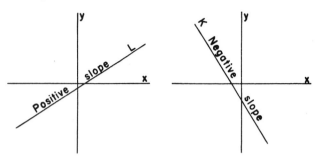

Lines that are parallel to the *x*-axis have a slope of 0. For example, the line through (2, ⁻3) and (5, ⁻3) has a rise of 0 and a run of 3 for these two points. Therefore, its slope is 0/3, or 0. Lines that are perpendicular to the *x*-axis, such as the line through (⁻2,1) and (⁻2,3), have a slope that is undefined. In this example the rise is 2 and the run is 0, but 2/0 is undefined. Between these two extremes the slopes of lines can be any positive or negative real number.

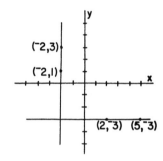

Graphs of Equations   The story is told that the idea of coordinates in geometry came to Descartes while he lay in bed and watched a fly crawling on the ceiling. Whether the story is true or not, it is useful in illustrating the relationship between curves and their equations. Each position of the fly can be given by two distances from the edges of the ceiling (in this example the *x* and *y* axes). Descartes discovered that these distances can be related by an equation. That is, each point on the curve has coordinates which satisfy an equation, and conversely, every two numbers *x* and *y* which satisfy the equation correspond to a point on the curve.

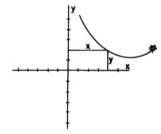

Descartes was interested primarily in beginning with a curve and finding its equation. Let's consider the circle with center at $(0,0)$ and radius of 5 units. This circle passes through 12 points whose x and y coordinates are whole numbers. Each of these coordinates satisfies the equation $x^2 + y^2 = 25$, which is the equation of this circle.

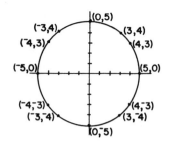

During the same period as Descartes, Pierre Fermat, a countryman, was working independently on coordinate geometry. Fermat was interested in beginning with equations and then finding their graphs. The most straightforward method of graphing equations is to select some values for x and compute the corresponding values for y. Consider the equation $y = x^2$. The following table contains a few of the numbers for x and y which satisfy this equation. Notice that each value of y is positive or zero. Also, for a given value of x and its negative, the values of y are equal. Because of these conditions the graph opens upward and is symmetric to the y-axis.

x	y
$^-3$	9
$^-2$	4
$^-1$	1
0	0
1	1
2	4
3	9

$$y = x^2$$

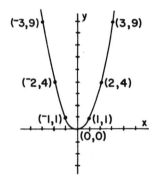

Linear Functions   The distance to an approaching thunderstorm can be determined by counting the seconds between a flash of lightning and the resulting sound of thunder. For every 3 seconds, sound travels approximately 1 kilometer. If you can count up to 6 seconds before hearing the thunder, the storm is approximately 2 kilometers away. Distance in this example is a function of time, and the graph of this function is a straight line. Since each y-coordinate (distance) is $1/3$ of each x-coordinate (time), the equation for this function is $y = (1/3)x$.

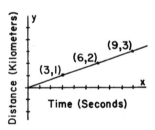

Sound travels faster in water than in air. In water it travels about 1.5 kilometers per second. The graph which shows this relationship between time and distance is linear and has the equation $y = 1.5x$, or $y = 3x/2$. The slope of this line is greater than the slope of the line for the speed of sound in air. Both of these lines have equations of the form $y = mx$, where m is the slope of the line.

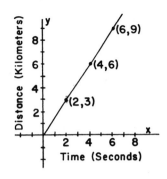

The speeds of sound in air and water are examples of special types of functions called *rates* (see page 347). Kilometers per hour, rent per month, and dollars per kilogram are three more examples of rates. These rates are linear functions, and as a rate increases the slope of the graph of its line also increases. For example, if a lawn mower costs $3 per hour to rent, it will cost $6 for 2 hours, $9 for 3 hours, etc. The graph of this function has a slope of 3 (see accompanying figure). If the rental rate is increased to $5 per hour, the graph of the function will have a slope of 5.

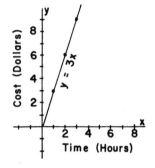

Suppose there is an initial fee of $2 for renting the lawn mower, in addition to $3 per hour. At time zero it costs $2; after the first hour the cost is $5; etc. The graph of this function is a line which intersects the y-axis at $(0,2)$ and has a slope of 3 (see accompanying figure). The equation of this line is $y = 3x + 2$. In general, equations for linear functions can be written in the form $y = mx + b$, where m is the slope of the line and b is the y-coordinate of the point where the line crosses the y-axis. You can see this by noticing that $(0,b)$ satisfies the equation $y = mx + b$. The variable b is called the *y-intercept* of the line and in the example shown here equals 2. When b equals 0, the equation of the line becomes $y = mx$, a line through the origin with a slope of m.

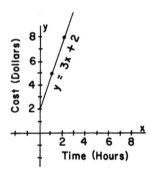

SUPPLEMENT *(Activity Book)*

Activity Set 8.1 Function and Coordinate Games (Bouncing ball and pendulum experiments; games of battleships with rectangular and polar coordinates, Hide-a-region; and What's My Rule?)

Additional Sources

Davis, R.B. *Discovery in Mathematics.* Reading, Mass: Addison-Wesley, 1964. pp. 79–124 (graphs and equations).

National Council of Teachers of Mathematics. *Experiences in Mathematical Ideas,* 2. Reston, Viginia: NCTM, 1970. "Graphs," pp. 96–123.

National Council of Teachers of Mathematics. *The Rational Numbers.* Reston, Virginia: NCTM, 1972. "Graphing," pp. 316–37.

School Mathematics Study Group. *Mathematics Through Science, Part III: An*

Experimental Approach to Functions, Teacher's commentary. Palo Alto: Stanford University, 1964.

Stertz, D.E. and J.L. Teeters. "Dual Concepts—Graphing with Lines (Points)." *The Mathematics Teacher,* **70** No. 8 (December 1977), 726–31.

Trask, III., F.K. "Circular Coordinates: A Strange New System of Coordinates." *The Mathematics Teacher,* **64** No. 5 (May 1971), 402–8.

EXERCISE SET 8.1

1. Polar grids are used for plotting information from radar data. There are several different radar systems for determining the heights of storms and the strengths and directions of winds. Radar echoes give indications of precipitation within a storm.

   The center of the polar grid represents the radar station. Each point on a polar grid has many different coordinates. Give another pair of coordinates for each of these points.

   ★   a. $(5, 60°)$ b. $(2, 200°)$

   ★   c. $(6, 280°)$ d. $(1, 340°)$

Plotting radar information on a polar grid

2. Electrical impulses that accompany the beat of the heart are recorded by an electrocardiograph. The electrocardiograph measures electrical changes in millivolts (1/1000 of a volt). The following graph shows the changes in millivolts (mV) as a function of time for a measure of a normal heartbeat.

   ★   a. How much time was required for this graph if each small space on the horizontal axis represents .04 second?

b. The tall rectangular part of the graph was caused by a 10-millivolt signal from the EKG machine. This is called a calibration pulse. What was the length of the time for this signal?

c. This graph shows ten heartbeats or pulses. Approximately how much time is there between each pulse (from the end of one pulse to the end of the next pulse)? At this rate how many pulses will there be per minute?

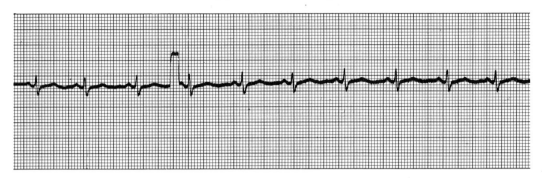

3. Great Britain's pound is worth about $1.70 in U.S. currency. That is , 1£ = $1.70, 2 £ = $3.40, etc. This is a linear function whose graph is shown here.

★ a. Use this graph to find the approximate value in dollars of these amounts in British currency:

£ 1.50   and   £ 2.50

★ b. Use the graph to find the approximate value in pounds of these amounts in U.S. currency:

$6   and $5.10

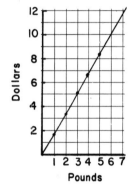

4. This graph shows the heights of a river for various amounts of water discharge. The points which are shown represent actual measurements of discharge and the corresponding heights of the river. The curve approximately fits the location of these points.

★ a. How many separate measurements were taken for this graph?

b. Use this graph to predict the river's height for a discharge of 5500 cubic meters of water per second. Predicting a value between two known values is called *interpolation*.

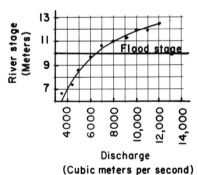

★   c. At 10 meters (flood stage) the river overflows its banks. At this point the height of the river increases less rapidly for large increases in discharge. Predict the river's height for a discharge of 13,000 cubic meters per second. Predicting a value beyond the known values is called *extrapolation*.

5. The cost of first class postage is a function of weight. The first ounce costs 15 cents and each additional ounce or fraction thereof costs 13 cents. For example, 1.4 ounces cost 28 cents and 2.7 ounces cost 41 cents.

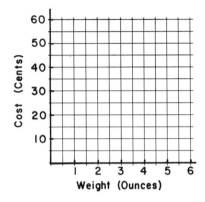

★   a. Find the costs for these weights:

    3.2 oz.   4.2 oz.   4.8 oz.

★   b. This postal rate holds for up to 13 ounces. Therefore, the domain of this function is all numbers from 0 to 13. There are 13 numbers in the range. List these numbers.

    c. Graph this function for weights less than or equal to 5 ounces.

★   d. Is this function one-to-one?

6. The lines on these coordinate systems are perpendicular. Compute the slope of each line. How are the slopes of perpendicular lines related?

a.
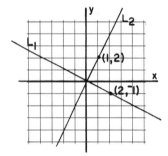

b.
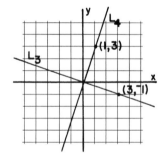

7. Graph the lines for these equations.

    I. $y = 3x$ II. $y = {}^{-}3x$ III. $y = {}^{-}3x + 5$ IV. $y = \dfrac{{}^{-}1}{3}x$

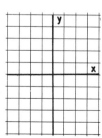
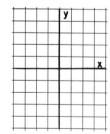
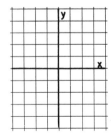
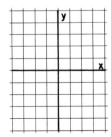

★   a. What is the slope of each line?

★   b. Which two of these lines are parallel?

        c. Which two of these lines are perpendicular?

8. The graphs of three lines and their equations are shown below.

★   a. What is the slope of each line?

★   b. Draw another line on the first coordinate system whose slope is greater than the given line. Is there any limit to how large the slope of a line can become?

        c. Draw a line on the third coordinate system whose slope is less than the slope of the given line. Is there any limit to how small the slope of a line can become? (*Hint:* The slope of a line can be negative.)

L_1 $y = 10x$ L_2 $y = x$ L_3 $y = \frac{1}{2}x$

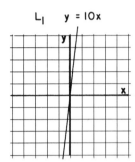 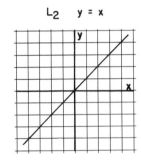 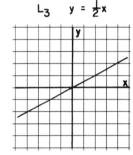

9. For these equations and values of x, compute the corresponding values of y. Graph these number pairs and connect the points with a curve. Which of these curves is symmetric to the y-axis?

★   a.

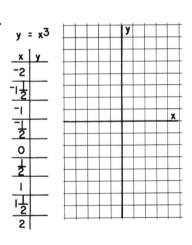

$y = x^3$

x	y
-2	
$-1\frac{1}{2}$	
-1	
$-\frac{1}{2}$	
0	
$\frac{1}{2}$	
1	
$1\frac{1}{2}$	
2	

      b.

$y = 1 - x^2$

x	y
-3	
-2	
-1	
0	
1	
2	
3	

10. Leaky Boat Rentals charges 1 dollar per hour for renting a canoe. If you are a member of their club there is no initial fee. Nonmembers who are state residents pay an initial fee of 2 dollars, and out-of-state people pay an initial fee of 5 dollars. The graphs of these rates are the lines shown in the accompanying figure.

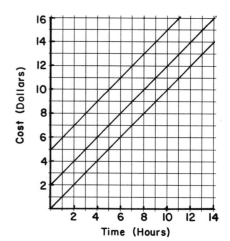

a. Label the graphs which correspond to: members; state residents who are nonmembers; and out-of-state residents. Are the slopes of these lines equal?

★ b. How much more will it cost an out-of-state resident than a club member to rent a canoe for 8 hours? Label the portion of the graph which corresponds to this difference.

c. Answer the question in part **b** for 11 hours.

11. The average annual cost of heating a home with solar energy is $100, with an initial investment of $8000. The average annual cost of heating with oil is $700 with an initial investment of $2000.

★ a. Which line shows the cost of oil heat and which one shows the cost of solar heat? Label the coordinates of two points on each of these lines.

b. Which line has the greater slope?

★ c. For the first few years it costs less to heat with oil than with solar energy. In how many years will these costs be equal? How can this question be answered by the graph?

d. How much more does it cost to heat with solar energy than with oil, for the first 4 years (including initial investments)? Mark the portion of the graph which shows this difference.

★ e. How much more does it cost to heat with oil than with solar heat for the first 12 years? Mark the portion of the graph which shows this difference.

12. In March of 1977 the Environmental Protection Agency (EPA) announced some startling new results concerning fuel economy and safety of heavy-duty trucks.

★   a. In a crash, the force of a truck (its kinetic energy) depends on the truck's mass and its speed. This force is computed by multiplying the mass by a speed factor. Use this table to find an equation for the speed factor for force as a function of the truck's speed.

Speed	Speed Factor for Force
30 ──────────▶	900
40 ──────────▶	1600
50 ──────────▶	2500
60 ──────────▶	3600

★   b. When the speed is doubled, how many times greater is the force of the truck?

c. The air resistance to a moving truck is also its mass times a speed factor, but in this case the speed factor is greater than that for force. Use this table to find an equation for the speed factor for air resistance as a function of speed.

Speed	Speed Factor for Air Resistance
30 ──────────▶	27000
40 ──────────▶	64000
50 ──────────▶	125000
60 ──────────▶	216000

d. As the speed is doubled, how many times greater is the air resistance to the truck?

13. *Calculator Exercise:* Some calculators have keys for the trigonometric functions. On such a calculator there is a key labelled "tan" for the tangent function. This key can be used to find the slope of a line, if we know the line's *angle of inclination* (the angle formed by the line and a horizontal line). In this figure the slope of $\overleftrightarrow{AB}$ is the tangent of 30°. The following calculator steps and displays show that the slope of $\overleftrightarrow{AB}$ to 10 decimal places is .5773502692. [The Degree (D) and Radian (R) switch on the calculator should be on D.]

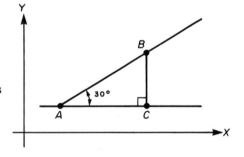

Steps	Displays
1. Enter 30	30.
2. $\boxed{\text{tan}}$	.5773502692

★ a. Find the tangent of the following angles. Make a conjecture about the tangents of angles as they get closer to 90°.

$$30° \quad 45° \quad 60° \quad 80° \quad 85° \quad 89° \quad 89.5° \quad 89.9°$$

★ b. Conversely, we can find a line's angle of inclination if we know its slope. This is done on a calculator by pressing the inverse key, $\boxed{\text{inv}}$, and then the tangent key, $\boxed{\text{tan}}$, as shown in the following steps. For a slope of .625, the angle of inclination is approximately 32°.

Steps	Displays
1. Enter .625	0.625
2. $\boxed{\text{inv}}$	0.625
3. $\boxed{\text{tan}}$	32.00538321

Find the angles of inclination to the nearest degree for lines with the following slopes. Make a conjecture about angles of inclination as the slopes of lines get closer to 0.

$$.424 \quad .306 \quad .231 \quad .158 \quad .070 \quad .017$$

c. When a carpenter speaks of the slope of a roof as being "10 inches," this means that for a run of 12 inches there is a rise of 10 inches. For such a roof, what is the angle of inclination between the roof and the floor of the attic?

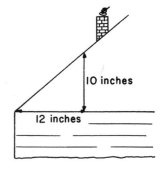

14. On the following page there are a photo of a radar view of a hurricane taken from an airplane and a drawing of a ship's sonar scope with the positions of two submarines.

★ a. Assume that the circles on the radar screen have radii of 25, 50, 75, and 100 kilometers, respectively. Using ray $\overrightarrow{OP}$ to indicate 0°, approximate the polar coordinates for the centers of clouds x, y, and z.

b. The angles on the sonar scope are measured clockwise, and 0° is at the top of the circle. The ray from the center of the circle to the 0° mark is always the direction the ship is headed. What are the polar coordinates of the centers of the two subs if the marks on the revolving cursor represent 10-km intervals?

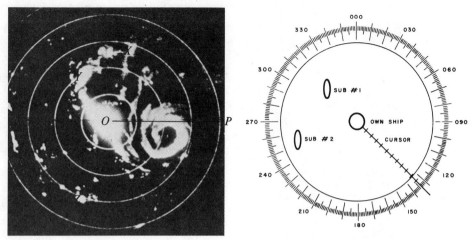

Airborne radar view of a hurricane

Ship's sonar scope with revolving cursor

15. *Slopes on Geoboards:* A rectangular geoboard has rows and columns of pegs or nails. Line segments and polygons can be formed on the geoboard by stretching rubber bands between the pegs. The line segment on this geoboard has a run of 3 and a rise of 2. Its slope is 2/3. How many line segments with different slopes (positive, negative, or zero) can be formed on this geoboard?

16. *Constellations:* Use these Cartesian coordinates to plot the star constellations. Plot each set of points on a separate coordinate system.

★   a. *Orion*   (10,4); (13,6); (12,9); (14,10); (16,11); (18,14); (16,4); (12,17).
      Orion's belt has three stars in a row. What are the coordinates of the points which represent these stars?

   b. *The Big Dipper and Pole Star*   (48,15); (51,23); (44,28); (51,7); (9,23); (48,19); (48,29); (49,11). Which point is the pole star?

Volcanic activity in Mount Etna

8.2 CONIC SECTIONS

The paths of hot lava being shot into the air by this explosion of Mount Etna form mathematical curves called parabolas. Two thousand years before Galileo made his important discoveries about the parabolic paths of projectiles, the Greeks had studied this curve as well as the ellipse and hyperbola. At that time these curves had no practical applications. By the seventeenth century they were required by scientists and engineers to describe the motions of planets, comets, and projectiles; to construct lenses; to study the curvature of light; and to design antennas and light reflectors.

Parabolas The paths of balls, bullets, or other objects which are thrown or shot into the air are parabolas. The distance an object travels varies with the angle it is aimed at, its *angle of elevation.* An object will travel its greatest distance away (horizontal distance) for a 45° angle of elevation. Using this condition, which waterspout on this fountain (see photo) appears to have an angle of elevation which is closest to 45° ? While parabolas have many different shapes, as shown by these paths of water, each parabola satisfies the following definition.

Definition: A *parabola* is the set of all points in a plane which are the same distance from a fixed point, called the *focus,* as from a fixed line, called the *directrix.* (The distance from a point to a line is the perpendicular distance.)

In both of the following figures the focus is labelled *F* and the directrix *D.* The line passing through *F* and perpendicular to line *D* is the *axis* of the parabola. Notice how the distances for the points *S, T,* and *R* on these curves satisfy the conditions of the definition. Select another point on these curves and use a piece of paper to mark off its distances to the focus and directrix. Compare these distances.

Fountain at Swirbul Library, Adelphi University

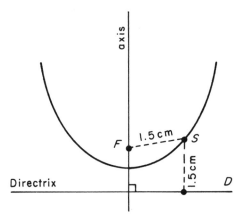

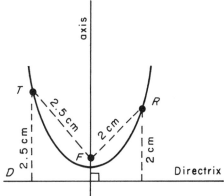

The equation of the parabola on this coordinate system is

$$y = \frac{x^2}{8}$$

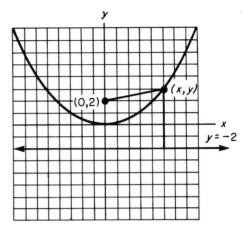

Any pair of numbers which satisfy this equation are the coordinates of points which are on this graph. As examples, (2,1/2), (‾2,1/2), (4,2), and (‾4,2) can all be used in this equation and you will find the corresponding points on the graph. Conversely, for every point on the graph, its coordinates will satisfy the equation $y = x^2/8$. The focus for this parabola is at (0,2), and the directrix is the line parallel to the x-axis and 2 units below it ($y = $‾2). Select a point on the graph and use a piece of paper to mark off its distances to the focus and directrix. Compare these measurements.

Parabolas have a very important reflection property which distinguishes them from other curves. Select any point P on a parabola and draw the tangent T to this point. (Intuitively, a tangent to a parabola is like a tangent to a circle—it touches the curve at one point but does not pass through it.) Lines from P to the focus and from P parallel to the axis of the parabola form equal angles (A and B) with the tangent.

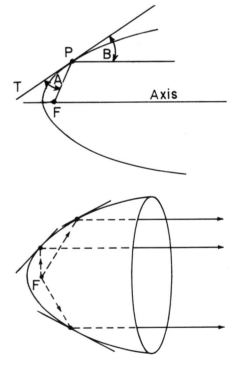

This property has applications in the *paraboloid*, a three-dimensional parabolic-shaped surface. A *paraboloid* is generated by rotating a parabola about its axis. These surfaces are used in searchlights, automobile headlights, and other reflectors. A light at the focus is reflected off the surface in rays which are parallel to the axis. The resulting beam of light is as much as 6000 times as bright as light emitted in the same direction by the light source alone.

The reflecting principle of the paraboloid is used in reverse in radio telescopes and radar antennas. In radio telescopes weak radio waves from space arrive in nearly parallel rays. These rays are reflected off the parabolic-shaped receiver and concentrated at the focus. Radar antennas use the reflecting principle for both receiving and transmitting radar signals. The radar antenna pictured here was used to transmit signals to the moon and to receive then moments later the echoes which bounced back.

Radio telescope,
Millstone Hill Radar Observatory

Ellipses   Religion and science did not develop as separate disciplines until Johannes Kepler's introduction of the laws of planetary motion. For 2000 years scientists thought that the planets moved in circular paths and at constant speeds. Then in 1609 Kepler announced his two laws: (1) the planets move about the sun in elliptical paths; and (2) the planets do not move at a constant speed. The earth, for example, moves in an elliptical path about the sun as shown in the accompanying diagram. If it moves from A to B in 1 month, then it will also take 1 month to move from C to D, provided the areas of regions ASB and CSD are equal.

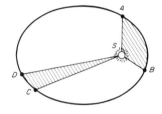

The location of the sun for the elliptical path of a planet is called the *focus*. Every ellipse has two focuses (foci).

Definition: An ellipse is the set of points in a plane such that the sum of the distances from each point to two fixed points, called *focuses,* is a constant.

The focuses for the ellipse shown here are labelled F_1 and F_2. For any point P which is chosen on this ellipse, the sum of the distances PF_1 and PF_2 will equal 5 centimeters. Select another point on this curve and mark off its distances to F_1 and F_2 on a piece of paper. Compare the total length to the 5-centimeter line segment below the ellipse.

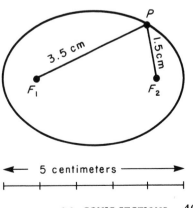

The equation of the ellipse on this coordinate system is

$$9x^2 + 25y^2 = 225$$

Every pair of numbers which satisfy this equation are the coordinates of a point on this curve, and conversely. As examples, the coordinates of points A, B, C, D are $(0,3)$, $(^-5,0)$, $(0, ^-3)$, and $(5,0)$, and each satisfies the equation of the ellipse. The focuses are at $(^-4,0)$ and $(4,0)$. The sum of the distances from point G to F_1 and F_2 is 10 units. Select any point on the ellipse and use a piece of paper to mark off the sum of its distances to F_1 and F_2. Compare this length to 10 units on this coordinate system.

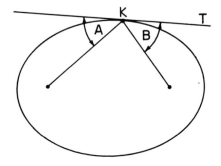

Every ellipse has a reflection property which involves its focuses. Select any point (K) on an ellipse and draw the tangent (T) to the ellipse at this point. Lines from the focuses to K form equal angles $(A$ and $B)$ with the tangent. This property has some interesting applications. Consider an elliptical-shaped pool table with a pocket at one focus. A ball shot from the other focus to any point on the ellipse will rebound into the pocket.

The three-dimensional counterpart to the ellipse is the *ellipsoid*. This figure is generated by revolving an ellipse about the line through its focuses. In a room with an ellipsoidal ceiling, a whisper at one focus can be clearly heard at the other focus. Statuary Hall in the Capitol at Washington, D.C., is a well-known example of a room with this whisper-chamber effect.

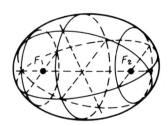

Hyperbolas   Hyperbolas are not seen as frequently in everyday life as parabolas and ellipses. One of the few times we see a hyperbola is by observing the shadow on a wall which is cast from a lamp with a cylindrical or a conical shade. The shadow above the lamp forms one branch of a hyperbola, and the shadow below forms the other branch. Hyperbolas also occur as the paths of comets. Comets which stay in the solar system follow elliptical paths. Those which enter the solar system and then leave again follow parabolic or hyperbolic paths.

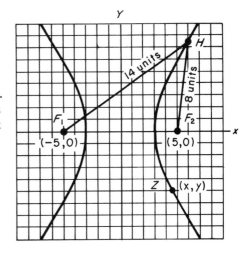

Every hyperbola has two focuses and two separate branches which satisfy the following definition.

Definition: A *hyperbola* is the set of all points in the plane such that the difference between the distances from any point to two fixed points, called *focuses*, is a constant.

The focuses for the hyperbola in the accompanying diagram are labelled F_1 and F_2. The difference between the distances PF_1 and PF_2 is 3 centimeters. Try computing a similar difference for the point K. For any point on the hyperbola, use a piece of paper to mark off two distances to the focuses. Compare the difference between these distances to the length of the 3-centimeter line segment below the hyperbola.

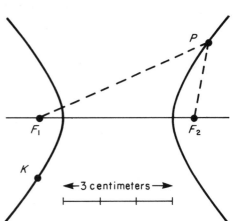

The equation of the hyperbola on this coordinate system is

$$16x^2 - 9y^2 = 144$$

Every point on this curve has coordinates which satisfy this equation, and conversely. For example, the points whose coordinates are ($^-3$,0) and (3,0) are on the curve, and it is easy to see that their coordinates satisfy the equation of this hyperbola. The difference between the distances from point H to F_1 and F_2 is $14 - 8$, or 6 units. Select another point and mark off its distances to F_1 and F_2. The difference should equal 6 units on this coordinate system.

The parabola, ellipse, circle, and hyperbola are called *conic sections* because each can be obtained from the intersection of a plane with a double cone, as shown below. A slightly tilted plane produces an ellipse; if the plane is horizontal, its intersection with the cone is a circle; a plane parallel to the sides of the cone results in a parabola; and the intersection of a vertical plane with the cone is two branches of a hyperbola.

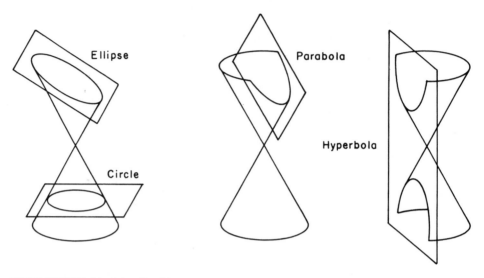

SUPPLEMENT *(Activity Book)*

Activity Set 8.2 Conic Sections (Drawings by string and paper-folding)

Just for Fun: Line Designs (Curves from sets of intersecting lines)

Additional Sources

Beck, A., M.N. Bleicher, and D.W. Crowe. *Excursions into Mathematics.* New York: Worth, 1969. pp. 248–54 (Cartesian coordinates, straight lines, and conics).

Gardner, M. "Diversions that involve one of the classic conic sections: the ellipse." *Scientific American,* **204** No. 2 (February 1961), 146–54.

_____. "On conic sections, ruled surfaces and other manifestations of the hyperbola." *Scientific American,* **237** No. 3 (September 1977), 24–42.

Menninger, K.W. *Mathematics in Your World.* New York: The Viking Press, 1962. "Curves, Visible and Invisible," pp. 123–34.

Ogilvy, C.S. *Through the Mathscope.* New York: Oxford University Press, 1956. "Cones and Conic Sections," pp. 60–73.

Von Baravalle, H. "Conic Sections in Relation to Physics and Astronomy." *The Mathematics Teacher,* **63** No. 2 (February 1970), 101–9.

EXERCISE SET 8.2

1. The solar furnace pictured here was
   constructed from the curved mirror
   of an old army searchlight. Instead of
   reflecting light outward from a central
   bulb, this mirror now reflects the sun's
   rays inward to a focal point, where temperatures reach as high as 3500° C.
   This effect accounts for the use of the
   word "focus," which in Latin means
   a hearth or burning-place.

★ a. What is the shape of the surface of
      the searchlight's reflector?

   b. Explain why this surface enables a
      searchlight to be converted to a
      solar furnace. (*Note:* The sun's
      rays are nearly parallel when they
      reach the earth.)

Solar furnace, constructed
from a searchlight

2. In 1594 Galileo discovered the laws
   of falling objects through his experiments at the Leaning Tower of Pisa
   in Italy. One of his discoveries was the
   formula for the distance in feet (d)
   travelled by an object falling from
   rest in a given number of seconds (t):
   $d = 16t^2$.

   a. Complete the table and plot the six
      points.

   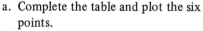

   b. Connect the points with a smooth
      curve. What would this curve look
      like if t were allowed to be negative
      as well as positive? What is the name
      of this curve?

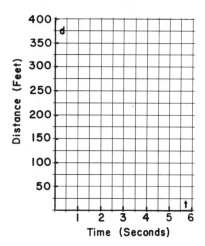

★ 3. When an object such as a bullet or an arrow is shot into the air, the horizontal distance it will travel depends on the initial force and the angle of elevation. If two objects are thrown into the air with the same amount of force and their angles are complementary, they will travel the same horizontal distance. Assuming the initial forces to be equal, which one of the following angles of elevation will produce the greatest distance? Which two will give the same distance?

    a. 35°   b. 45°   c. 55°   d. 65°

4. If a ball is thrown into the air at a speed of 96 feet per second, the distance, d, that it travels in t seconds is $d = 96t - 16t^2$. Complete the table for this equation and sketch the curve.

★   a. What type of curve is it?

      b. How high will the ball go?

★   c. How long will the ball be in the air?

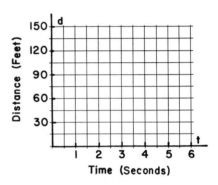

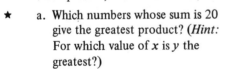

t	0	1	2	3	4	5	6
d							

5. There are an infinite number of pairs of numbers whose sum is 20: $(1 + 19, 12.5 + 7.5,$ etc.$)$. Some of these pairs of numbers have relatively small products, such as $.1 \times 19.1 = 1.91$, and others have much larger products, such as $12 \times 8 = 96$. If we let x and $(20 - x)$ be the two numbers whose sum is 20, their product is $y = x(20 - x)$ or $y = 20x - x^2$. For each value of x in the table find the values of y from this equation, and sketch the curve.

★   a. Which numbers whose sum is 20 give the greatest product? (*Hint:* For which value of x is y the greatest?)

      b. What happens to the product when one number gets very close to 0 and the other number is close to 20?

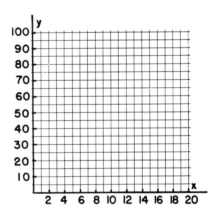

x	0	2	4	6	8	10	12	14	16	18	20
y											

6. On the following ellipses try the coordinates in the equations in order to match each equation to its curve.

★ a. $\dfrac{x^2}{16} + \dfrac{y^2}{9} = 1$

b. $\dfrac{x^2}{4} + \dfrac{y^2}{25} = 1$

c. $\dfrac{x^2}{36} + y^2 = 1$

I.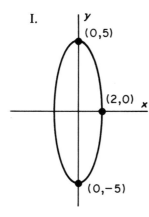

II.

III.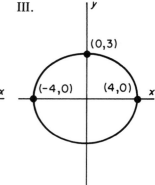

7. It is possible to roughly sketch an ellipse which is centered at the origin by knowing only its *x* and *y intercepts*. The intercepts are the four points where the ellipse intersects the coordinate axes. The coordinates of these points will have either *x* or *y* equal to 0. Find the missing coordinates for these points and write them in the tables. Plot these points and sketch each ellipse.

★ a. $\dfrac{x^2}{25} + \dfrac{y^2}{16} = 1$

b. $\dfrac{x^2}{9} + \dfrac{y^2}{36} = 1$

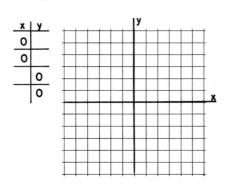

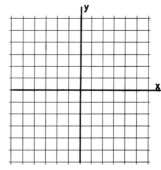

8. The area of an ellipse is πmn, where m is half the length of the major axis and n is half the length of the minor axis. (The major axis is the longer of the two axes.) Compute the areas of these two ellipses.

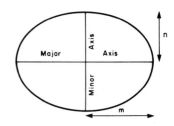

★ a.

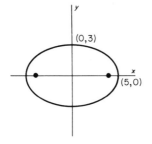

b.

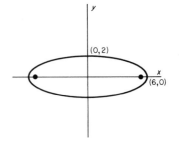

c. If an ellipse is almost circular, its major and minor axes will be almost equal. Use this observation to explain how the formula for the area of a circle is a special case of the formula for the area of an ellipse.

9. The *eccentricity* of an ellipse is the ratio c/a where c is the distance from the origin to the focus and a is the distance along the major axis from the origin to the ellipse. This number varies between 0 and 1. If it is close to 0, the ellipse is almost circular; and if it is close to 1, the ellipse is elongated (eccentric).

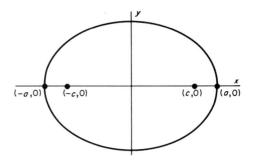

★ a. What happens to the eccentricity as the focuses move toward $(0,0)$? What does this do to the shape of the ellipse?

b. What happens to the eccentricity as the focuses move toward $(a,0)$ and $(^-a,0)$? What does this do to the shape of the ellipse?

10. When x and y are related by the equation $y = 1/x$, they are said to be *inversely related*. That is, as x increases, y decreases; and as x decreases, y increases. Complete the table for this equation and sketch its graph.

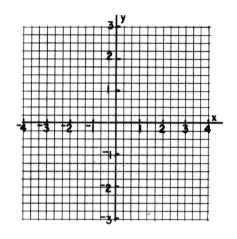

x	4	-3	-2	-1	$\frac{-1}{3}$	$\frac{1}{3}$	1	2	3	4
y										

★ a. Explain what happens to y when x is positive and gets close to 0.

b. What happens to y when x is negative and gets close to 0?

★ c. What is the name of this curve?

11. A rectangle with an area of 144 cm² can have an infinite number of different widths (x) and lengths (y). These widths and lengths are inversely related to each other by the equation $xy = 144$ or $y = 144/x$. Complete the following table and sketch the graph of this equation.

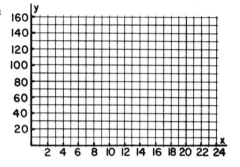

x	24	18	16	12	8	6	4	2	1
y									

★ a. What happens to the values of y as x increases?

b. What happens to the values of y as x gets close to 0?

★ c. Can x be equal to 0?

d. If x and y were allowed to take on negative values in this equation, the graph would be one of the conic sections. Which one?

Daniel Webster Hoan Memorial Bridge,
Milwaukee, Wisconsin

8.3 MATHEMATICAL CURVES

Today's technology requires a wide range of curves. The designing of bridges, roads, and buildings are only a few of the many applications. Each of the curves in these photographs can be represented by equations and graphs using coordinate geometry. For each curve there is an equation, and for each equation there is a corresponding curve.

Kresge Auditorium, Massachusetts
Institute of Technology

Network of roads outside Empire State
Plaza, Albany, New York

Here are three of the better-known curves and their equations.

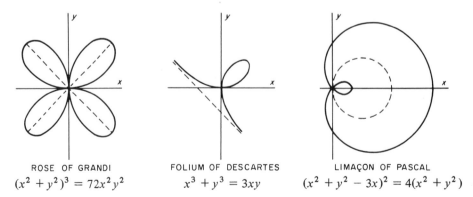

ROSE OF GRANDI
$$(x^2 + y^2)^3 = 72x^2y^2$$

FOLIUM OF DESCARTES
$$x^3 + y^3 = 3xy$$

LIMAÇON OF PASCAL
$$(x^2 + y^2 - 3x)^2 = 4(x^2 + y^2)$$

S-Shaped and Exponential Curves The graphs of growth rates for sizes, weights, populations, etc., of living things have similar shapes. If growth can be measured in the early and late stages of the life of an organ (heart, liver, brain, etc.), organism (plant, animal, insect, etc.), or population (bacteria, animal, insect, etc.), then the graph of the growth rate has three phases: the *lag phase,* the *exponential phase,* and the *stationary phase.* These phases are shown on the accompanying graph of a 5-day experiment in the growth of mold. This is the typical *Sigmoid curve* or *S-shaped curve* of growth. Two more illustrations of S-shaped curves are contained in the graphs that follow.

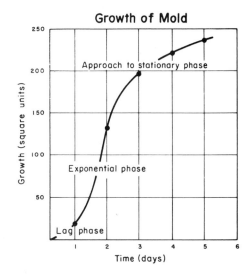

Growth of Mold

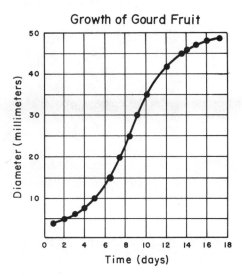

Growth of Gourd Fruit

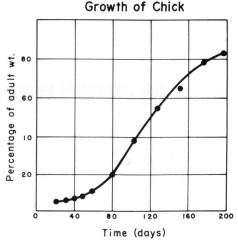

Growth of Chick

Theoretically, during the exponential phase of growth, 1 cell produces 2 cells, then 4, 8, 16, and so on, in a geometric sequence. This is represented by the exponential equation, $y = 2^x$. When x takes on the values 0, 1, 2, 3, 4, . . . , from the domain, y equals 1, 2, 4, 8, 16, . . . , in the range. When x is negative, the values of y are less than 1. For example, if $x = {}^-3, y = 2^{-3} = 1/2^3 = 1/8$.

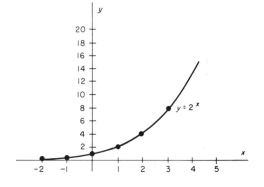

The values of 2^x when x is any real number (fraction or irrational number) can be approximated by connecting the points shown in the preceding graph to form a smooth curve. In general, the equation $y = a^x$ (where a is any positive real number) is called an *exponential equation with base a*. The domain of the function for this equation is the set of all real numbers (positive and negative) and the range is the set of all positive real numbers.

There is one value in particular for the base a in the equation $y = a^x$, which occurs so often in the physical sciences that it is called the "natural base" and denoted by the letter e. This is an irrational number which to the first 5 decimal places is 2.71828. Since e is greater than 2 and less than 3, the graph of $y = e^x$ is between the graphs of $y = 2^x$ and $y = 3^x$. The equation $y = e^x$ is so important it is called the *exponential function.*

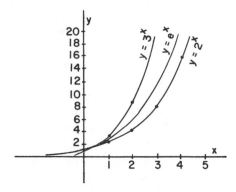

Spirals   The spiral is a common shape in plants. In daisies and sunflowers there are two sets of intersecting spirals of florets winding clockwise and counterclockwise. The heads of some young ferns, such as the cinnamon and Christmas ferns, are called fiddle heads because their tips are spiraled and resemble the curved heads of violins and fiddles. The Sago Fern or Silver Tree Fern shown here is a New Zealand tree fern which is cultivated for its handsome spiral-shaped crown.

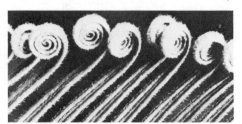

Sago Fern, also called Silver Tree Fern, from New Zealand

The frequent occurrence of spirals in living things can be explained by different growth rates. Living forms curl so that the faster growing or longer surface lies outside and the slower growing or shorter surface lies inside. Consider, for example, the chambered nautilus. As it grows it lives in successively larger compartments. The outer surfaces of these compartments have faster growing rates than the inner surfaces. Similarly, in the ram's horns the leading edges grow more than the trailing edges and so the horns curl back.

Chambered nautilus.
Courtesy of the American Museum of
Natural History.

Dall sheep.
Courtesy of the American Museum of
Natural History.

The mathematical curve called a *spiral* is a plane curve
which winds around a fixed point while receding from it
(see figure below). If this curve rises from the plane into
three dimensions, it is called a *helix*. The common type of
spring is a *cylindrical helix*. If the helix rises up to a
point as illustrated in this picture of the shell of a snail,
it is a *conical helix*.

Caribbean land snail

There are several types of spirals. One of these was
described by Archimedes in his treatise, *On Spirals,*
and is now called the *Archimedean spiral*. This curve
has a complicated equation in rectangular coordinates
but can be described very simply in polar coordinates.
Starting at the center and moving counterclockwise,
the distances from the center increase by fixed amounts
(arithmetic sequence) as the angles increase by fixed
amounts. For each 20-degree increase in the angles of
this Archimedean spiral the points on the curve are
half a unit further from the center. Here are the
coordinates of five points which lie on this spiral:
$(0,0°)$, $(1/2,20°)$, $(1,40°)$, $(1\ 1/2,60°)$ and $(2,80°)$.

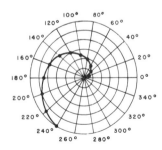

Sine Curves We live in an "ocean of air," and just as the ocean it is filled with waves. These waves are produced by compressions of air. If, for example, you strike the prong of a tuning fork, its vibrations cause the air around it to be repeatedly compressed. These successive compressions of air travel outward much like waves when a stone is dropped in a pool of water. Like water waves, mathematicians and scientists picture sound waves as having high points and low points. The high points occur at each vibration when the air is compressed (see accompanying figure). The beginning to the end of each wave (*a* to *b*, *b* to *c*, *c* to *d*) is called a *cycle*. The number of waves or cycles in a given length of time, usually a second, is called the *frequency* of the sound.

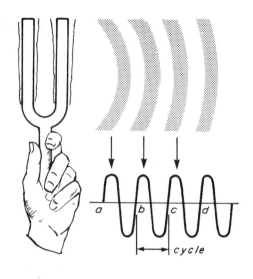

Each sound has its own frequency and, thus, its own unique curve. The greater the frequency, the higher the pitch. Curve A in this example has a higher pitch than Curve B. It has about two cycles for every one of the other curve. The loudness of a sound is related to the *altitude* or *height* of its curve. The altitude is half the distance between the high and low points of the curve. The louder the sound, the greater the altitude. In this example curve B corresponds to a louder sound than curve A.

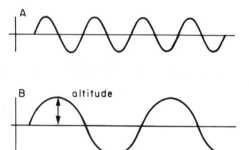

Mathematical curves which have the shapes of sound waves are called *sine curves.* The graph of the sine curve on the following page was plotted by using the circle at the right. For each number of degrees, *d,* from 0 to 360, the vertical distance from the circle to the horizontal diameter is the *sine of d.* For example, the sine of 60° is the vertical distance marked *y,* and this same distance is the height above 60 on the *x*-axis of the following curve. These *y* values are all less than 1 because the circle has a

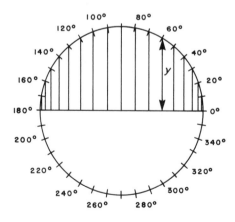

radius of 1 unit. For each of the degrees marked on this circle, the vertical distance to the diameter of the circle is the *y* value shown on the graph. The negatives of these distances are used from 180 to 360 degrees, which completes one cycle of the graph of the equation *y* = sine of *x*.

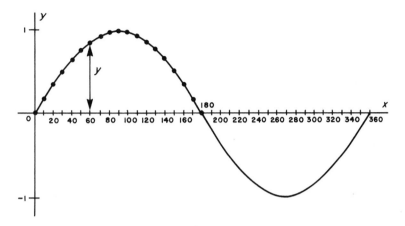

This curve continues without end to the right of 360 and to the left of the *y*-axis, repeating the basic cycle over and over. The vertical distances to this curve are obtained from the same circle by passing around the circle more than once for angles greater than 360 degrees and by passing around the circle clockwise for negative angles, just as in the case of polar coordinates. This means that the domain of the function for *y* = sine of *x* is all real numbers (positive and negative), and the range is all real numbers from negative 1 to positive 1.

SUPPLEMENT *(Activity Book)*

Activity Set 8.3 Curves and Moving Points (Spirals, cycloids, Aristotle's wheel, and sine curves)

Just for Fun: Curve Stitching (Cardioids and cusped curves)

Additional Sources
_____

Gardner, M. *The Unexpected Hanging.* New York: Simon and Schuster, 1969. "Spirals," pp. 103–13.

Linn, C.F. *The Golden Mean.* New York: Doubleday, 1974. pp. 79–87 (curves and music).

Maor, E. "The Logarithmic Spiral." *The Mathematics Teacher,* **67** No. 4 (April 1974), 321–27.

School Mathematics Study Group. *Mathematics and Living Things,* Teacher's commentary. Palo Alto: Stanford University, 1965. pp. 101–22 (examples of curves of growth).

Steinhaus, H. *Mathematical Snapshots.* New York: Oxford University Press, 1950. "Squirrels, Screws, Candles, Tunes, and Shadows," pp. 192–205.

"Before a symphony can be played by an orchestra, there must be collaboration of many parties—a composer, the makers and players of many instruments, and the conductor of the orchestra. All are, or have been, at work to produce—just a curve."

Sir James Jeans

EXERCISE SET 8.3

1. Stockholm's newly designed Sergel's Square contains oval rings of restaurants and shops located on curves which designer Piet Hein calls super-ellipses.* These curves are flatter on the ends than ellipses and more suited to the flow of traffic than an elliptical-shaped traffic circle. The coordinates of several points have been labelled on the superellipses in graphs I, II, and III. Try these coordinates in the equations in order to match equations in parts **a**, **b**, and **c** to their respective curve.

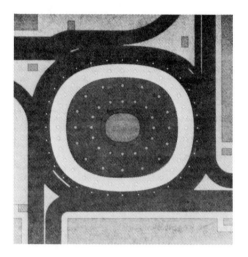

★ a. $\dfrac{x^4}{81} + \dfrac{y^4}{16} = 1$ b. $16x^4 + y^4 = 256$ c. $\dfrac{x^6}{64} + y^6 = 1$

I.

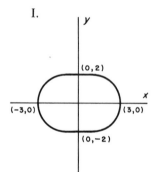

II.

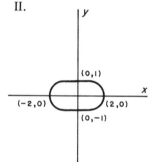

III.

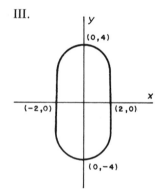

*M. Gardner, *Mathematical Carnival* (New York: Knopf, 1975), pp. 240–54.

★ 2. The intensity of light varies inversely as the reciprocal of the square of its distance from the source. For example, if the amount of light is 10 watts per square centimeter at a distance of 1 meter, it will be $10/2^2$ or 2.5 watts per square centimeter at a distance of 2 meters. For a distance of d meters and an intensity of I watts per square centimeter, the formula for this example is $I = 10/d^2$. Complete this table and sketch the graph of this equation on the coordinate axes that follow.

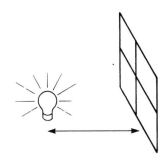

d	1	2	3	4	5	6
I	10	2.5				

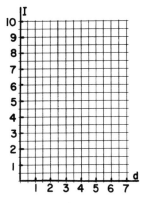

★ a. What happens to the graph as the distance becomes less than 1 meter?

b. What happens to the values of I as d increases?

★ 3. Complete the tables for $y = 2^x$, $y = 3^x$, and $y = 4^x$, and graph these equations on the same coordinate system. For positive values of x, what can you conclude about the graphs of exponential equations for increasing bases?

$y = 2^x$

x	-3	-2	-1	0	1	2	3
y							

$y = 3^x$

x	-3	-2	-1	0	1	2	3
y							

$y = 4^x$

x	-3	-2	-1	0	1	2	3
y							

★ 4. Complete the tables for each equation and plot these curves on the same coordinate system. What general statement can be made about the graphs of $y = a^x$ and $y = 1/a^x$?

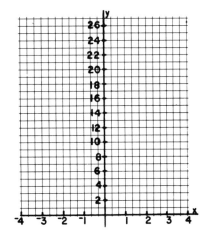

$y = 3^x$

x	-3	-2	-1	0	1	2	3
y							

$y = \dfrac{1}{3^x}$

x	-3	-2	-1	0	1	2	3
y							

5. The *Escherichia coli* is a bacterium which under ideal conditions doubles every 20 minutes. Beginning with a single bacterium, there will be 8 at the end of the first hour. Compute the numbers of these bacteria for each of the hours from 2 to 6 and graph the results. Connect these points with a smooth curve. What is the equation of this curve?

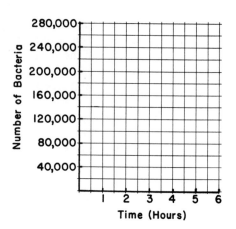

6. This graph shows the population growth in the United States since 1660. (The population for 1980 is estimated.)

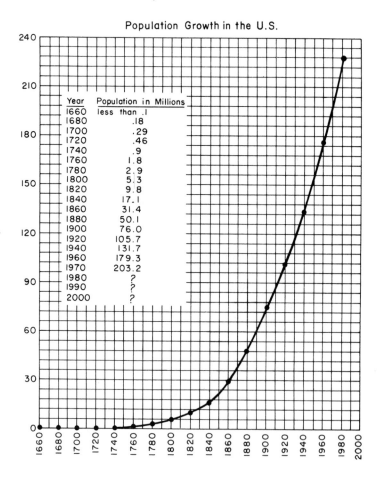

Population Growth in the U.S.

Year	Population in Millions
1660	less than .1
1680	.18
1700	.29
1720	.46
1740	.9
1760	1.8
1780	2.9
1800	5.3
1820	9.8
1840	17.1
1860	31.4
1880	50.1
1900	76.0
1920	105.7
1940	131.7
1960	179.3
1970	203.2
1980	?
1990	?
2000	?

★ a. Assuming that we continue in the exponential phase of this curve for the next 10 years, predict the population in the year 1990.

b. Predict the population in the year 2000 if the exponential phase continues that long.

c. If the 1980s mark the beginning of the stationary phase, estimate the upper limit of population for this country.

7. Graph the following coordinates, where the first number represents time in weeks and the second number represents weight, in grams, of a corn plant. Connect these points with an S-shaped curve. Mark the approximate locations of the lag, exponential, and stationary phases of this curve.

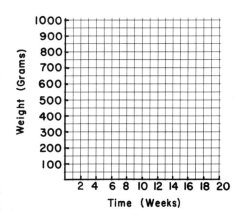

(1,10); (2,30); (3,50); (4,90); (5,140); (6,210); (7,280); (8,370); (9,450); (10,560); (11,640); (12,710); (13,760); (14,800); (15,840); (16,860); (17,880); (18,900)

★ a. How many weeks was this plant in its exponential growth phase?

b. How much weight did it increase during the exponential phase?

8. The numbers in this table are the polar coordinates of points on an Archimedean spiral. Continue the patterns of angles and distances until reaching a unit distance of 7. (The circles on the graph have increasing radii from 1 to 7.) Graph these points and sketch the spiral.

Angles	0°	20°	40°	60°	80°	100°	120°	140°	160°
Unit Distances	0	.2	.4	.6	.8	1	1.2	1.4	1.6

★ a. Approximately how many revolutions (turns through 360°) are there on this graph?

★ b. If you trace this curve at a constant speed, beginning at the center, what can be said about the time needed for each succeeding revolution?

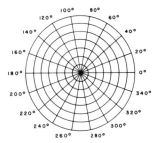

9. When two or more notes are combined, the curves of the individual notes merge into one curve. The five top curves (A, B, C, D, E) combine to form curve F below. This is what Sir James meant by the production of *just one curve* (see beginning of Exercise Set 8.3).

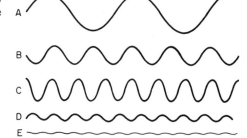

★ a. Which of the first five curves has the note with the highest pitch?

b. Which of the first five curves has the loudest note?

★ c. Which of the first five curves has the same number of cycles per second as curve F?

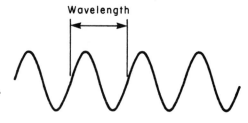

10. About 100 years ago the island of Krakatoa in the South Pacific was destroyed by a tremendous volcanic explosion which sent sound waves completely around the earth. The sound was heard in India and China, more than 5000 kilometers away.

★ a. Sound travels 335 meters per second. How long did it take the sound from Krakatoa to travel 5000 kilometers? Compute the answer to the nearest hour.

b. As a sound travels away from its source, what happens to the altitude of its curve?

★ c. Does the frequency of a curve change as its sound travels away from its source?

★ 11. The length of a wave is the horizontal distance from the beginning to the end of one cycle. You can figure out the wavelength of a sound if you know its frequency. Middle C on a piano has 262 vibrations per second. In 1 second, sound travels 335 meters. Therefore, the length of each wave or cycle is 335/262 or 1.28 meters. Compute each wavelength for the notes of the middle octave of the piano.

Wavelength

Note	C	D	E	F	G	A	B	C
Frequency	262	294	330	349	392	440	494	524
Wavelength (meters)	1.28							

★   a. How many times greater is the wavelength of middle C than that of the next higher C?

b. What is the relationship between the frequencies of these two C notes?

12. Most people can hear sound waves that range in wavelength from about 1 centimeter to 16 meters. The length of a wave divided into the distance that sound travels in 1 second (335 m), gives the frequency of the sound (see Exercise 11).

a. Draw a few cycles for a sound whose wavelength is 1 centimeter. How many of these cycles will occur in 1 second?

b. What is the frequency of sound whose wavelength is 16 meters?

★   c. About how many times greater is the frequency of the highest pitch we can hear as compared to the frequency of the lowest pitch we can hear?

13. Use this circle whose radius is 1 unit to approximate the sines of the following angles to the nearest tenth of a unit. To measure these distances use the scale on the radius of the circle. The sines of angles between 180° and 360° are negative.

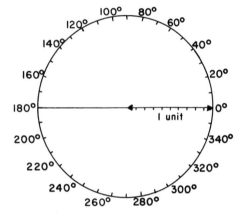

★   a. 30°   b. 60°

★   c. 150°   d. 240°

★   e. 260°   f. 320°

g. Compare the sines of some angles which are less than 180° with the sines of their supplements. What do you notice?

14. Plot the sines of the angles for each multiple of 30°: (0, 30, 60, 90, etc.) using the circle in Exercise 13. Sketch the graph of one cycle of the sine function.

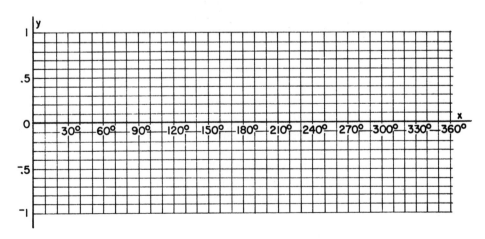

★ 15. *Calculator Exercise:* Calculators which have keys for the trigonometric functions have a key labelled "sin" for the sine function. The following steps show how to find the sine of 40°. (Before carrying out these steps, the Degree (D) and Radian (R) switch should be on D.)

Steps	Displays

    1. Enter 40   40.

    2. | sin | .64278760

★   a. Find the sine of the following angles. What happens to the sine of an angle as it gets close to 90°? Check your answers to this question by looking at the circle in Exercise 13.

$$20° \quad 40° \quad 60° \quad 80° \quad 86° \quad 89° \quad 89.5°$$

★   b. The sums of the following pairs of angles are 180°. Find the sines of these angles. Make a conjecture about the sines of supplementary angles.

$$40°,140° \quad 55°,125° \quad 10°,170°$$

★   c. The following pairs of angles differ by a multiple of 360°. Find the sines of these angles. Make a conjecture about the sines of angles which differ by a multiple of 360°.

$$10°,370° \quad 30°,390° \quad 90°,450° \quad 270°,630°$$

16. In the early 1970s a group of **MIT** scientists created a computer model acclaimed as able to forecast the world population growth. This World Model takes into account such factors as pollution, food supplies, and natural resources.*

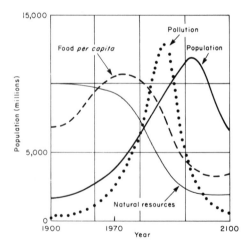

★   a. What is this graph's domain? What part of this domain represents prediction of the future?

    b. In 1900 all but one of these curves were increasing. Which curves are increasing and which are decreasing in the year 2000?

★   c. What are the approximate intervals of years during the exponential phases when the population and pollution graphs are increasing? What is happening to the food supply during these intervals?

*Donnella H. Meadows et al., *Limits to Growth* (New York: Universe Press, 1972).

d. What year does the population begin to decline? According to this graph, what causes this decline?

17. *Fraser Spiral:* The background in this photo produces an illusion called the Fraser spiral. Can you explain what is wrong with this "spiral"? (*Hint:* Try tracing it.)

18. *Calculator Exercise:* The exponential equation $y = e^x$ and variations of this equation occur frequently in analyzing rates of growth and decay. On a calculator which has a key for the exponential function, e^x can be found by two steps. The second step here shows e^2 to 9 decimal places.

Steps	Displays
1. Enter 2	2.
2. $\boxed{e^x}$	7.389056099

★ a. The number e is irrational. Use your calculator to find the first few decimal places in e by evaluating e^1 (e to the first power).

★ b. For negative exponents, e^x is less than 1. What are the first few decimal places in e^{-3}? (*Hint:* $e^{-3} = 1/e^3$.)

c. Find e^x for each of the values of x which are marked on the x-axis. Round off these numbers to the nearest tenth. Plot the pairs (x,y) and connect these points to form the graph of $y = e^x$.

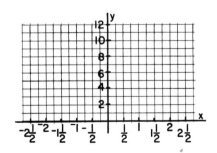

CHAPTER NINE

MOTIONS IN GEOMETRY

Mathematics is an obscure field, an abstruse science, complicated and exact; yet so many have attained perfection in it that we might conclude almost anyone who seriously applied himself would achieve a measure of success.

Cicero

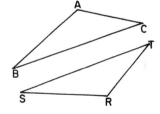

"WE'RE HERE TO FIX THE COPIER!"

9.1 CONGRUENCE MOTIONS

There is an old belief that everyone has a "double"—someone who looks exactly like him or her, somewhere in the world. Each man in this cartoon has two doubles, which seems appropriate for a team of copier repairmen. Copy machines have the ability to quickly reproduce words and figures which have the same *size* and *shape* as the original. Such figures are said to be *congruent*. Intuitively, we say that two plane figures are *congruent* if one can be *moved* onto the other so that they coincide. The idea of motion or movement is one of the more important concepts in mathematics. The particular motions that are associated with congruence will be studied in this section.

Mappings   If triangle *ABC* is traced on paper and flipped over, it can be placed on triangle *RST* so that the points of each triangle coincide. The correspondence of point *A* with *R*, *B* with *S*, and *C* with *T* is indicated by

$$A \leftrightarrow R, \quad B \leftrightarrow S, \quad C \leftrightarrow T$$

By placing triangle *ABC* onto triangle *RST*, each point on the first triangle corresponds to exactly one point on the second triangle. This one-to-one correspondence of points is a special type of function. In Section 8.1 there are functions of numbers in which each number in one set corresponds to one and only one number in a second set. Similarly, there are functions of points for which each point in one set corresponds to a unique point in a second set. In geometry, functions are called *mappings* or *transformations*. The mapping of triangle *ABC* to triangle *RST* is denoted by $\triangle ABC \leftrightarrow \triangle RST$. This notation indicates that the following sides and angles are matched with each other.

Corresponding Sides	Corresponding Angles
$\overline{AB} \leftrightarrow \overline{RS}$	$\angle ABC \leftrightarrow \angle RST$
$\overline{BC} \leftrightarrow \overline{ST}$	$\angle BCA \leftrightarrow \angle STR$
$\overline{AC} \leftrightarrow \overline{RT}$	$\angle CAB \leftrightarrow \angle TRS$

These pairs of sides and angles are called *corresponding sides* and *corresponding angles* for the mapping $\triangle ABC \rightarrow \triangle RST$. For each point in triangle ABC, the point it is *mapped to* is called its *image*. This terminology will be used to examine three mappings and their corresponding motions.

Translations   A translation is a special kind of mapping which can be described by a sliding motion. Each point is moved the same distance and in the same direction. The translation on the right maps A to A', B to B', C to C', $\overline{BC}$ to $\overline{B'C'}$, and pentagon K to pentagon K'. This translation is completely determined by the point A and its image A'. That is, given any point X we can find its image X' by moving in the *same direction* as from A to A' and the *same distance* as AA'.

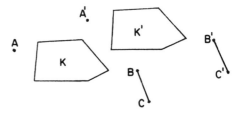

Translations occur for space figures as well as for plane figures. Just as in the case of two-dimensional figures, a translation in three dimensions is described by a sliding motion of every point in the same direction and for the same distance. This photo shows a sliding motion of the earth's crust, which geologists call a block fault. The arrow points to one side of the fault along which the earth's crust has been displaced.

Fault line showing displaced rock

Reflections   A reflection about a line is a mapping which can be described by folding. If this page is folded about line L, each point will coincide with its image. E will be mapped to E', F to F', $\overline{EF}$ to $\overline{E'F'}$ and figure M to figure M'. Since point S is on L, it does not move for this mapping. S and all other points on L are called *fixed points* for the reflection about L.

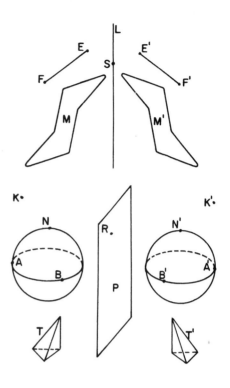

Reflections in space take place about planes. Each point to the left of plane P has a unique image on the right side of P. The sphere is mapped to the sphere, point K to K', tetrahedron T to tetrahedron T', etc. Point R and all other points on the plane are *fixed points* of the mapping. That is, each point on the plane is its own image. Reflections in space can be illustrated by mirrors. If plane P is replaced by a mirror, so that the figures to the left of the mirror are reflected, their images will appear to be in the positions of the figures on the right side of the mirror.

Surprisingly clear reflections can be created by mirror images from pools. Pick out some points on the building shown in this photo and their images.

Model of the New Delhi United States Embassy, 1959

In the previous mappings about line L and plane P, each point and its image are on lines which are perpendicular to the line or plane of reflection. For example, $\overline{EE'}$ is perpendicular to L, and $\overline{NN'}$ is perpendicular to P. Furthermore, each point is the same distance from the line or plane as is its image. These two conditions hold for all reflections.

Rotations   The third type of mapping is a rotation. To illustrate a 90° rotation about O, place a piece of paper on this page and trace $\overline{FG}$ and quadrilateral $ABCD$. Hold a pencil at point O and rotate the paper 90° in a clockwise direction. (A 90° rotation can be determined by placing the edges of the paper parallel to the edges of this page.) Each of

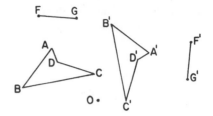

the points you trace will coincide with its image after this rotation. Quadrilateral $ABCD$ is mapped to quadrilateral $A'B'C'D'$, and $\overline{FG}$ is mapped to $\overline{F'G'}$. The only fixed point for this mapping is point O.

Space figures are rotated about lines. If the sphere shown here is rotated 90° about the vertical axis through N and S, H will be mapped to H' and B to B'. Each point will be mapped to a new location except for the points N and S, which remain fixed.

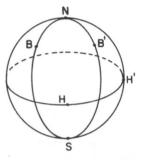

The restaurant and observation deck at the top of the 60-story Space Needle in Seattle, Washington, rotates once every 60 minutes. Each point on this moving structure traces out a circular path during one complete revolution. These moving points are constantly changing their locations and being mapped to each other. For example, consider the locations of points at 10:00 A.M. and again at 10:15 A.M. During this 15-minute interval, each point rotates 90° and finishes in a position which was previously occupied by another point of the structure.

Space Needle, Seattle, Washington

Composition of Mappings   This wood-engraving by M.C. Escher combines translations and reflections. The white swan W is mapped onto the black swan B by a translation followed by a reflection. This mapping can be carried out by tracing swan W and its center line on a piece of paper. Then slide the paper diagonally to swan B so that the two center lines coincide. The swan which was traced can how be made to coincide with swan B by a reflection about the center line. A translation followed by a reflection is called a *glide reflection.*

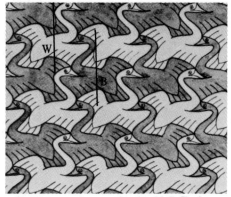

Swans, wood-engraving by M.C. Escher

When one mapping is followed by another, it is called a *composition* or *product* of mappings. Any combination of translations, reflections, or rotations can be used. In the figure shown here, a 90° clockwise rotation about point O is followed by a translation of each point three spaces to the right. Triangle ABC is mapped to triangle $A'B'C'$ by the rotation, and then the translation maps triangle $A'B'C'$ to triangle $A''B''C''$. The composition of the rotation and translation is the mapping which takes triangle ABC to triangle $A''B''C''$.

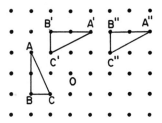

Two-dimensional patterns such as those on wallpaper and tiled floors are systematically created by translations, reflections, rotations, and compositions of these mappings. This is accomplished by beginning with a basic figure, such as the mushroom-shaped figure in the upper left square of this grid. To obtain the position of the figure in each square of the rows, the mushroom will be rotated 180° about the center of the edge of an adjacent square. To go down the columns from square to square, each mushroom will be reflected about the common edge of the square.

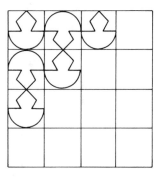

The Moors of Spain used combinations of mappings to generate two-dimensional patterns. In commenting about their work, Escher said, "What a pity that Islam did not permit them to make 'graven images.' They always restricted themselves, in their massed tiles, to designs of an abstract geometrical type."* The following patterns from the Alhambra were sketched by Escher. The first pattern was obtained by merely translating

*M.C. Escher, *The Graphic Works of M.C. Escher* (New York: Random House, 1967).

the basic figure (shown above the pattern) across the rows and down the columns. In the second pattern the basic figure is rotated 90° and then translated both down and to the right to the adjacent figures. The third pattern is generated by the same transformations as used in the rows and columns of the mushroom pattern, described in the preceding paragraph.

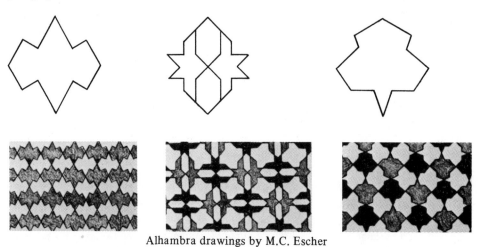

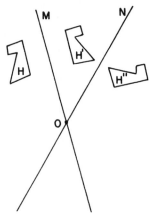

Alhambra drawings by M.C. Escher

The composition or product of mappings is similar to an operation on numbers. For example, the product of two integers is always another integer (closure property), and the composition of two mappings is always another mapping. The composition of some pairs of mappings is quite easy to determine. A 30° rotation followed by a 45° rotation can be replaced by a 75° rotation. Two

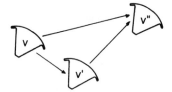

translations can always be replaced by one translation, as illustrated by the figure shown above. The first translation maps V to V' and the second maps V' to V''. The result of these two translations can be accomplished by the single translation which maps V to V''.

The situation for reflections is more interesting as shown by the figure on the right. For a reflection about line M, figure H is mapped to H'. Then H' is mapped to H'' by a reflection about line N. These two reflections can be replaced by a single rotation about point O which maps H to H''.

Compositions of different types of mappings, such as a rotation followed by a translation, or a reflection followed by a rotation, can also be replaced by a single mapping. It can be proven that the composition of any two of the three mappings (rotation, translation, or reflection) is a rotation, translation, reflection, or glide reflection.

Congruence   Translations, reflections, and rotations all have something very important in common. If A and B are any two points and A' and B' are their respective images, then the distance between A and B is the same as the distance between A' and B'. That is, for these mappings the lengths of line segments are the same as the lengths of their images. Such mappings are called *distance preserving* or *isometric*. What this means intuitively is quite simple: As figures are rotated, translated, or reflected, their size and shape do not change.

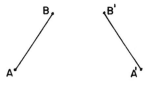

Up to this point we have thought of two figures as being congruent if they have the same size and shape or if one can be made to coincide with the other. For example, figure W is congruent to figure Z because W can be placed on Z so that they coincide. Thinking of congruence in terms of coinciding figures is fine for an introductory notion, but this concept can be defined more explicitly in terms of mappings. In this example figure W is congruent to figure Z because W can be mapped to Z by a glide reflection (a translation to the right and a reflection about line L).

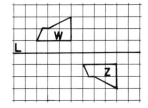

The need for a more careful definition of congruence becomes evident when we consider congruence in three dimensions. For example, we need a way of defining congruence for these two kitchen grinders, but it does not make sense to say that they coincide.

A suitable definition of congruence for both plane and space figures can be given in terms of mappings.

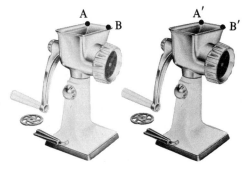

Definition: Two geometric figures are congruent if and only if there exists a mapping (translation, reflection, rotation, or glide reflection) of one figure onto the other.

This definition says that for each of these four mappings a figure is congruent to its image. Conversely, if two figures are congruent, one can always be mapped to the other by one of these mappings. This definition gives us a way of viewing congruence of both plane and space figures. The plane figures W and Z shown above are congruent because there is a glide reflection mapping one figure to the other. The two grinders are congruent because there is a translation which maps one to the other. Each point on the left grinder is mapped to a corresponding point on the right grinder. The distances between any two points on the left grinder, such as points A and B, and their images, A' and B', are equal. Defining congruence in terms of distance-preserving mappings is the mathematical way of saying that two objects have the same size and shape.

Mapping Figures onto Themselves   Michael
Holt and Zoltan Dienes in their book
Let's Play Math describe the following
scheme for coloring pictures of a house.*
Cut out a square and color the corners
with four different colors. Both the front
and back sides of each corner should have
the same color. Place the square on a
piece of paper and draw a frame around it.
At each corner of the frame write (or
draw pictures for) the words "wall,"
"roof," "door," and "window." This is
called the Rainbow Toy. The different
positions in which the square can be placed

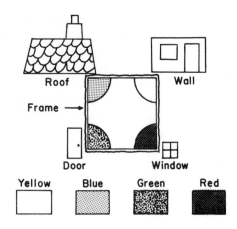

on the frame determine different arrangements of colors for the wall, roof, door, and
window of the house. For the position of the square which is shown here, we get the colors
for house 1. Color schemes for houses 2, 3, and 4 are obtained by rotating the square into
three different positions. By flipping the square over, there are four more positions for
the color schemes for houses 5 through 8.

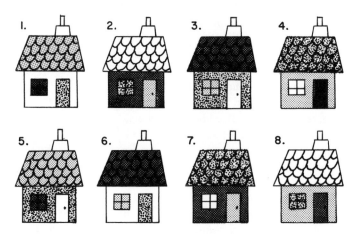

*M. Holt and Z. Dienes, *Let's Play Math* (New York: Walker and Company, 1973), pp. 88–94.

The Rainbow Toy is an elementary way of illustrating the eight different mappings of a square back onto itself. Four of these are rotations of 90°, 180°, 270°, and 360°. The other four mappings are reflections. The reflections about D_1 (the diagonal from upper left to lower right) interchanges corners b and d but leaves corners c and a in the same location. There is also a reflection in diagonal D_2 (the diagonal from upper right to lower left) and reflections about the horizontal and vertical lines, H and V.

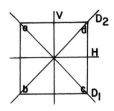

If we cut out a square and draw a frame about it, as for the Rainbow Toy, some interesting observations can be made about the composition of the eight mappings. Let's label the corners of the square and the frame by a, b, c, and d. The position shown here, with a in the a corner of the frame, b in the b corner, etc., will be called the Initial Position of the square. Now, if the square is rotated 90° clockwise and then reflected about the horizontal axis, b and d will end up in the same corner in which they started, but c and a will be interchanged.

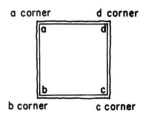

a corner   d corner

b corner   c corner

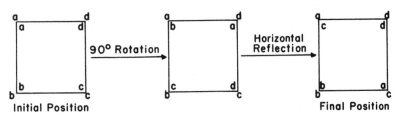

Initial Position   90° Rotation   Horizontal Reflection   Final Position

We can get the same Final Position by beginning with the Initial Position and reflecting about D_2 (the diagonal from upper right to lower left). That is, the composition of a 90° rotation followed by a horizontal reflection is the same as a reflection about D_2.

Something surprising happens when we reverse the order of these two mappings. The following figures show that by first carrying out the horizontal reflection and then a 90° rotation, we do not get the same Final Position as we did in the previous example.

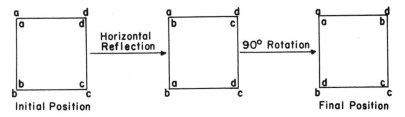

Initial Position   Horizontal Reflection   90° Rotation   Final Position

This time c and a are in the same position in which they started, but b and d have changed places. The composition of these two mappings is the same as a reflection about D_1 (the diagonal from upper left to lower right). These examples show that the order in which mappings are carried out is significant. In other words, the composition of mappings is not commutative.

SUPPLEMENT *(Activity Book)*

Activity Set 9.1 Translations, Rotations, and Reflections (Illustrations of mappings and their compositions by geometric models and circular geoboards)

Additional Sources

Bell, S. *Transformations and Symmetry.* Boston: Houghton Mifflin, 1968.

Bruni, J.V. *Experiencing Geometry.* Belmont, California: Wadsworth, 1977. "Geometric Transformations," pp. 185–209.

Bruni, J.V. and H.J. Silverman. "Making patterns with a square." *The Arithmetic Teacher,* 24 No. 4 (April 1977), 265–72.

Haak, S. "Transformation Geometry and the Artwork of M.C. Escher." *The Mathematics Teacher,* 69 No. 8 (December 1976), 647–52.

Jackson, S. B. "Applications of Transformations to Topics in Elementary Geometry: Part I." *The Mathematics Teacher,* 68 No. 7 (November 1975), 554–62.

Johnson, M.L. "Generating patterns from transformations." *The Arithmetic Teacher,* 24 No. 3 (March 1977), 191–94.

Troccolo, J. "A Strip of Wallpaper." *The Mathematics Teacher,* 70 No. 1 (January 1977), 55–58 (mappings are used to construct wallpaper).

EXERCISE SET 9.1

1. This silhouette of lines and angles was produced by joining six photographs of a construction staging, side by side.

   ★ a. Are these six photographs congruent?

   b. Are the six photographs of the staging congruent?

   ★ c. What mapping is suggested by this picture?

2. The design on this nineteenth century quilt also occurs in the fourteenth century Moorish palace Alhambra. Beginning with the black figure in the upper left corner, the top row can be generated by a sequence of 180-degree rotations.

   a. Locate the centers of rotation for these mappings.

   ★ b. Beginning with the figure in the upper left corner, what mapping can be carried out to generate the left column of figures?

Arabic latice patchwork quilt, ca. 1850

Collections of Greenfield Village and the Henry Ford Museum, Dearborn, Michigan

3. The importance of triangles to architecture is due to a basic mathematical
fact, which is not true for polygons
with more than three sides. The following questions will help you to see
why triangles are common in
construction.

a. Construct two noncongruent quadrilaterals whose sides are congruent
to the following four line segments.

——————————————
——————————————
——————————————
—————————

★ b. How many noncongruent quadrilaterals can be constructed whose
sides are congruent to the four line
segments in part **a**? (*Hint:* Imagine
changing the shape of a quadrilateral
formed by linkages; see page 166.)

Construction of Lundolm Field House,
University of New Hampshire

c. Construct a triangle using segments that are congruent to these line segments.

———————————————   —————————   ———————————————

★ d. Can you construct another triangle from the line segments in part **c** which is not
congruent to your first triangle?

e. A triangle is the basic figure for strengthening buildings, bridges, and other structures because triangles hold their shape better than polygons with four or more
sides. Use your results from parts **a** through **d** to explain why this is true.

4. If a triangle is constructed from three
given line segments, it will have a
unique size and shape. That is, another
triangle which uses the same line seg-
ments will be congruent to the first one. This is not true for triangles that are constructed from two given line segments and an angle. Construct two noncongruent
triangles which contain two sides and an angle congruent to those given here.

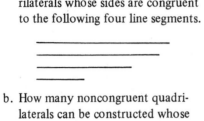

5. For the translation which takes A to A', sketch the
image of the hexagon.

★ a. The line through point D and its image is parallel
to $\overleftrightarrow{AA'}$. Is this true for every point on the hexagon
and its image?

b. If E' and G' are the respective images of E and
G, how does the length $E'G'$ compare with the
length EG ?

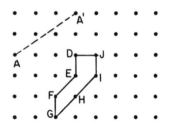

★   c. Compare the area of the hexagon with the area of its image.

6. Sketch the image of the pentagon for a reflection about line L.

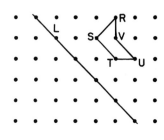

★   a. If R' is the image of R, what is the measure of the angles formed by the intersection of $\overleftrightarrow{RR'}$ and L?

    b. If U' is the image of U, how does the distance from U to L compare with the distance from U' to L?

★   c. What are the fixed points for this mapping?

7. Sketch the image of the quadrilateral for the 90° rotation about O which takes A to A'. (Trace the figure onto a piece of paper and rotate.)

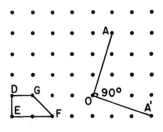

★   a. If E' is the image of E, what is the measure of $\angle EOE'$?

    b. If G' is the image of G, how does the length of EG compare with the length of $E'G'$?

    c. Are there any fixed points for this mapping?

8. Map quadrilateral $ABCD$ to quadrilateral $A'B'C'D'$ by the translation which moves point A to A'. Then map quadrilateral $A'B'C'D'$ to quadrilateral $A''B''C''D''$ by the translation which takes A' to A''.

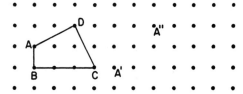

★   a. The translation which maps A to A' can be described by "over 4 and down 1." The translation from A' to A'' is "over 2 and up 2." Describe the translation which maps A to A''.

    b. What is the image of quadrilateral $ABCD$ for the composition of the following two mappings: the translation which takes A to A'; followed by the translation which takes A' to A?

9. Reflect pentagon *RSTUV* about line *M* and then reflect its image about line *N*.

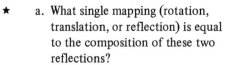

★ a. What single mapping (rotation, translation, or reflection) is equal to the composition of these two reflections?

b. Let R' be the image of R for the reflection about M, and R'' be the image of R' for the reflection about N. Compare the distance from R to R'' with the distance from line M to line N. What relationship do you find? Will this hold for other points and their images?

10. The hexagons in the accompanying sketch can be mapped to each other by compositions of reflections about lines *M* and *N*.

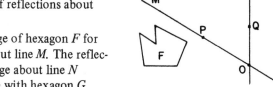

a. Sketch the image of hexagon *F* for a reflection about line *M*. The reflection of this image about line *N* should coincide with hexagon *G*.

★ b. There is a rotation about point *O* which will map figure *F* to figure *G*. How is the number of degrees in this rotation related to $\angle POQ$?

11. There are two pairs of congruent polygons in the next diagram.

a. Describe a single mapping (reflection, translation, or rotation) which will map one parallelogram to the other.

b. Describe two mappings whose composition will map one hexagon to the other. Show all lines of reflection and points of rotation.

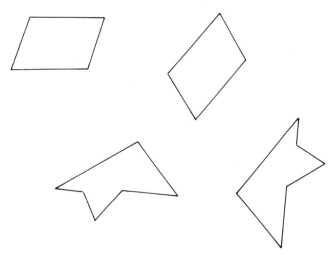

12. Complete the patterns for the grids by carrying out the mappings on the basic figure.

a.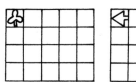

b.

★   a. Rows: Rotate 180°.
Columns: Reflect about edge of square.

b. Rows: Reflect about edge of square.
Columns: Reflect about edge of square.

★ 13. Mark off a square region which can be used as a basic figure to generate this wallpaper pattern. Describe the mappings for obtaining the rows and columns.

14. A translation maps P with coordinates ($^-2,1$) to P' with coordinates (3,2). Graph these points and draw an arrow from P to P' to indicate the length and direction of the translation.

a. Sketch the image of $\triangle ABC$ for this translation.

★   b. If A', B', and C' are the images of A, B, and C, what are their coordinates?

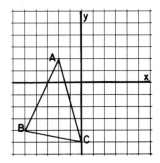

15. Draw the image of quadrilateral $DEFG$ for a reflection about the x-axis. Label the images of these vertices as D', E', F', and G'.

a. What are the coordinates of D', E', F', and G'?

b. Reflect quadrilateral $D'E'F'G'$ about the y-axis and label its vertices by D'', E'', F'', and G''. What are the coordinates of the vertices of this figure?

c. What single rotation will map quadrilateral $DEFG$ to quadrilateral $D''E''F''G''$?

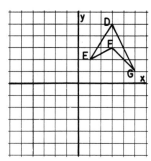

16. Mappings are sometimes given by describing what will happen to the coordinates of each point. For example, the mapping $(x,y) \rightarrow (x + 2, y - 3)$ is a translation which maps each point 2 units to the right and down 3 units. To find the image of a particular point, such as (1,7), substitute these values for x and y into $(x + 2, y - 3)$. In this case, (1,7) maps to (3,4). Find the image of each figure for the following mapping. Label each mapping as a translation, rotation, reflection, or glide reflection.

★   a. $(x,y) \rightarrow (x - 2, y + 1)$ b. $(x,y) \rightarrow (x + 1, {}^-y)$ c. $(x,y) \rightarrow ({}^-x, {}^-y)$

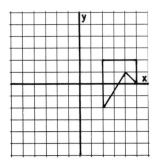 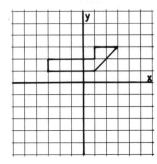 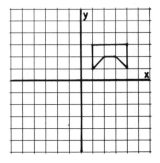

17. *Tic-Tac-Toe:* Just about everyone is familiar with the game of Tic-Tac-Toe. The first player to place three X's or O's in a row, column, or diagonal of a 3 by 3 grid wins the game.

a. There are nine ways to place the first X (or O) on the grid. The two top grids shown here are congruent. How many noncongruent ways are there of making the first move? Sketch them.

b. After the first move (an X in the upper left corner in this example) there are only eight ways to place the O, but some of these, such as the two lower grids, are congruent. Sketch all the noncongruent moves for the second turn.

c. Only one of the moves in part **b** can prevent the X player from winning. Which one?

18. *Isolations:* * The following types of number puzzles are called *isolations*. The object is to place consecutive whole numbers $(1, 2, 3, \ldots, n)$ in the circles so that no two consecutive numbers are in circles that are connected by a line. To put this another way, consecutive numbers should be *isolated*. Find a solution for the following isolations.

★   a.
Four-in-a-row

b.
Five-in-a-row

*More of these puzzles are described by D. T. Piele, "Isolations," *The Mathematics Teacher* 67 No. 8 (December 1974), 719-22.

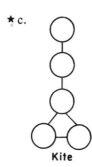

★ c.

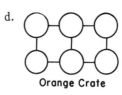

d.

Orange Crate

Kite

e. For each figure in parts **a**, **b**, **c**, and **d**, a reflection about lines of symmetry will change the locations of the numbers. Find another solution for figures **b, c,** and **d** which cannot be obtained from the above solutions by a reflection or a rotation. (*Note:* There are 1 solution for Four-in-a-row; 7 solutions for Five-in-a-row; 2 solutions for the Kite; and 6 solutions for the Orange Crate, if reflections and rotations are not allowed.

19. *Puzzle:* This arrangement of eight circles is another isolation number puzzle. It is from *An Unexpected Hanging,* by Martin Gardner. If we agree that all solutions which can be obtained from each other by rotations and reflections of this diagram are the same, then there is only one solution. Find this unique solution.

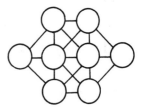

Model of Fort Wayne, Indiana, in environmental wind tunnel

9.2 SIMILARITY MOTIONS

Our intuitive notion of similar figures is "same shape" with no special requirement on size. Such figures occur in a wide variety of ways. An indispensable application in the design of large objects is the use of models. Research in ship and plane design, for example, is routinely carried out by testing models in ship-model towing tanks and wind tunnels. The scale model of Fort Wayne, Indiana, shown in the previous photo, is being tested in an environmental wind tunnel in the Fluid Dynamics and Diffusion Laboratory at Colorado State University.

Virtually all manufactured objects, whether large or small, are designed and constructed from scale drawings. A drawing or plan may be scaled down or scaled up, depending on the size of the original object. This picture shows a computer-on-a-chip and its scaled-up enlargement.

All of our familiar optical instruments involve similar figures. When you look through a magnifying glass or a microscope you see an image that is similar to the original but scaled up. Telescopes, binoculars, cameras, and even our eyes all produce images that are similar to objects before them.

Enlargement of computer-on-a-chip, actual chip on finger

Shadows and Mappings   Similar figures can be created by lights and shadows. Hold a flat object perpendicular to a flashlight's rays, and the light will produce a shadow similar to that of the original figure. Because the light is from a small bulb and spreads out in the shape of a cone, the shadow is larger than the object. The sun, on the other hand, is so large that its rays are almost parallel as they strike the earth. If a flat object is placed perpendicular to the sun's rays and its shadow is cast onto a surface that is also perpendicular to the sun's rays, the shadow will be congruent to the original figure.

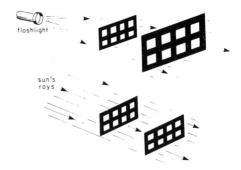

Light rays and shadows are like mappings and their images. For each point on the object there is a corresponding "shadow point" which is its image. The rays of light are like lines projecting from a central source to the object and then to its image. This type of mapping is illustrated here by quadrilateral *ABCD* and its image. Point *O*, which is taking the place of the light source, is called a *projection point*, and the mapping of one figure to the other is called a *similarity mapping*. Each point on quadrilateral $A'B'C'D'$ is twice as far from point *O* as is its corresponding point on quadrilateral *ABCD*. For instance, OA' is twice *OA*, OB' is twice *OB*, etc. Because of this relationship between points and their images, this mapping is said to have a *scale factor* of 2.

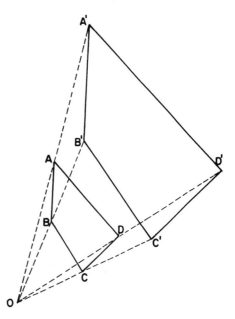

Scale Factors   If the scale factor for a similarity mapping is greater than 1, the image is an enlargement of the original figure. The following mapping from figure *XZW* to figure $X'Z'W'$ has a scale factor of 3. Each image point of figure $X'Z'W'$ is 3 times further from *O* than its corresponding point on figure *XZW*. That is, $OX' = 3$ times *OX*, $OZ' = 3$ times *OZ*, etc.

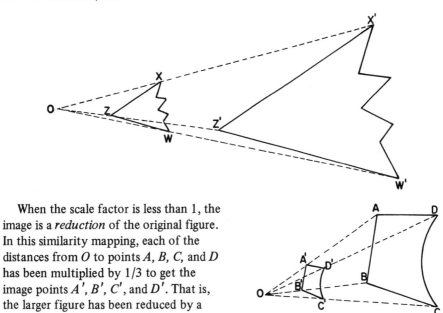

When the scale factor is less than 1, the image is a *reduction* of the original figure. In this similarity mapping, each of the distances from *O* to points *A, B, C,* and *D* has been multiplied by 1/3 to get the image points $A', B', C',$ and D'. That is, the larger figure has been reduced by a scale factor of 1/3.

It is even possible for a scale factor to be negative. In this case the original figure and its image will be on opposite sides of the projection point, and the image will be "upside down" from the original figure. The larger flag shown here is projected through point O to the smaller flag by a scale factor of $^-1/2$. In particular, A is mapped to A', and G is mapped to G'. As in the previous examples, the scale factor determines the size of the image. For a

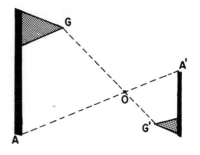

scale factor of $^-1/2$, each image point is half as far from the projection point as is its corresponding point (and these points are on opposite sides of the projection point). For example, OG' is half of OG, and OA' is half of OA.

The lenses of our eyes and of cameras invert the images of scenes much like a similarity mapping with a negative scale factor. These lenses are the projection points, and the scene is produced upside down on the retinas of our eyes and the film of a camera.

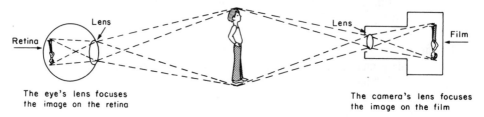

The eye's lens focuses the image on the retina

The camera's lens focuses the image on the film

Similarity Mappings   The pantograph is a mechanical device for enlarging and reducing figures which employs the principle of a similarity mapping. It can easily be constructed from four strips of wood which are hinged at points A, B, C, and P so that they move freely. Point O is the projection point and should be held fixed. As point P traces the original figure, a pencil at P' (its image) traces the enlargement. Since P' is twice as far from the projection point O as point P, the scale factor for this mapping is 2. In order to reduce a figure, the pencil is held at P, and P' is moved around the original figure. In this case the scale factor will be $1/2$. The pantograph can be changed for different scale factors by adjusting the settings at points B and C.

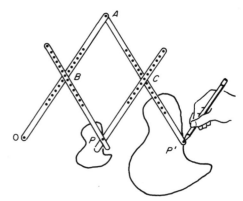

Pantograph, for reproducing similar figures

The ability of the pantograph to reproduce figures with the same shape is due to two important properties of similarity mappings. First, the sizes of angles do not change. Consider this similarity mapping of triangle *DEF*. Angle *EDF* is congruent to $\angle E'D'F'$, and $\angle DEF$ is congruent to $\angle D'E'F'$. In general, all angles are congruent to their images under a similarity mapping.

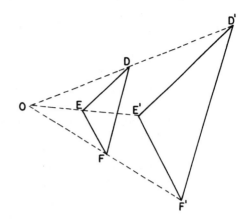

Second, the lengths of line segments all change by the same factor. Each side of triangle $D'E'F'$ is twice as long as its corresponding side in triangle *DEF*. In general, if k is the scale factor of a similarity mapping, each line segment will have an image which is k times as long. Another way of stating this second condition is to say that similarity mappings preserve ratios. For instance, the ratio DE/EF is equal to $D'E'/E'F'$. This is true because both $D'E'$ and $E'F'$ are twice as long as DE and EF.

In our study of congruence the mappings we used (rotations, translations, and reflections) preserved size and shape. Similarity mappings, on the other hand, do not preserve size but they do preserve shape. What we have been intuitively referring to as the "same shape" can now be made more precise by the following definition of similar figures.

Definition: Two geometric figures are similar if and only if there exists a similarity mapping of one figure onto the other.

This definition also holds for space figures. These two boxes are similar because the smaller one can be mapped onto the larger one by using projection point *O*, inside the small box, and a scale factor of 3. The larger box has a width, length, and height which are each three times the corresponding dimension of the smaller box. As with plane figures, a scale factor greater than 1 produces an enlargement, and a scale factor between *O* and 1 reduces the original figure.

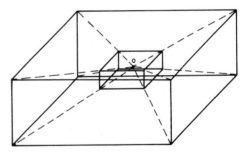

Similar Polygons   There are many applications of
similar figures for which it is inconvenient
or even impossible to set up similarity mappings by using projection points. A solution
to this problem is to produce similar figures
from measurements of angles and distances.
The construction of maps and charts is one
example. The polygon on this chart connects
5 points and is similar to the large polygon
over the water which connects the actual
landmarks. These positions on the chart
could have been plotted by measuring the
5 vertex angles and 5 distances between
these islands. The following theorem
shows that these measurements are all
that is necessary to obtain similar polygons.

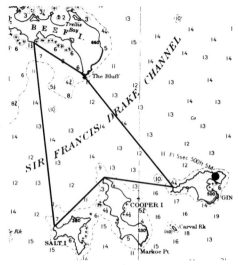

Virgin Islands, West Indies,
scale factor 139,000

Theorem: **Two polygons are similar if there exists a one-to-one correspondence between
their vertices such that:**

(1) Corresponding angles are congruent.
(2) Lengths of corresponding sides have the same ratio.

Both conditions (1) and (2) are usually
necessary for two polygons to be similar.
A square and a rectangle have congruent
angles (all 90°) but their sides do not
satisfy condition (2). Therefore, they are
not similar. The rectangle and the parallelogram satisfy condition (2) because
the sides of one are equal to the corresponding sides of the other, but the
angles in the parallelogram are not congruent to those in the rectangle. Therefore, these two figures are not similar.

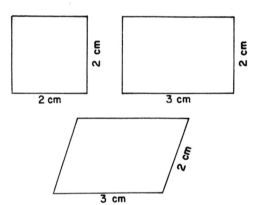

When the two polygons are triangles they will be similar if either condition (1) or condition (2) of the preceding theorem is satisfied. Triangles *KLM* and *K'L'M'* satisfy condition (2) because the sides of the larger triangle are two times longer than the corresponding sides of the smaller triangle. Therefore, we can conclude that the two triangles are similar and that the vertex angles at *K*, *L*, and *M* are congruent to the corresponding angles at *K'*, *L'*, and *M'*.

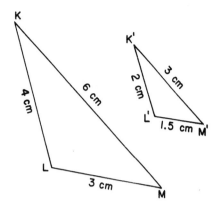

On the other hand, if two triangles satisfy condition (1), they are similar. For example, these two triangles are similar because the angles of one are congruent to the angles of the other. In fact, because there are 180° in every triangle, it is necessary only to find two angles from one triangle which are congruent to two angles from the other. Why? When this condition is satisfied we can conclude that the lengths of the corresponding sides have the same ratio.

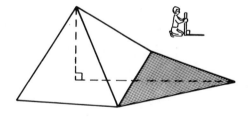

These facts about similar triangles have many applications. In particular, they form the basis of indirect measurement.

Indirect Measurement   The Greek mathematician Thales (ca. 600 B.C.) used the shadow of a staff to measure the heights of the Egyptian pyramids. This is the earliest known example of determining the heights of objects by indirect measurement. By holding a stick vertically to the ground, the stick and its shadow form a small right triangle that is similar to a right triangle formed by the pyramid and its shadow.

Here is an example of how this method can be used to find the height of a tree. Both the stick and the tree in the next figure form a right angle with the ground. Furthermore, ∠*CAB* and ∠*STR* are both congruent because they are formed by the sun's rays. With this information about the angles, we know that the two triangles are similar, with $\overline{AC}$ corresponding to $\overline{TS}$, $\overline{AB}$ to $\overline{TR}$, and $\overline{CB}$ to $\overline{SR}$. Therefore, the ratios of the corresponding sides are equal. In particular, $TR/AB = SR/CB$. Three of these distances, *TR*, *AB*, and *CB*, can be measured directly and are given in the diagram. The height of the tree, *SR*, can be computed from these ratios.

$$\frac{35}{200} = \frac{SR}{80}$$

This equation is satisfied for $SR = 14$. Therefore, the height of the tree is 14 meters.

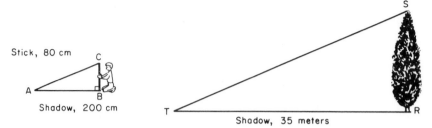

Stick, 80 cm

Shadow, 200 cm

Shadow, 35 meters

Scale Factors—Area and Volume

The smaller of the two knives pictured here is a regular-size Scout knife, whose length is about equal to the width of the palm of your hand. This newspaper clipping says that the bigger knife is "three times larger" than the conventional Scout knife. Does this mean that the length of the larger knife is three times greater, or that its surface area, or volume is three times greater? Phrases such as "twice as large" or "three times bigger" can be misleading. They often refer to a comparison of linear dimensions, as in the case of these knives (compare their lengths). The "three" in this example refers to the scale factor. It means that the dimensions of length, width, and height of the big knife are three times greater than the corresponding dimensions of the smaller knife. As another example, when scientists say that a

Prepared for anything

What could be the world's largest Scout-type knife is ready for the world's largest potato. Wayne Goddard, a professional knife-maker who works at his home at 473 Durham St., Eugene, turned this one out for Dennis and Raymond Ellingsen, Eugene knife collectors. Completely functional, the knife is 24½ inches long when opened. It weighs 4¼ pounds and is three times larger than the conventional Scout knife.

microscope enlarges an object 1000 times, they are referring to the linear dimensions and not an increase in area. On the other hand, when a realtor speaks of one plot of land as being twice as large as another, he/she is referring to the area and not its length or width. If a farmer wants a silo which is twice as large, this refers to the volume and not the height or surface area of the silo. In the following paragraphs you will see the effect that scale factors have on surface area and volume.

Area—These two rectangles are similar. The scale factor from the smaller to the larger is 3. That is, the length and width of one are 3 times greater than the length and width of the other. However, the area of the larger rectangle is 9 times greater. (The smaller rectangle has an area of 8 square units, and the larger rectangle has an area of 72 square units.) In general, if one figure is similar to another by a scale factor of k, where k is any positive number, then one will have an area which is k^2 times the area of the other.

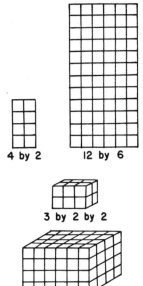

4 by 2   12 by 6

3 by 2 by 2

6 by 4 by 4

The relationship between the scale factor and the surface area of a three-dimensional figure is the same as that for plane figures. Consider the two boxes shown here. The scale factor from the smaller box to the larger is 2. That is, the length, width, and height of the larger box are each two times greater than the corresponding dimensions of the smaller box. Let's compare the areas of the sides of these boxes. The front side of the smaller box has an area of 6, and the front side of the larger box has an area of 24. The larger area is 2^2 or 4 times greater than the smaller area. If a similar comparison is made between each face of the smaller box and the corresponding face of the larger box, you will see that the larger area is 4 times greater than the smaller area. In general, the surface areas of two similar figures are related by the *squares of their scale factors*. If the scale factor is k, where k is any positive number, then one figure will have a surface area which is k^2 times the surface area of the other figure.

Volume—There is also a relationship between the volumes of two similar space figures. There are 12 cubes in the smaller box and 96 cubes in the larger one. The larger box has 8 (or 2^3) times more cubes. In general, the volumes of two similar figures are related by the *cubes of their scale factors*. If two figures are similar and their scale factor is any positive number k, the volume of one will be k^3 times the volume of the other.

We are now prepared to examine the relationships between the areas and volumes of the two knives shown in the news clipping. The scale factor from the smaller knife to the larger is 3. Therefore, the larger knife has a surface area that is 3^2 or 9 times greater than the surface area of the smaller knife. The volume of the larger knife is 3^3 or 27 times greater than the volume of the smaller knife.

Let's apply the relationships between scale factor and area and volume to another example. Pictured here is a nineteenth century scale model of a cookstove. This miniature stove was used by a travelling salesperson as a sample and has all the features of the life-size appliance. The scale factor from this miniature to the life-size stove is 5. Therefore, the larger stove has a surface area which is 5^2 or 25 times greater than that of the smaller stove. Since the surface area of the top of the smaller stove is 300 square centimeters, the surface area of the top of the big stove

Nineteenth century scale model of cookstove

is 25 × 300, or 7500 square centimeters. The increase in volume between the two stoves is determined by the cube of the scale factor. The volume of the life-size stove is 5^3 or 125 times greater than the volume of the miniature stove. The oven of the small stove has a volume of 1000 cubic centimeters. Therefore, the volume of the oven in the large stove is 125 × 1000, or 125,000 cubic centimeters.

Sizes and Shapes of Living Things   The relationship between the surface area and volume of similar figures has some important and interesting applications in nature. For every type of animal there is a most convenient size and shape. A flea cannot have the shape of a hippopotamus, and an elephant cannot be smaller than a rabbit. One factor which governs the size and shape of a living thing is the ratio of its surface area to its volume. All warm-

blooded animals at rest lose the same amount of heat for each unit area of skin. Small animals have too much surface area for their volumes, and so a major reason for eating is to keep warm. For example, 5000 mice weigh as much as a person, but their combined surface area and food consumption are about 17 times greater! At the other end of the scale, large animals tend to overheat because they have too little surface area for their volumes. This is one reason why you do not see elephants and other large animals with fur.

Let's take a closer look at the relationship between surface area and volume as the size of an object increases. This table contains the surface areas and volumes of four different-size cubes. As the sizes of the cubes increase, both the surface areas and the volumes increase, but the volumes

Box Size	Surface Area	Volume	Area / Volume
2 by 2 by 2	24	8	3
3 by 3 by 3	54	27	2
4 by 4 by 4	96	64	1.5
10 by 10 by 10	600	1000	.6

increase at a faster rate. One way of viewing this change is to form the ratio of surface area to volume. This ratio is 3 for a 2 by 2 by 2 cube, and as the dimensions of the cubes increase, the ratio decreases. For a 10 by 10 by 10 cube it is less than 1.

Another way to compare the change in surface area and volume as the size of a cube increases is by graphs. For a cube whose length, width, and height are x, its area is $6x^2$ and its volume is x^3. The graphs of $y = 6x^2$ and $y = x^3$ show that for $x < 6$ the area is greater than the volume; for $x = 6$ the area equals the volume; and for $x > 6$ the volume is greater than the area.

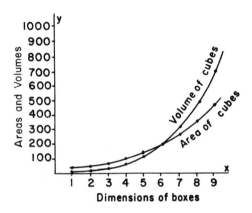

To relate these changes in surface area and volume to the problem of maintaining body temperatures, assume that the three cubes shown here are animals and that the ideal ratio between surface area and volume is 2. In this case the 2 by 2 by 2

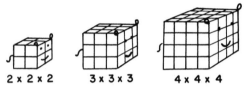

animal has too much surface area (it would tend to be too cold) because its area-to-volume ratio is 3; the area-to-volume ratio for the 3 by 3 by 3 animal is 2, which is "just right"; and the 4 by 4 by 4 animal has too little surface area (it would tend to be too hot) because its ratio of area to volume is 1.5.

SUPPLEMENT *(Activity Book)*

Activity Set 9.2 Devices for Indirect Measurement (Stadiascope, clinometer, hypsometer, transit, and plane table)

Additional Sources

Bruni, J.V. *Experiencing Geometry.* Belmont, California: Wadsworth, 1977. "Congruence and Similarity," pp. 125–48.

Haldane, J.B.S. "On Being the Right Size," from J. Newman, *The World of Mathematics,* 2. New York: Simon and Schuster, 1956, pp. 952–57.

Hogben, L. *Mathematics in the Making.* New York: Doubleday, 1960. pp. 17–25 (ratio and indirect measurement).

Menninger, K.W. *Mathematics in Your World.* New York: The Viking Press, 1962. "Why the Elephant is not a Flea," pp. 15–26.

NCTM. *Experiences in Mathematical Ideas,* 2. Reston, Virginia: NCTM, 1970. "Ratio," pp. 69–95.

EXERCISE SET 9.2

1. Engineers use models of planes to gain information about wing and fuselage (central body) designs. The scale factor from this model of the Boeing B52 to the full-size plane is 100.

Boeing B52 being adjusted for wind tunnel test

★ a. The lift of an airplane depends on the surface area of its wings. How many times greater is the surface area of the B52 than the surface area of its model?

   b. The weight of a plane depends on its volume. How many times greater is the volume of the B52 than the volume of its model?

★ c. In actual flight the wings of the B52 flap up and down a distance of 6 meters. How much wing flap can be expected in a wind tunnel test of the model?

2. For each figure and the given scale factor, sketch a similar figure.

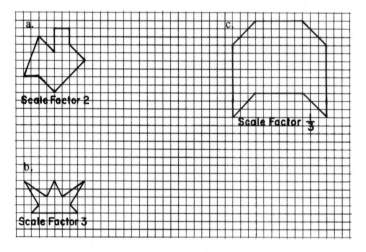

3. Use O as a projection point and find the images of triangle T for scale factors of 2, 3, and 1/2. Look for a relationship between the coordinates of the vertices of T and the coordinates of their images. Triangle T has an area of 16 square units. What are the areas of its three images?

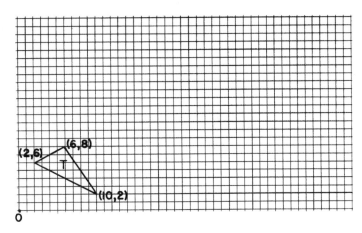

4. The similarity mappings in parts **a**, **b**, and **c** describe what will happen to each point on the given figures by relating the coordinates of each point to the coordinates of its image. The mapping in part **a** *doubles the coordinates* of each point. For example, point ($^-$1,1) gets mapped to ($^-$2,2). Sketch the images of the given figures. What is the scale factor for each mapping?

★ a. $(x, y) \rightarrow (2x, 2y)$ b. $(x,y) \rightarrow \left(\dfrac{x}{2}, \dfrac{y}{2}\right)$ c. $(x,y) \rightarrow (^-3x, ^-3y)$

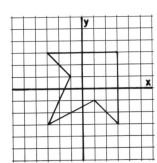
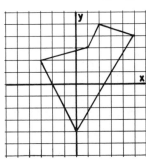
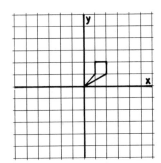

5. A pinhole camera can easily be made from a box. When the pinhole is uncovered, rays of light from the object strike a light-sensitive film. These rays of light travel in straight lines from the object to the pinhole of the camera, like lines through a projection point. Without a lens to gather light rays and increase their intensity, it takes from 60 to 75 seconds for enough light to pass through the pinhole to record the image.

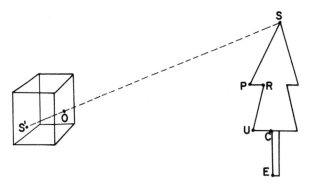

a. Draw lines from the letters of this tree to point O and label their images on the back wall (film) of the camera.

b. Sketch the complete image of the tree.

★ c. What is the scale factor for this projection?

6. Using projection points K and M and a scale factor of 2, sketch the two images of the pentagon.

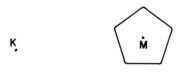

★ a. Is each image similar to the original pentagon?

★ b. Are the two images congruent to each other?

c. Explain why the size of the image of a similarity mapping depends only on the scale factor and not on the location of the projection point.

★ 7. *Diagonal Test:* There is an easy way to test for similar rectangles. If one rectangle is placed on the other with two of their right angles coinciding, as shown in these figures, then the rectangles are similar if their diagonals lie on the same line. For example, rectangle $AEFG$ is similar to rectangle $ABCD$ but the two rectangles in the lower figure are not similar.

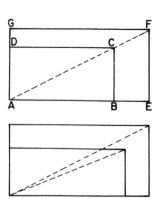

★ a. Explain why rectangle $ABCD$ is similar to rectangle $AEFG$.

★ b. Use the diagonal test to find out which two of the following rectangles are similar.

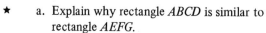

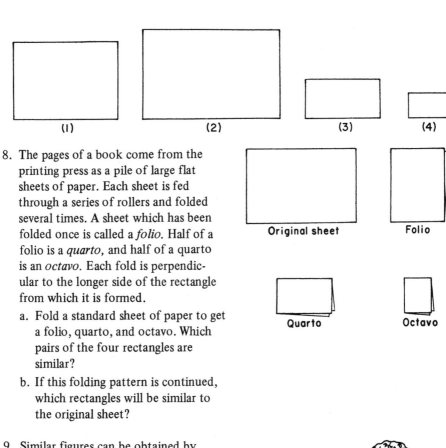

(1)   (2)   (3)   (4)

8. The pages of a book come from the printing press as a pile of large flat sheets of paper. Each sheet is fed through a series of rollers and folded several times. A sheet which has been folded once is called a *folio*. Half of a folio is a *quarto*, and half of a quarto is an *octavo*. Each fold is perpendicular to the longer side of the rectangle from which it is formed.

Original sheet   Folio

Quarto   Octavo

a. Fold a standard sheet of paper to get a folio, quarto, and octavo. Which pairs of the four rectangles are similar?

b. If this folding pattern is continued, which rectangles will be similar to the original sheet?

9. Similar figures can be obtained by reproducing a figure from one grid to another grid of a different size. This practice is common for enlarging patterns in quilting and sewing. The patchwork doll pattern shown here was enlarged by a scale factor of 3 to get the larger doll.

Reproduce the figure from Grid A (see next page) onto Grids B and C by copying a square at a time. (The figure will become larger on Grid B and smaller on Grid C.)

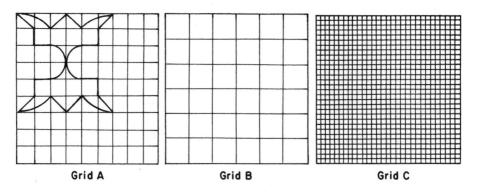

| | Grid A | Grid B | Grid C |

★   a. What is the scale factor from Grid A to Grid B? Grid B to Grid C?

      b. The scale factor from Grid A to Grid C is 3/10. How can this scale factor be obtained from the two scale factors in part **a**?

★   c. How many times greater is the area of the figure on Grid B than the area of the figure on Grid A?

10. On the Xerox 7000 any one of five switches may be selected to determine the size of the reproductions.

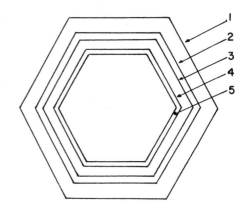

If switch 1 is used, the reproduction is congruent to the original. The remaining switches reduce the original to smaller sizes. For example, switch 2 was used to reduce the largest hexagon shown here to hexagon 2; switch 3 reduced the largest hexagon to hexagon 3; etc.

★   a. Measure the sides of each hexagon in millimeters and record them in this table.

Hexagon	1	2	3	4	5
Length of Sides	27	23			

b. What is the scale factor for each switch on the Xerox 7000?

Switch	1	2	3	4	5
Scale Factor	1	$\frac{23}{27}$			

c. If switch 5 is used to reduce a figure, its area will be approximately what fraction of the area of the original figure: 1/2, 1/3, or 1/4?

★ 11. The larger box shown here has linear dimensions that are twice those of the smaller box. Doubling the linear dimensions produces a similar box with a scale factor of 2, and tripling the dimensions is equivalent to a scale factor of 3. For the scale factors in the following table, record the dimensions of the boxes which are similar to the small box. Compute the areas and volumes of these boxes.

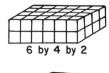

6 by 4 by 2

3 by 2 by I

	Scale Factors				
	I	2	3	4	5
Dimensions	3 x 2 x I	6 x 4 x 2			
Surface Area	22 cm²				
Volume	6 cm³				

12. These two figures are similar. Each dimension of the larger figure is twice the corresponding dimension of the smaller figure.

★ a. The smaller figure has a volume of 7 cubic units. What is the volume of the larger figure?

b. Sketch an enlargement of the next figure for a scale factor of 3. Find the surface areas and volumes of both figures.

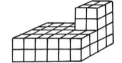

13. The setting pictured here appears to be life-size except for the nickel on the table. These pieces of furniture and dishes were handcrafted in the early 1800s for a doll-house collection. The scale factor from the life-size objects to these antiques is 1/8.

★ a. The nickel and the base of the candleholder both have the same diameter. What is the diameter of a nickel?

b. Use your answer in part **a** and the scale factor to find the diameter of the base of the life-size candleholder.

★ c. How many times smaller is the area of the miniature rug than the area of the similar life-size rug?

d. The miniature cream pitcher holds one milliliter of cream. How much cream will a similar life-size pitcher hold?

e. If you were to make a doll similar to yourself for this setting, what would be the height of the doll?

14. This is not an example of trick photography. There are two average-size adults sitting in this chair, and there is room for several more. The scale factor from the small chair to the large chair is 8.

★ a. The small chair is 40 cm tall and its seat is 24 cm wide. What is the height of the large chair and the width of its seat?

b. How many times more paint will the large chair require than the small chair?

★ c. The small chair weighs about 1 kilogram. Both chairs are made of the same kind of wood. What is the weight of the large chair?

15. In *Gulliver's Travels,* by Jonathan Swift, Gulliver went to the kingdom of Lilliput where he found that he was 12 times taller than the average Lilliputian.

a. The Lilliputians computed Gulliver's surface area to make him a suit of clothes. About how many times more material would be needed for his suit than one of theirs?

★ b. They computed Gulliver's volume to determine how much he should be fed. About how many times more food would he require than a Lilliputian?

c. Tiny people like the Lilliputians can exist only in fairy tales because the ratio of their volume to surface area would not enable them to maintain the proper body temperature. Would they be too warm or too cold?

16. *Similarity Stretcher:* Loop or knot two rubber bands together and hold one end fixed at point *O*. Stretch the bands so that as the knot at point *P* traces one figure, a pencil at point *P'* traces an enlargement. Use this method to enlarge the map of the United States. Assume the distance from *O* to *P'* is 2.3 times the distance from *O* to *P*. How many times greater will the distance from Denver to Kansas City be on the enlarged map than on the smaller map?

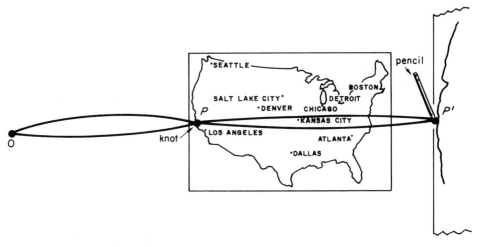

17. *Paper Chase Game:* Player 1 and Player 2 each choose a beginning point near the center of a grid. For the example shown here these points are *A* and *B*. Each player will draw a connected path of line segments by drawing only one line segment on each turn. The line segments may be drawn in a horizontal, vertical, or diagonal direction. Play begins with Player 1 drawing a line segment from *A*. Then Player 2 draws a line segment from point *B*, which is twice as long and in the same direction. Play continues in this manner

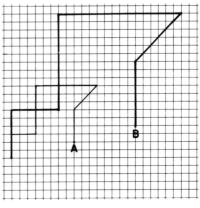

with both players beginning at the ends of their previous line segments and Player 2 always drawing a segment which is twice as long as Player 1's and in the same direction.

The object of the game is for Player 1 to move so that when he/she finishes a turn, Player 2 finishes his/her turn at the same point. When the game is over, the players' paths will be similar figures. The point at which the game ends has a very special relationship to the starting points. Describe the location of this point.

Anamorphic painting of H.M.S. *Victory* by James Steere

9.3 TOPOLOGICAL MOTIONS

Topology has been called "the mathematics of distortion." It is a special kind of geometry which involves analyzing figures and surfaces when they are twisted, bent, stretched, shrunk, or, in general, distorted from one shape to another. These descriptions are all examples of topological motions. A figure may be changed so much by a topological motion that it bears little resemblance to its original shape. The above painting of the H.M.S. *Victory* by James Steere is a topological distortion called anamorphic art. The reflection of this painting in the cylindrical mirror which has been placed at the center of the picture brings the picture into clear

H.M.S. *Victory* by Joseph Marshall

perspective. It shows the stern of the ship from the painting by Joseph Marshall. You may wonder how mathematicians find anything left to study when figures can be so drastically changed. In the following paragraphs we will consider some of the properties of figures which remain unchanged by topological motions.

Topological Mappings   Topology is sometimes referred to as rubber-sheet geometry because topological mappings in the plane can be performed as if the plane were a rubber sheet. As the rubber sheet in these figures is stretched, the face is distorted into several shapes. In each picture the original circle has changed shape, but it is still a simple closed curve.

Similar distortions can take place in space by thinking of an object as made of an elastic material. In the following sequence a donut-shaped object, called a *torus,* has been deformed in three steps into the shape of a cup. Since topologists view each of these as being "the same" in the sense that they have the same topological properties, a topologist is sometimes described as a person who doesn't know the difference between a donut and a cup of coffee.

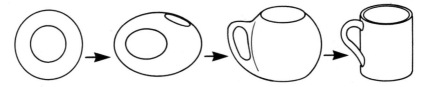

These stretching and bending motions from one object to another are like mappings, with each point having a corresponding image point. The definition of a topological mapping for both plane and space figures contains two conditions.

Definition: A mapping from one set of points to another is a *topological mapping* if:

(1) There is a one-to-one correspondence between the two sets.

(2) The mapping is continuous.

If there is a topological mapping from one set to another, the two sets are said to be *topologically equivalent.*

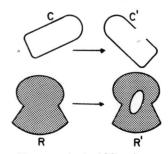

The basic idea of continuity, in condition (2) of the definition, is that points "close together" cannot be cut or torn apart and separated by the mapping. As examples, it is not permissible to deform a curve by cutting it or to deform a region by punching a hole in it. In this figure the simple closed curve C is not topologically equivalent to curve C', and region R is not topologically equivalent to R'.

Non-topological Mappings

There is, however, one exception to this restriction on cutting. Cutting is permissible as long as the points along the cut are matched back together again. For example, the solder wire in the form of a simple closed curve in Figure A has been cut at point K; knotted in Figure B; and resoldered at point K in Figure C. The curves formed by the solder in Figures A and C are topologically equivalent.

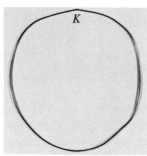

Figure A

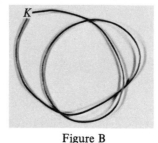

Figure B

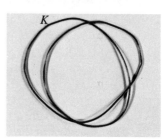

Figure C

All simple closed curves are topologically equivalent. This includes circles, squares, triangles, and irregular simple closed curves such as the one shown on the right. All line segments and simple nonclosed curves such as the following figures are topologically equivalent. If we think of these figures as rubber string, each could be bent or stretched into the shape of another without cutting or separating points.

In three dimensions spheres (balls), cubes (boxes), ellipsoids (footballs), and cylinders (cans) are all topologically equivalent. If we think of these as being made of clay, any one of them could be deformed into the shape of another without punching holes or separating points.

Topological Properties   In Section 9.1 we saw that the size and shape of an object remain the same for congruence mappings. To state this another way, these mappings preserve the distance between points and their images. In Section 9.2, we saw that an object changes its size but not its shape for a similarity mapping. In this case, the mapping preserves measures of angles, and ratios of the lengths of line segments, but does not preserve distances between points. Now, with topological mappings both the size and the shape can vary greatly between an object and its image. In other words, topological mappings do not preserve distances between points, measures of angles, or ratios of the lengths of line segments. In spite of this freedom there are properties of figures which are preserved by topological mappings. One of these is the number of sides and edges of a surface.

A sheet of paper has two sides and one edge, in the sense that an ant crawling on one side must cross over an edge to get to the other side. Deform the sheet by crumpling it in your hands, and it still has two sides and one edge. The two-sidedness and one-edgedness are properties that are preserved under a topological mapping.

Flat sheet of paper

Two sides and one edge

A cylindrical band of paper has two sides and two edges. A surface must be crossed to get from edge to edge, and an edge must be crossed to get from surface to surface.

In the nineteenth century the German mathematician Ferdinand Moebius (1790–1868) made a surprising discovery by putting a half-twist in a strip of paper and fastening the edges. The resulting surface has only one

Cylindrical band

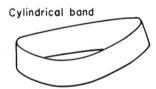

Two sides and two edges

edge and one side! This means that an ant crawling on this surface can reach any other spot without crawling over an edge. This surface is called a Moebius band or Moebius strip. While the Moebius band is only a slight variation of a cylindrical band, these two figures cannot be topologically equivalent because of their different number of sides.

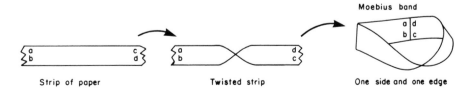

Strip of paper   Twisted strip   One side and one edge

The Moebius band has fascinated people for years and has been a frequent subject in literature and in art. It is pictured here in Escher's 1963 woodcut, *Moebius Strip II*. In at least one case the Moebius band has been put to practical use. The Sandia Corporation has invented a Moebius band nonreactive resistor which has been patented by the U.S. Atomic Energy Commission. This resistor has several desirable electrical properties which ordinary bands do not have. The following sketch is from this patent.

Metal Moebius band electrical resistor

Moebius Strip II, 1963
woodcut by M.C. Escher

Another property that is preserved by topological mappings can be roughly described by speaking of the number of "holes" in an object. For example, the four objects pictured next are all topologically different. Mathematicians describe their differences by referring to the genus of a surface. The *genus* is the maximum number of circular cuts (simple closed curves) that can be made without dividing the surface into two separate pieces. For example, two cuts can be made (as shown) on the two-holed surface without separating its surface into disconnected pieces. Similarly, the torus can be cut once without separating its surface into disjoint pieces. Any circular cut on the surface of a sphere will divide its surface into two pieces. The numbers under these figures are their *genus numbers.*

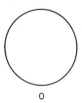
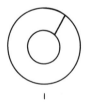
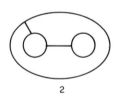
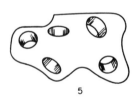

0   1   2   5

Networks   In the eighteenth century in the town of Königsberg, Germany, a favorite pastime was walking along the Pregel River and crossing the town's seven bridges. During this period a natural question arose: Is it possible to take a walk and cross each bridge only once? This question was solved by the Swiss mathematician Leonhard Euler. His solution was the beginning of network theory which is an important branch of topology.

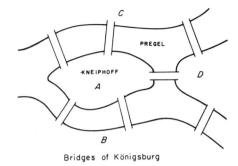

Bridges of Königsburg

Euler represented the four land areas of Königsberg (*A, B, C,* and *D* in the previous figure) by 4 points, and the 7 bridges by 7 lines joining these points. For example, the island of Kneiphoff has 5 bridges, and in the diagram at the right there are 5 lines from point *A*. The 3 lines from point *D* represent 3 bridges, etc. This kind of a diagram is called a *network.* Notice that Euler was concerned not with the size and shape of the bridges and land regions but rather with how the bridges were connected.

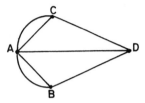

A *network* is a collection of points, called *vertices,* and a collection of lines connecting these points, called *arcs.* A network is *transversible* if you can trace each arc exactly once by beginning at some point and not lifting your pencil from the paper. The problem of crossing each bridge exactly once reduces to one of transversing the network for these bridges.

Euler made a remarkable discovery about networks which depends on the number of odd vertices. In the Königsberg network there are an odd number of arcs at point *A* and so *A* is called an *odd vertex.* If the number of arcs meeting at a point is even, it is called an *even vertex.* Euler found that the only transversible networks are those which have either no odd vertices or exactly 2 odd vertices! Since the Königsberg network has 4 odd vertices, it is not transversible.

Here are some examples of transversible and nontransversible networks.

Transversible
(2 odd vertices)

Transversible
(No odd vertices)

Nontransversible
(4 odd vertices)

Nontransversible
(6 odd vertices)

Four-Color Problem   If you try to color this map of 8 states with only 3 colors so that no 2 states with a common boundary have the same color, you will find it cannot be done. It can, however, be colored with 4 colors.

For the past 125 years mathematicians have been trying to prove that 4 colors are enough to color any map no matter how many regions it contains or how they are situated. The only condition which must be satisfied is that regions with a common boundary must be colored differently. It is agreed that if 2 regions meet only at a point, then they do not share a common boundary. Since no one was able to produce a map which required more than 4 colors, it was conjectured that 4 colors were sufficient to color any map. This simple appearing conjecture became one of the most famous unsolved problems in mathematics.

Then on July 22, 1976, two University of Illinois mathematicians, Kenneth Appel and Wolfgang Haken, announced they had proven that 4 colors are all that is necessary to color any map. The problem was solved by representing map regions as points and boundaries as arcs connecting these points. Using this network approach they were able to reduce the problem to the examination of 1936 basic map forms. All other maps are topologically equivalent to these basic forms or else they differ in some insignificant way. Then they fed their map forms into the computer, and after 1200 hours the computer determined that each map could be colored with 4 colors or less.

SUPPLEMENT *(Activity Book)*

Activity Set 9.3 Topological Entertainment (Moebius bands, topological tricks, Gale Game, and game of "Sprouts")

Additional Sources

Bergamini, D. *Mathematics.* New York: Time-Life Books, 1963. "The Mathematics of Distortion," pp. 176–91.

Euler, L. "The Seven Bridges of Königsberg," from J. Newman, *The World of Mathematics,* **1.** New York: Simon and Schuster, 1956. pp. 571–80.

Glenn, W.H. and D.A. Johnson. *Exploring Mathematics on Your Own.* New York: Doubleday, 1961. "Topology: The Rubber-Sheet Geometry," pp. 197–234.

Hayter, R.J. "New Light on the Four Colour Conjecture." *Mathematics in Schools,* **6** No. 2 (March 1977), 3–5.

Newman, J.R. and E. Kasner. *Mathematics and the Imagination.* New York: Simon and Schuster, 1940. "Rubber-Sheet Geometry," pp. 265–98.

Steinhaus, H. *Mathematical Snapshots.* New York: Oxford University Press, 1950. "Platonic Bodies Again, Crossing Bridges, Tying Knots, Coloring Maps, and Combing Hair," pp. 214–40.

Struble, M. *Stretching a Point.* Philadelphia: Westminster Press, 1971 (topics in topology).

EXERCISE SET 9.3

1. Each Figure in parts **a** through **g** is topologically equivalent to the sphere in Figure 1, or the torus in Figure 2, or the two-holed object in Figure 3. Match the correct number with each one.

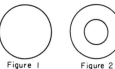

Figure 1   Figure 2   Figure 3

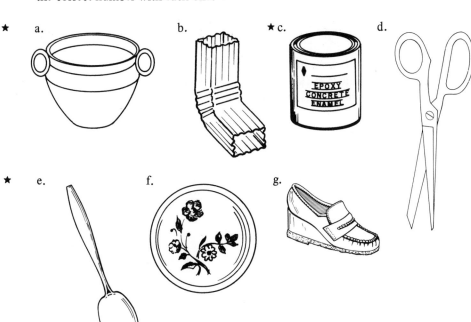

★ a.   b.   ★ c.   d.

★ e.   f.   g.

2. According to the Swiss psychologist Jean Piaget, children's first discoveries of geometry are topological, and by the age of three they readily distinguish between open and closed figures.* Assuming that a child "thinks topologically," match up four pairs of figures which he/she would consider to be "the same."

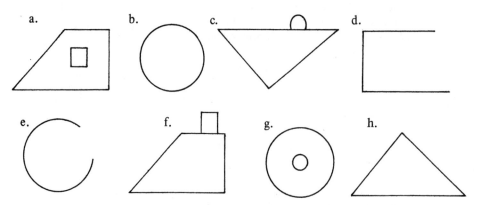

3. Two of these networks are transversible. For each transversible network, show a beginning point and an ending point, and mark the route with arrows.

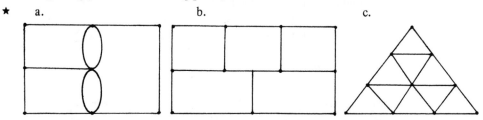

★ 4. Each of these networks has 2 odd vertices and is transversible. Show a beginning point and an ending point for each. What will always be true about the beginning and ending points of networks with exactly 2 odd vertices?

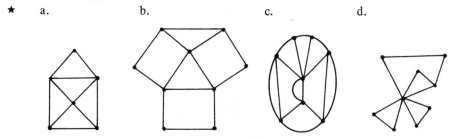

*J. Piaget, "How Children Form Mathematical Concepts," *Scientific American*, **189** No. 5 (1953), 74–79.

★ 5. The vertices and edges of polyhedra are three-dimensional networks.

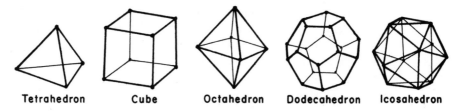

Tetrahedron   Cube   Octahedron   Dodecahedron   Icosahedron

★   a. Which one of the five regular polyhedra is transversible?

★   b. What is the least number of diagonals that need to be drawn on the faces of the cube before the network will be transversible?

★ 6. Many years after Euler proved that it was impossible to take a walk in which each of the 7 bridges of Königsberg is crossed over exactly once, an eighth bridge was built. Sketch a network with 4 vertex points for the land areas *A, B, C,* and *D,* and 8 arcs for the bridges. Is this network transversible?

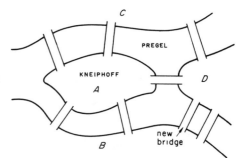

7. This is a sketch of 2 islands (*A, B*), 4 land regions (*C, D, E, F*), and 15 bridges. Is it possible to plan a walk in which each bridge is crossed exactly once? Explain the reason for your answer.

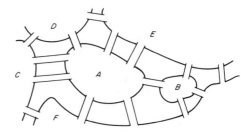

★ 8. This museum floor plan has 5 rooms and 16 doors. Is it possible to plan a walk in which each door is passed through exactly once? (*Hint:* Form a network by connecting the 5 inside points and the outside point *P,* with arcs which run through each of the doors.)

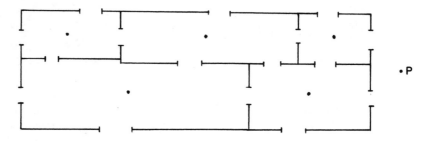

9. Figure A has 4 regions and 10 line segments. A type of puzzle which is similar to planning the museum walk in Exercise 8 requires that a continuous arc be drawn which passes through each line segment of the figure exactly once. The solution in this example is shown in Figure B. Try drawing such an arc for each of the following figures. (*Note:* If you start inside the figure it is not necessary to end inside, and if you start outside the figure it is not necessary to end outside.)

Figure A

Figure B

★ a.

b.

c.

d.

★ e. Figure A has 2 regions with an even number of sides and 2 regions with an odd number of sides. How is the number of regions with an odd number of sides related to solving this type of puzzle? Check your conclusion in parts **a** through **d**.

10. The curves shown next are simple closed curves. The Jordan curve theorem (page 165) says that if a point inside a simple closed curve is connected to a point outside the curve, the connecting arc will intersect the curve. In these examples, *B* is outside. Can you draw an arc from *A* to *B* which does not intersect the curve? In other words, is *A* inside or outside?

a.

b.

•*B*

★ c. Draw line segments $\overline{AB}$ for the curves in parts **a** and **b**. Count the number of times that $\overline{AB}$ intersects each curve. How can this number be used to tell when a point is inside or outside a simple closed curve? Check your answer on a few curves.

11. This simple closed curve is from the NCTM publication *Puzzles and Graphs* by John Fujii. Are the pairs of points which are given in parts **a, b, c,** and **d** on the same side of the curve? (*Hint:* Connect pairs of points by a line segment and count the number of intersections of the line segment and the curve.)

★   a. *A,B*   b. *B,C*   c. *D,C*   d. *B, D*

12. To color maps it is desirable for regions with a common boundary to have a different color.

★   a. What is the minimum number of colors required to color any map, regardless of the number of regions?

       b. What is the minimum number of colors required for this map of Europe?

13. Each of the maps in parts **a** through **d** can be colored with 2 or 3 colors, so that regions with a common boundary are colored differently. Which ones require 3 colors?

a.   b.   c.   d.

   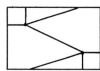

★   e. Think of each map in parts **a** through **d** as a network. Compare the number of even and odd vertices inside each rectangle (omit vertices on the rectangle) with the number of colors required. How are they related? Draw some more network type maps and test your conjecture.

14. Anamorphic art is a special type of topological mapping. One example of this art is the *slant picture,* which is designed to be viewed at an angle. The technique for drawing such pictures is to use grids as we did for producing similar figures in Exercise 9, Exercise Set 9.2. However, this time the grid is distorted along one axis, such as shown in Figure 2. Finish copying the design in Figure 1 onto the corresponding regions of the trapezoidal grid in Figure 2. The completed figure should look normal if viewed by placing your eye to the right of this grid and a little above the surface of this page.

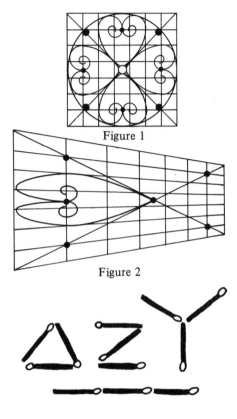

Figure 1

Figure 2

15. *Matchstick Problem:* ** The 3 matchstick patterns shown here are the only topologically different patterns that can be formed with 3 matches, subject to the condition that the matches meet only at their end points with no overlapping. (The line segment with 3 matches in a row is topologically equivalent to the Z-shaped pattern.)

    a. Sketch the 5 topologically distinct patterns that can be formed with 4 matches.

★   b. Form 10 topologically distinct patterns with 5 matches.

*For more details on slant drawings as well as techniques for sketching cylindrical and conical anamorphic pictures, see M. Gardner, "The Curious Magic of Anamorphic Art," *Scientific American,* **232** No. 1 (1975), 110–14.

**M. Gardner, *The Unexpected Hanging* (New York: Simon and Schuster, 1969), pp. 79–81.

CHAPTER TEN

PROBABILITY AND STATISTICS

"Statistical thinking will one day be as necessary for efficient citizenship as the ability to read and write."

— H.G. Wells

10.1 PROBABILITY

Probability, a relatively new branch of mathematics, was started in Italy and in France during the sixteenth and seventeenth centuries to provide strategies for gambling games. The Italian mathematician Jerome Cardan made the first notable contribution to probability in *The Book on Games of Chance.* In this book Cardan analyzed the probability of obtaining a sum of 7 on a roll of two dice. Cardan also included tips on how to cheat and how to detect cheating. In the seventeenth century the mathematicians Blaise Pascal and Pierre Fermat worked together to develop what is considered to be the first theory of probability. Once again, it was questions about dice games that provided the initial incentive.

From these beginnings, probability has evolved to where it has applications in all walks of life. Life insurance companies use probability to estimate how long a person is likely to live. Each different age has a different probability, called *life expectancy.* Scientists use rules of probability for interpreting statistics and estimating values for experimental data. The outcomes of athletic events are predicted in terms of odds, which is another way of stating a probability. If the odds are high, the probability is close to 1, and if the odds are low, the probability is close to 0. According to Lucy, the odds of an involuntary muscle spasm occurring at a given moment are 1 to 10 billion which is very close to 0.

Theoretical Probability (Computing Probability in Advance)   Before the theory of probability was developed, the chance of an event happening was determined solely by experience. Pascal and Fermat, on the other hand, were interested in determining the probabilities of events without relying on experiments. For example, on the roll of an ordinary die any one of six faces has an equally likely chance of turning up. Since there are six possible outcomes, the *probability* of rolling any particular number, such as 5, is 1/6. This is written as

$$P(5) = \frac{1}{6}$$

and is read as, "the probability of 5 equals 1/6." Similarly, the probability of rolling either a 3 or a 5 is 2/6 because two out of six faces have the desired outcomes.

$$P(3 \text{ or } 5) = \frac{2}{6}$$

In general, when each of the outcomes is equally likely, we have the following definition.

Definition: Probability of event = $\dfrac{\text{Number of favorable outcomes}}{\text{Total number of outcomes}}$

Because this probability is computed theoretically and does not rely on experiments, it is called *theoretical probability*. Since the number of favorable outcomes is always less than or equal to the total number of outcomes, this definition shows that the probability of an event is some fraction from 0 to 1. If there are no favorable outcomes, such as rolling a die and getting an 8, the probability is 0. If every outcome is a favorable outcome, the probability is 1. For example, the probability of rolling a die and getting a number less than 7 is 1.

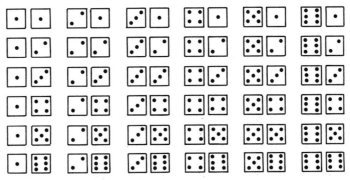

Thirty-six possible outcomes for two dice

When two dice are rolled there are 36 equally likely outcomes. Find the pairs of dice in this array whose numbers have a sum of 7. You will see something special about their locations. Since six of these pairs have a sum of 7, the probability of rolling a 7 is 6/36 or 1/6.

$$P(\text{sum of 7}) = \frac{1}{6}$$

Locate the dice whose numbers have a sum of 3. What is the probability of rolling a sum of 3?

Many questions in probability can be answered by listing all possible outcomes and then determining the number of favorable outcomes. To consider another illustration of this approach suppose that the four playing cards facing down are numbered 2, 3, 4, and 5. If two cards are selected, what is the probability of getting the cards with 2 and 3? In all, there are six different ways that pairs of cards can be selected (see next page). These six pairs of cards represent equally likely outcomes. Since only one pair, namely, the cards with 2 and 3, is a favorable outcome, the probability of selecting a 2 and a 3 is 1/6.

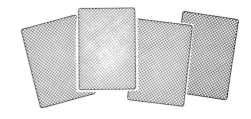

Empirical Probability (Computing Probability from Experience)

When a probability is determined from experience or by observing the results of experiments, it is called *empirical probability*. Suppose we were to flip a bottle cap instead of a coin. Because of its construction we cannot assume it is equally likely to land right side up as upside down. If we flip the bottle cap a large number of times and find that it lands right side up 20% of the time, we would conclude that the probability of landing upside down is greater than the probability of landing right side up. If many repeated experiments yield similar results we will conclude from experience that the probability of landing right side up is about 20/100 or 1/5.

The empirical approach to probability is the only means of determining probability in many fields. Insurance companies measure the risks against which people are buying insurance in order to set premiums. A person's age and life expectancy are important factors. To compute the probability that a person 20 years old will live to be 65 years old, insurance companies gather birth and death records of large numbers of people and compile mortality tables. One such table indicates that for every 10 million births, 6,800,531 will live to age 65 (see table on page 482). Thus, the probability that a newborn baby will live to be 65 is 6,800,531/10,000,000 or about .68 (68 percent). According to the table, 9,664,994 people will live to age 20. Therefore, the probability of a baby living to age 20 is about .966, or 96.6 percent. We can compare the numbers from the table for two different ages to determine the probability of living from one age to another. The probability that a person of age 20 will live to be 65 is 6,800,531/9,664,994 or approximately .7 (70 percent).

Odds

Racetracks state probabilities in terms of odds. Suppose that the odds against Blue Boy winning are 4 to 1. This means that for every dollar you bet on Blue Boy, the racetrack management will match it with $4. For each win, you receive the money you bet plus the money matched by the racetrack. That is, for $1 you receive $5, for $2 you receive $10, etc. With 4-to-1 odds, the racetrack management expects Blue Boy to lose four out of every five races. Thus, the probability of Blue Boy losing the race is 4/5 and the probability of his winning is 1/5.

"BLUE BOY SEEMS TO BE HOLDING BACK A BIT."

This example suggests a procedure for converting odds to probabilities. If the *odds in favor* of an event are m to n, then the probability of the event is $m/(m + n)$. In this case, the *odds against* this event will be n to m, and the probability of the event not happening will be $n/(m + n)$. These relationships will be illustrated in the following examples.

In a deck of 52 cards there are four aces. The probability of drawing an ace is 4/52.

$$\frac{\text{Number of favorable outcomes}}{\text{Total number of outcomes}} = \frac{4}{52} \quad \text{or} \quad \frac{1}{13}$$

The probability of not drawing an ace is 48/52.

$$\frac{\text{Number of unfavorable outcomes}}{\text{Total number of outcomes}} = \frac{48}{52} \quad \text{or} \quad \frac{12}{13}$$

The odds of drawing an ace are 1 to 12.

Number of favorable outcomes to *Number of unfavorable outcomes* = 4 to 48   (1 to 12)

The odds of not drawing an ace are 12 to 1.

Number of favorable outcomes to *Number of unfavorable outcomes* = 48 to 4   (12 to 1)

These examples show the close relationship between odds and probability. Both are just different ways of presenting the same information. This relationship can be clarified further by a diagram. The bar in this sketch has 13 equal parts, 12 to represent unfavorable

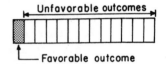
Unfavorable outcomes
Favorable outcome

outcomes and 1 representing a favorable outcome. The *probability* of an event is the ratio of the number of favorable outcomes to the total number of outcomes. This is illustrated in the diagram by comparing the number of shaded parts to the total number of parts, 1 to 13 or 1/13. The *odds* of an event is a ratio between the number of favorable outcomes and the number of unfavorable outcomes. This is illustrated by comparing the number of shaded parts of the bar to the number of unshaded parts, 1 to 12.

Mathematical Expectation   To evaluate the fairness of a game we must consider the prize or monetary gain as well as the probability of winning. The probability of winning may be fairly small, but if the prize is large enough the game may be a good risk. We know, for example, that the probability of rolling a sum of 7 on two dice is 1/6. This means that on the average a sum of 7 will come up once on every six rolls. If it costs a dollar every time you roll the dice and the payoff is 6 dollars for rolling a 7, then over a period of time you could expect to break even. The probability of winning multiplied by the value of the prize is called the *mathematical expectation*. In this game the mathematical expectation is 1 dollar: $1/6 \times 6 = 1$. Since this is also the cost of playing each game, this game is considered to be *fair*.

The mathematical expectation for roulette shows that this game favors the "house." A roulette wheel has 38 compartments: Two are numbered 0 and 00 and are colored green; the remaining ones are numbered 1 through 36, and half are red and half are black. One way of playing this game is to bet on the red or black colors. The probability of a red is 18/38 or 9/19 (number of red compartments/total number of compartments). If you bet 1 dollar on the red and win, you are paid 2 dollars. This game has a mathematical expectation of 9/19 × $2, or approximately 95¢. Since this is less than a 1-dollar bet the game is unfair.

Sometimes there is more than one category for winning. Several states, for example, have sweepstakes and for each ticket there are several categories of prizes. One type of ticket which is sold in New Hampshire has five-digit numbers, with 0 allowed for any of the digits. After the drawing you can win $5000 if you have the same number; $200 if you have the first four digits or the last four; and $20 if you have the first three digits, the middle three digits, or the last three. The probabilities of winning these amounts are 1/100,000, 18/100,000 and 261/100,000, respectively, and the mathematical expectation is

$$\frac{1}{100,000} \times \$5000 \;+\; \frac{18}{100,000} \times \$200 \;+\; \frac{261}{100,000} \times \$20 \;=\; 14 \text{ cents}$$

In such cases where there are several categories of prizes and different probabilities of obtaining each prize, the *mathematical expectation* is computed by multiplying the different probabilities of winning times the respective amounts to be won and then adding these products.

SUPPLEMENT *(Activity Book)*

Activity Set 10.1 Probability Experiments
Just for Fun: Buffon's Needle Problem

Additional Sources

Huff, D. and I. Geis. *How to Take a Chance.* New York: W.W. Norton, 1959.

Menninger, K.W. *Mathematics in Your World.* New York: The Viking Press, 1962. "The Realm of Lady Luck," pp. 155–77, and "Life and Death by Mathematical Analysis," pp. 247–64.

National Council of Teachers of Mathematics. *Experiences in Mathematical Ideas,* **2**. Reston, Virginia: NCTM, 1970. "Dealing with Uncertainty," pp. 174–207.

Starr, N. "A Paradox in Probability Theory." *The Mathematics Teacher,* **66** No. 2 (February 1973), 166–68.

"The greatest advantage in gambling comes in not gambling at all."

Jerome Cardan

EXERCISE SET 10.1

1. Out of 36 possible ways two dice can come up, six ways produce a sum of 7. Compute the probability of rolling each of the other sums. (You may want to use the array of dice on page 475.)

Sum	2	3	4	5	6	7	8	9	10	11	12
Probability						$\frac{6}{36}$					

★   a. What is the probability of a sum greater than or equal to 8?

★   b. What is the probability of a sum greater than 4 and less than 8?

2. The dice game called "craps" is played by the following rules. If the player rolling the dice gets a sum of 7 or 11, he/she wins. If this player rolls a sum of 2, 3, or 12, he/she loses. If the first sum rolled is a 4, 5, 6, 8, 9, or 10, the player continues rolling the dice. The player now wins if he/she can obtain the first sum rolled before rolling a 7.

a. What is the probability of rolling a 7 or an 11? (See table in Exercise 1.)

★ b. What is the probability of losing on the first roll?

c. Suppose a player rolls an 8 on the first turn. Which is the better probability on the second turn: rolling a 7 or an 8?

3. A chip is to be drawn from a box containing the following numbers of colored chips: 8 orange; 5 green; 3 purple; and 2 red. Determine the probabilities of selecting the following chips.

a. A purple chip.

★ b. A green or a purple chip.

c. A chip that is not orange.

★ 4. The five regular polyhedra, called Platonic solids, are pictured next. These are the only polyhedra which can be used for fair dice. The regular polyhedron with 20 faces (see upper right corner) was used as a die by the Egyptians more than 2000 years ago. The faces of these five polyhedra are labelled with consecutive whole numbers beginning with 1. For example, the tetrahedron has numbers 1, 2, 3, 4; and the icosahedron has numbers from 1 to 20. Determine the probabilities of rolling the numbers in the following table for each of these "dice."

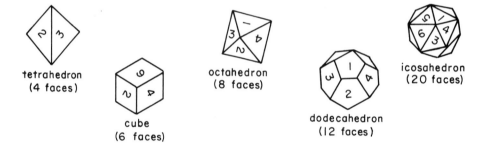

tetrahedron
(4 faces)

cube
(6 faces)

octahedron
(8 faces)

dodecahedron
(12 faces)

icosahedron
(20 faces)

	Tetrahedron	Cube	Octahedron	Dodeca-hedron	Icosahedron
★ a. A number less than 3					
★ b. An even number					
★ c. The number 2					

5. A deck of 52 cards contains 13 cards in each of four suits: spades, hearts, clubs, and diamonds. The clubs and spades are black and the diamonds and hearts are red. Each suit has three face cards: jack, queen, and king. Compute the probabilities and odds of drawing each of the following cards.

		Prob.	Odds
★	a. An ace		
	b. A face card		
★	c. A diamond		
	d. A black face card		

6. When the probability that an event will happen is added to the probability that the event will not happen, the sum is 1. This fact and your answers in Exercise 5 will be helpful in computing the following probabilities.

★   a. Not drawing an ace.   b. Not drawing a face card.

★   c. Drawing a club, heart, or spade.   d. Not drawing a black face card.

7. For each of the given events and probabilities, determine the odds in favor of each event.

★   a. The probability of living to age 65 is 7/10.

    b. The probability of selecting a person with type O blood is 3/5.

★   c. The probability of winning a certain raffle is 1/500.

    d. The probability of rain on Monday is 80%.

8. Gambling syndicates predict the outcomes of sporting events in terms of odds. Convert each of the following odds to probabilities.

★   a. The odds on the Packers over the Rams are 4 to 3.

    b. The odds on the University of Michigan defeating Ohio State are 7 to 5.

    c. The odds on the Yankees winning the pennant are 10 to 3.

9. The mortality table on the next page is based on the lives and deaths of policyholders in several large insurance companies. Use this information to answer the following questions.

★   a. What is the empirical probability that a child will live to be 1 year old?

    b. What is the empirical probability of a person living to age 50?

★   c. What is the empirical probability that a person of age 34 will live to age 65?

d. What is the empirical probability that a 60-year-old person will live to be 65?

★ e. If an insurance company has 7000 policyholders at age 28, how many death claims must they be prepared to pay before these people reach age 29?

Age	Number Living	Number Dying	Age	Number Living	Number Dying
0 ...	10,000,000	70,800	35 ...	9,373,807	23,528
1 ...	9,929,200	17,475	36 ...	9,350,279	24,685
2 ...	9,911,725	15,066	37 ...	9,325,594	26,112
3 ...	9,896,659	14,449	38 ...	9,299,482	27,991
4 ...	9,882,210	13,835	39 ...	9,271,491	30,132
5 ...	9,868,375	13,322	40 ...	9,241,359	32,622
6 ...	9,855,053	12,812	41 ...	9,208,737	35,362
7 ...	9,842,241	12,401	42 ...	9,173,375	38,253
8 ...	9,829,840	12,091	43 ...	9,135,122	41,382
9 ...	9,817,749	11,879	44 ...	9,093,740	44,741
10 ...	9,805,870	11,865	45 ...	9,048,999	48,412
11 ...	9,794,005	12,047	46 ...	9,000,587	52,473
12 ...	9,781,958	12,325	47 ...	8,948,114	56,910
13 ...	9,769,633	12,896	48 ...	8,891,204	61,794
14 ...	9,756,737	13,562	49 ...	8,829,410	67,104
15 ...	9,743,175	14,225	50 ...	8,762,306	72,902
16 ...	9,728,950	14,983	51 ...	8,689,404	79,160
17 ...	9,713,967	15,737	52 ...	8,610,244	85,758
18 ...	9,698,230	16,390	53 ...	8,524,486	92,832
19 ...	9,681,840	16,846	54 ...	8,431,654	100,337
20 ...	9,664,994	17,300	55 ...	8,331,317	108,307
21 ...	9,647,694	17,655	56 ...	8,223,010	116,849
22 ...	9,630,039	17,912	57 ...	8,106,161	125,970
23 ...	9,612,127	18,167	58 ...	7,980,191	135,663
24 ...	9,593,960	18,324	59 ...	7,844,528	145,830
25 ...	9,575,636	18,481	60 ...	7,698,698	156,592
26 ...	9,557,155	18,732	61 ...	7,542,106	167,736
27 ...	9,538,423	18,981	62 ...	7,374,370	179,271
28 ...	9,519,442	19,324	63 ...	7,195,099	191,174
29 ...	9,500,118	19,760	64 ...	7,003,925	203,394
30 ...	9,480,358	20,193	65 ...	6,800,531	215,917
31 ...	9,460,165	20,718	66 ...	6,584,614	228,749
32 ...	9,439,447	21,239	67 ...	6,355,865	241,777
33 ...	9,418,208	21,850	68 ...	6,114,088	254,835
34 ...	9,396,358	22,551	69 ...	5,859,253	267,241

10. The odds which are posted by racetracks are the "odds against" a particular horse, dog, or car entered in the event. For each of the following bets and odds, determine how much money the bettor receives back on a win.

	a.	★b.	c.
Bet	$2	$6	$10
Odds	3 to 2	4 to 1	7 to 5

11. Suppose that there are 3 red, 4 blue, and 5 green chips in a bag and that you win by selecting either a red or a green chip. If you get a red chip you win 3 dollars, and a green chip pays 2 dollars.

★ a. What is the mathematical expectation?

b. If it costs $1.50 to play this game, would you expect to win or to lose in the long run?

12. One type of bet that can be made in roulette is to place a chip on a single number. If the ball lands in the compartment with your number, the "house" pays you 36 chips.

★ a. What is the probability that the ball will land on 13?

b. If each chip is worth $1, what is the mathematical expectation in this game?

★ c. Is the mathematical expectation for playing a color (as computed on page 478) greater than, less than, or equal to that for playing a particular number?

13. There are six digits under the six squares on this sweepstakes ticket. If the same digit occurs five times you win $10,000, and so forth, as shown on the ticket. Assuming that only the digits 1, 2, 3, 4, 5, 6, 7, 8, and 9 are used, compute the following probabilities.

★ a. A 3 on the first square.

★ b. Two 3s on the first two squares.

★ c. Two numbers of the same kind on the first two squares.

d. Determine the mathematical expectation for the following outcomes and hypothetical probabilities.

Outcome	Probability	Prize
5 of a kind	$\dfrac{1}{100,000}$	$10,000
4 of a kind	$\dfrac{1}{5000}$	$1,000
3 of a kind	$\dfrac{1}{20}$	$5
2 of a kind	$\dfrac{1}{10}$	$2

14. In the Nineteenth Big Giveaway by the Publishers Clearing House each person receives 14 prize numbers. These numbers are part of a lot of 1540 numbers. If a person returns these numbers, he/she has 14 chances out of 1540 of getting the lucky number in his/her lot.

★ a. What is the probability of getting the lucky number in a given lot?

b. If you get the lucky number from your lot, you go into an End-of-Giveaway drawing of 60,000 lucky numbers. Assuming that this happens, what is the mathematical expectation of getting the Grand Prize of $100,000?

★ c. Suppose that the probabilities of winning a 5th, 6th, 7th, 8th, 9th, or 10th prize are 1/500, 1/250, 1/100, 1/50, 1/25, and 1/10, respectively. What is the mathematical expectation for winning one of these prizes?

10 5th PRIZES—$500 each
25 6th PRIZES—$250 each
100 7th PRIZES—$100 each
500 8th PRIZES—$50 each
2,500 9th PRIZES—$25 each
5,000 10th PRIZES—$10 each

and thousands of specially minted Bicentennial medallions

Photo from NASA's Synchronous Meteorological Satellite-1

10.2 COMPOUND PROBABILITIES

Meteorologists use computers and probability to analyze weather patterns. In recent years meteorological satellites have improved the chances of accurate weather forecasting. NASA's Synchronous Meteorological Satellite-1 returned the above photograph of the Western Hemisphere on May 28, 1974. Four storms can be seen across the top of the picture from Western Canada to the Atlantic Ocean. The small-scale structure of cumulus clouds over Florida and the Caribbean Sea allows meteorologists to infer wind speed and direction. Weather forecasts are usually stated in terms of probability: "There is a 20% chance of rain." The probability we are given is determined by a combination of probabilities. For example, there may be one probability of a cold front approaching from the north and a different probability for a storm approaching from the west. The probability of two or more events happening is called a *compound probability*.

Many of the questions which led to the development of probability contained two or more events. For example, if one die is to be rolled twice, what is the chance of getting an even number on each roll? This table shows that there are 36 possible outcomes for these two events and that 9 of these contain pairs of even numbers (circled in table). Therefore, the probability of both numbers being even is 9/36 or 1/4. Find all the pairs of numbers which have at least one 3. What is the probability of rolling a die twice and getting at least one 3?

Second Toss

	·	··	∴	::	∵:	:::
·	1,1	1,2	1,3	1,4	1,5	1,6
··	2,1	(2,2)	2,3	(2,4)	2,5	(2,6)
∴	3,1	3,2	3,3	3,4	3,5	3,6
::	4,1	(4,2)	4,3	(4,4)	4,5	(4,6)
∵:	5,1	5,2	5,3	5,4	5,5	5,6
:::	6,1	(6,2)	6,3	(6,4)	6,5	(6,6)

First Toss

Independent Events   In the preceding examples, the first toss of the die had no effect on the outcome of the second toss. Two or more events that have no influence on each other are called *independent events*. The probability of several independent events all happening can be computed by the following formula.

Theorem: The probability of independent events is the product of the separate probabilities of each event.

Let's apply this formula to one of the preceding examples. The probability of rolling an even number on a die is 1/2. Therefore, the probability of an even number on both tosses of the die is $1/2 \times 1/2 = 1/4$.

Here is another example of independent events. What is the probability of drawing a 4 from the five cards shown here and also rolling a 5 on the die? The probability of drawing a 4 is 3/5 (3 of the 5 cards have 4s), and the probability of rolling a 5 is 1/6. By the formula for the probability of independent events, the probability of drawing a 4 and rolling a 5 is $3/5 \times 1/6 = 3/30$ or 1/10. We can see that this probability is correct by listing all possible outcomes (as shown in the table) and then counting those which consist of a 4-card and a toss of 5 (circled pairs).

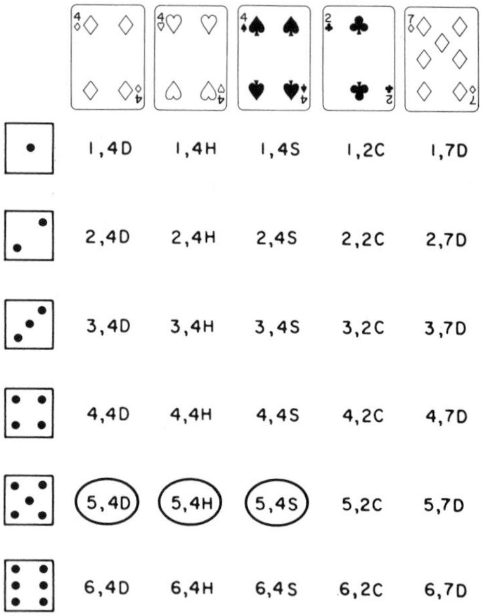

The formula we are using shows that the probability of two events occurring in succession is much poorer than the probability of either event. Consider the chance of drawing two aces in a row from a deck of 52 cards. The probability of selecting an ace on the first draw is 4/52 or 1/13. If the card selected is then replaced, the probability of drawing an ace the second time is also 1/13. These draws are independent of each other, so the probability of selecting two aces is $1/13 \times 1/13$, or 1/169.

die	4♦	4♥	4♠	2♣	7♦
1	1,4D	1,4H	1,4S	1,2C	1,7D
2	2,4D	2,4H	2,4S	2,2C	2,7D
3	3,4D	3,4H	3,4S	3,2C	3,7D
4	4,4D	4,4H	4,4S	4,2C	4,7D
5	(5,4D)	(5,4H)	(5,4S)	5,2C	5,7D
6	6,4D	6,4H	6,4S	6,2C	6,7D

Dependent Events   Suppose that the first card selected in the previous example is not replaced for the second drawing. Then the chance of drawing an ace on the second draw depends on the outcome of the first. If the first card is an ace, then only three of the remaining 51 cards are aces. In this case the probability of an ace on the second draw is 3/51. When one event affects the outcome of another, they are called *dependent events*. As with independent events, the probability of two dependent events can be computed by multiplying the probabilities of each event. In this example, the probability of drawing two aces is $1/13 \times 3/51 = 3/663$ or 1/221. As you would expect, this probability is smaller than getting two aces when the first card is replaced (1/169).

In most card games the different outcomes are dependent events. What is the probability of being dealt two hearts from the following simplified deck of cards: 9 of hearts, 8 of hearts, 2 of hearts, 9 of clubs, and 5 of spades? The probability of a heart on the first card is 3/5, and if the first card is a heart, the chance that the second card will be a heart is 2/4. The probability of getting two hearts is 3/5 × 2/4 = 6/20 or 3/10. This can be verified by comparing the pairs of cards with both hearts (circled pairs) to the total number of pairs in the table.

	9H,8H	9H,2H	9H,9C	9H,5S
8H,9H		8H,2H	8H,9C	8H,5S
2H,9H	2H,8H		2H,9C	2H,5S
9C,9H	9C,8H	9C,2H		9C,5S
5S,9H	5S,8H	5S,2H	5S,9C	

In most familiar card games there is such a variety of combinations that one rarely gets the same hand twice. The total number of different five-card poker hands is 2,598,960, and there are 635,013,559,600 different bridge hands. The chance of getting

Poker hand – a flush

any particular hand involves dependent events. Consider, for example, the chance of receiving a flush in poker (all five cards of the same suit). The first card dealt could be any suit. Suppose it were a club. The second card must also be a club and the probability of getting it would be 12/51. The probabilities that the third, fourth, and fifth cards are clubs are 11/50, 10/49, and 9/48, respectively. Thus, the probability of getting five clubs (or five cards of any one suit) is

$$\frac{12}{51} \times \frac{11}{50} \times \frac{10}{49} \times \frac{9}{48} = \frac{11,880}{5,997,600} \quad \text{or about} \quad \frac{1}{505}$$

Binomial Probability According to birth records there are more boys born than girls. However, since each sex is approximately 50% of the total, it is customary to assume that the probability of a girl is 1/2 and that for a boy is 1/2. If we also assume that having two or more children are independent events, then the probability of two girls is 1/2 × 1/2 = 1/4. Similarly, the probability of having three boys is 1/2 × 1/2 × 1/2 = 1/8. The probability of several repeated independent events in which there are two outcomes for each event is called a *binomial probability*.

The Harrison family from Tennessee had 13 boys. The probability of this happening is

$$\frac{1}{2} \times \frac{1}{2} \times \frac{1}{2} \times \frac{1}{2} \times \frac{1}{2} \times \frac{1}{2} \times \frac{1}{2} \times \frac{1}{2} \times \frac{1}{2} \times \frac{1}{2} \times \frac{1}{2} \times \frac{1}{2} \times \frac{1}{2} = \frac{1}{8192}$$

The chance of 14 consecutive boys is 1/16,384. Since the probability of this happening is very small, you might expect that if a couple had 13 boys, the next child would likely be a girl. However, since each event is independent, the probability of another boy is still 1/2.

Here is another example that is contrary to our intuition. It seems reasonable to expect that if a couple is planning to have 4 children, the probability of having 2 boys and 2 girls is 1/2. To determine the correct probability, consider the combinations of 4 children which are listed in the right-hand column of this tree diagram. The various paths of this diagram show that there are 16 combinations and that only 6 of these paths have 2 boys and 2 girls. Therefore, the probability of 2 boys and 2 girls is 6/16 (or 3/8) rather than 1/2.

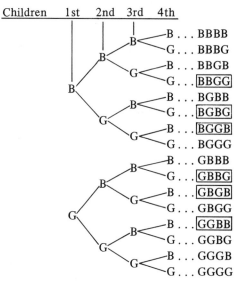

Complementary Probabilities   The probability that an event will happen and the probability that it will not happen are called *complementary probabilities*. The sum of two complementary probabilities is always 1. The probability of not getting a sum of 7 on a roll of two dice is 5/6, because the probability of getting a 7 is 1/6. The 5/6 and 1/6 are complementary probabilities.

There is a particular type of problem in which the use of complementary probabilities is especially convenient. Consider the following example:

What is the probability of getting at least *one 6 in four rolls of a die?*

One way to handle this problem is to find the probability of not getting a 6 on any roll. There is a 5/6 chance of not getting a 6 each time the die is rolled, and so the probability of no 6s in four rolls is

$$\frac{5}{6} \times \frac{5}{6} \times \frac{5}{6} \times \frac{5}{6} = \frac{625}{1296} \qquad \text{or approximately 48\%}$$

The complement to this probability, 52%, is the probability of getting at least one 6.

In this example, finding the probability of rolling at least one 6 without using complementary probability is much more difficult. This is because of the many ways of succeeding. Rolling a 6 four times in succession or getting two 6s in four rolls, etc., are all ways of satisfying the condition of "at least one 6."

The Birthday Problem—This famous problem has a solution that is unexpected and difficult to believe.

What is the smallest arbitrary group of people for which there is better than a 50% chance that at least two of them will have a birthday on the same day of the year?

Surprisingly, the answer is only 23 people. In fact, for 30 people the probability of two birthdays on the same day is 70%, and for 50 people it is about 97%. This graph shows that for more than 50 people we can be almost certain of two birthdays on the same day.

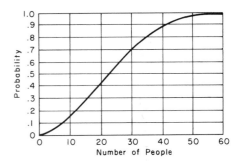

The Birthday Problem can be solved by complementary probabilities. That is, we will determine the probability that all 23 people have birthdays on different days, and then subtract this probability from 1. To begin with, consider the problem for just two people. It doesn't matter what date the first person's birthday is on; there is a probability of 364/365 that the second person's birthday *will not* be on the same day. When a third person joins this group the probability that his/her birthdate will differ from those of the other two people is 363/365. Therefore, the probability that three people will not share a birthdate is 364/365 × 363/365.

Similarly, the probability that all 23 people will have different birthdays is

$$\frac{364}{365} \times \frac{363}{365} \times \frac{362}{365} \times \cdots \times \frac{344}{365} \times \frac{343}{365} \quad \text{or approximately 49\%}$$

Therefore, the probability that there will be two or more birthdays on the same day, for a group of 23 people, is 100% − 49% or 51%. Make a prediction next time you're in a group of 23 or more people. The odds will be in your favor that there will be at least two birthdays on the same day.

SUPPLEMENT *(Activity Book)*

Activity Set 10.2 Compound Probability Experiments

Just for Fun: Trick Dice

Additional Sources

Bergamini, D. *Mathematics.* New York: Time-Life Books, 1963. pp. 126–47.

Fey, J.T. "Probability, Integers, and Pi." *The Mathematics Teacher,* **64** No. 4 (April 1971), 329–32.

Gardner, M. *The Scientific American Book of Mathematical Puzzles and Diversions.* New York: Simon and Schuster, 1959. "Probability Paradoxes," pp. 47–54.

Newman, J.R. and E. Kasner. *Mathematics and the Imagination.* New York: Simon and Schuster, 1940. "Chance and Chanceability," pp. 223–64.

Ogilvy, C.S. *Through the Mathscope.* New York: Oxford University Press, 1956. "What are the Chances," pp. 29–37.

Rudd, D. "A Problem in Probability." *The Mathematics Teacher,* **67** No. 2 (February 1974), 180–81.

If you had it all to do over, would you change anything? "Yes, I wish I had played the black instead of the red at Cannes and Monte Carlo."

Winston Churchill

EXERCISE SET 10.2

1. Find the following roulette wheel probabilities. (There are 18 red, 18 black, and 2 green compartments.)

   ★ a. Winning twice in a row on red.

   b. Losing twice in a row if you bet on red each time. (*Hint:* You lose if the ball lands in a green or a black compartment.)

   ★ c. Winning three times in a row on black.

   d. Winning 26 times in a row on black. Indicate your answer, but don't multiply it out. While the chance of black coming up 26 times is very small, it has happened before as related by Darrell Huff in *How To Take A Chance.* * On August 18, 1913, at a casino in Monte Carlo, black came up 26 times in a row on a roulette wheel. After about the 15th occurrence of black, there was a near-panic rush to bet on red. People kept doubling their bets in the belief that, after black came up the 20th time there was not a chance in a million of another repeat. The casino came out ahead by several million francs.

★ 2. If you flipped a fair coin 9 times and got 9 heads, what is the probability of a head on the next toss?

3. Classify the following events as *dependent* or *independent* and compute their probabilities.

   ★ a. Tossing a coin 3 times and getting 3 heads in a row.

   ★ b. Drawing 2 aces from a complete deck of 52 playing cards. (The first card selected is not replaced.)

   c. Rolling 2 dice and getting a sum of 7 twice in succession.

   d. Selecting 2 green balls from a bag of 5 green and 3 red. (The first ball selected is not replaced.)

*D. Huff, *How to Take a Chance* (New York: W.W. Norton, 1959), pp. 28–29.

4. Alice and Bill make one payment each week and it is determined by the "debits spinner." Assume each outcome on the spinner is equally likely.

⭐ a. What is the probability of making a fuel payment 2 weeks in a row?

b. What is the probability they will not make an electricity payment this week?

⭐ c. If they don't make an electricity payment within the next 3 weeks, their lights will be shut off. What is the probability they will lose their lights?

"O.K., Alice, spin the wheel and let's see who gets paid this week."

5. The paths of this tree diagram show the various combinations of 3 children. There are 3 ways of having 2 girls (BGG, GBG, and GGB); there is 1 way of having no girls, that is, all 3 children are boys; etc.

a. Complete the following table with the numbers of combinations for having 0, 1, 2, or 3 girls in a family of 3 children.

⭐ b. What is the probability of having 1 girl and 2 boys?

c. If a family has 2 girls, what is the probability that the next child will be a boy?

1st	2nd	3rd	
B	B	B	... BBB
		G	... BBG
	G	B	... BGB
		G	... BGG
G	B	B	... GBB
		G	... GBG
	G	B	... GGB
		G	... GGG

Number of Girls	0	1	2	3
Combinations	I		3	

6. A couple is planning to have 4 children.

a. Determine the numbers of combinations for which there will be 0, 1, 2, 3, or 4 girls. (*Hint:* See the tree diagram on page 488.)

⭐ b. The numbers in Pascal's triangle (page 7) can be used to find the numbers of combinations in part **a** of Exercises 5 and 6. Explain how.

Number of Girls	0	1	2	3	4
Combinations					

```
        I
      I   I
     I  2  I
    I  3  3  I
   I  4  6  4  I
  I  5  10 10  5  I
```

c. Use your observations in part **b** to determine the numbers of combinations for having 0, 1, 2, 3, 4, or 5 girls in a family of 5 children. Write them in the table.

Number of Girls	0	1	2	3	4	5
Combinations						

★ d. What is the probability of having 1 girl and 4 boys in a family of 5 children?

7. A bureau drawer contains 10 black socks and 10 brown socks. Assuming the socks are to be selected in the dark, find the probabilities of the following events.

★ a. Selecting 2 black socks.

b. Selecting a black sock and then a brown sock.

★ c. Selecting 3 socks and having 2 of the same color.

8. On a test of 10 true or false questions what is the probability of getting them all correct by guessing?

© 1973 United Feature Syndicate, Inc.

9. In an experiment designed to test estimates of probability, people were given the three alternatives in parts **a** and **b**. In each case which response seems most preferable? Check your answers by determining the probabilities.

★ a. 1) Drawing once for a winning ticket in a box of 10.

2) Drawing for a winning ticket in a box of 5, two different times, and winning both times.

3) The chances are the same in both cases.

b. 1) Drawing once for a winning ticket in a box of 10.

2) Drawing 2 times for a single winning ticket in a box of 20. (*Hint:* There are 190 different combinations of 2 tickets and 19 winning pairs.)

3) The chances are the same in both cases.

10. Determine the probabilities of the following events. (*Hint:* Use complementary probabilities.)

★ a. Getting a sum of 7 at least once on four rolls of a pair of dice.

b. Getting at least one 6 on four rolls of one die.

★ c. Getting at least one sum of 7 or 11 on three rolls of a pair of dice.

11. What is the probability that the next stranger you meet will have a birthday on the same day as yours?

12. Mr. and Mrs. Petritz of Butte, Montana, have 5 children who were all born on April 15. In the following questions, assume that it is equally likely that a child will be born on any of the 365 days of the year.

a. After the first Petritz child was born, what was the probability the second child would be born on April 15?

★ b. The third and fourth Petritz children are twins. What was the probability that the second and third children would be born on April 15?

c. If a couple has a child on April 15, what is the probability that their next 3 children will be born on April 15, if there are no multiple births?

13. The typical slot machine has three wheels which operate independently of one another. Each wheel has six different kinds of symbols which occur various numbers of times, as shown in the chart. If any one of the winning combinations appears in a row, the player wins money according to the payoff assigned to each combination. Find the probabilities of the following events.

	Wheel I	Wheel II	Wheel III
Cherries	7	7	0
Oranges	3	6	7
Lemons	3	0	4
Plums	5	1	5
Bells	1	3	3
Bars	1	3	1
	20	20	20

★ a. A bar on wheel I.

b. A bar on all three wheels.

★ c. Bells on wheels I and II and a bar on wheel III.

d. Plums on wheels I and II and a bar on wheel III.

14. In Sweden a motorist was accused of overparking in a restricted time zone. A police officer testified that this particular parked car was seen with the tire valves pointing to one o'clock and to six o'clock. When the officer returned later (after the allowed park-
ing time had expired), this same car was there with its valves pointing in the same directions—so a ticket was written. The motorist claimed that the car had been driven away from that spot during the elapsed time, and upon returning later the tire valves coincidentally happened to come to rest in the same positions as before.

★   a. Using only the twelve hour-hand positions, what is the probability that two given tire valves of a driven car would return to the same position? (Assume that due to variations in tire sizes and turning corners, the tires will not turn the same amount.)

     b. The driver was acquitted but the judge remarked that if the position of the tire valves of all four wheels had been recorded and found to point in the same direc- tion, the coincidence claim would be rejected as too improbable.* What is the probability of all four tire valves returning to the same position?

*See F. Mosteller et al., *Statistics: A Guide to the Unknown* (New York: Holden-Day, 1972), p. 102.

"Hello? Beasts of the Field? This is Lou, over in Birds of the Air. Anything funny going on at your end?"

10.3 DESCRIPTIVE AND INFERENTIAL STATISTICS

The employees of the Animated Animal Company keep records of the number of toys each machine produces and the number of breakdowns. The table shows that machine III outproduced machines I and II, and machine II had the most problems. Numerical information such as this is often called statistics. More technically,

	M	T	W	Th	F	Break-downs
Machine I	165	158	98	125	260	13
Machine II	117	82	46	6	30	24
Machine III	182	243	196	305	261	4

Weekly record of birds produced

statistics is the branch of mathematics which deals with the analysis of numerical data. There are two main branches: descriptive statistics and inferential statistics. The processes of describing and interpreting data is called *descriptive statistics.* The average number of birds produced weekly by Machine I is an example of descriptive statistics. On the other hand, the processes for using data to predict future outcomes is called *inferential statistics.* For example, sampling is an important part of inferential statistics. If the manager of Birds of the Air randomly selects 100 birds off the assembly line and finds that three are defective, he can estimate that the number of bad birds in a batch of 5000 will be 150.

Statistics had its beginning in the seventeenth century through the work of the English businessman John Graunt. Graunt used a publication called "Bills of Mortality," which listed births, christenings, and deaths. Here are some of his conclusions: The number of male births exceeds the number of female births; there is a higher death rate in urban (as opposed to rural) areas; and more men die violent deaths than women. Graunt used this source to publish his book *Natural and Political Observations of Mortality.* In his work he summarized great amounts of data to make it understandable (descriptive statistics) and made conjectures about large populations based on small samples (inferential statistics).

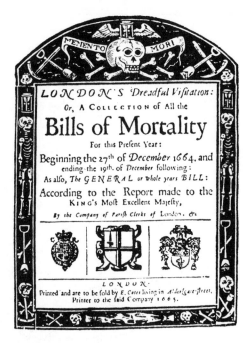

Descriptive Statistics   The table on the next page lists the statistics for the Miss America winners from 1946 to 1978. By a quick glance at these columns of numbers it is difficult to know which measurements occur most often and which occur least often. There are several ways of summarizing such lists of data. One method is a *frequency distribution,* as shown on the next page. This table contains the heights of the Miss Americas to the nearest half-inch and the number having each height. The heights of 5 feet 6 inches, 5 feet 7 inches, and 5 feet 8 inches occurred most frequently. Miss America winners with heights below 5 feet 5 inches and above 5 feet 9 inches do not occur very often. The bar graph and frequency polygon are pictorial methods of illustrating frequency distributions. Notice how these graphs are high near the middle portion of the scale and decrease in height towards the ends of the scale.

Miss America Winners

		Height	Wgt.	Age
1946	Marilyn Buferd, Los Angeles, Calif.	5-8	123	21
1947	Barbara Walker, Memphis, Tennessee	5-7	130	21
1948	BeBe Shopp, Hopkins, Minnesota	5-9	140	18
1949	Jacque Mercer, Litchfield, Arizona	5-4	106	18
1951	Yolande Betbeze, Mobile, Alabama	5-5½	119	21
1952	Colleen Kay Hutchins, Salt Lake City, Utah	5-10	143	25
1953	Neva Jane Langley, Macon, Ga...............	5-6¼	118	19
1954	Evelyn Margaret Ay, Ephrata, Pa.	5-8	132	20
1955	Lee Meriwether, San Francisco, Calif.	5-8½	124	19
1956	Sharon Ritchie, Denver, Colarado	5-6	116	18
1957	Marian McKnight, Manning, S.C.	5-5	120	19
1958	Marilyn Van Derbur, Denver, Colorado	5-8½	130	20
1959	Mary Ann Mobley, Brandon, Miss..............	5-5	114	21
1960	Lynda Lee Mead, Natchez, Miss.	5-7	120	20
1961	Nancy Fleming, Montague, Michigan	5-6	116	18
1962	Maria Fletcher, Asheville, N.C.	5-5½	118	19
1963	Jacquelyne Mayer, Sandusky, Ohio	5-5	115	20
1964	Donna Axun, El Dorado, Arkansas	5-6½	124	21
1965	Vonda Kay Van Dyke, Phoenix, Ariz.	5-6	124	21
1966	Deborah Irene Bryant, Overland Park, Kansas	5-7	115	19
1967	Jane Anne Jayroe, Laverne, Oklahoma	5-6	116	19
1968	Debra Dene Barnes, Morin, Kansas	5-9	135	20
1969	Judith Anne Ford, Belvidere, Ill.	5-7	125	18
1970	Pamela Anne Eldred, Birmingham, Mich.	5-5½	110	21
1971	Phyllis Ann George, Denton, Texas	5-8	121	21
1972	Laurie Lea Schaefer, Columbus, Ohio	5-7	118	22
1973	Terry Anne Meeuwsen, DePere, Wisconsin	5-8	120	23
1974	Rebecca Ann King, Denver, Colorado	5-9	125	23
1975	Shirley Cothran, Fort Worth, Texas	5-8	119	21
1976	Tawney Elaine Godin, Yonkers, N.Y............	5-10½	128	18
1977	Dorothy Kathleen Benham, Edina, Minn.	5-7½	120	20
1978	Susan Perkins, Columbus, Ohio	5-6	105	23

Frequency Distribution

Height (Feet-Inches)	5-4	5-4½	5-5	5-5½	5-6	5-6½	5-7	5-7½	5-8	5-8½	5-9	5-9½	5-10	5-10½
Frequency	1	0	3	3	5	2	5	1	5	2	3	0	1	1

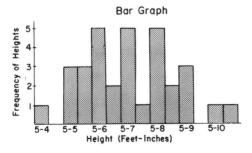

Bar Graph

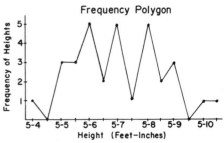

Frequency Polygon

Sometimes the data are spread over a wide range, and there may be only a few or even no numbers occurring with the same frequency. In this case it is convenient to group the data in intervals. The numbers of days for the gestation periods of animals have been grouped in 30-day intervals in the frequency distribution that follows. This table shows that two animals have a gestation period of less than 30 days and the most common gestation periods fall in intervals of 30 to 59 days, 60 to 89 days, and 90 to 119 days.

Animal	Days
Ass	365
Baboon	187
Badger	60
Bat	50
Bear	
Black	219
Grizzly	225
Polar	240
Beaver	122
Buffalo	278
Camel	406
Cat (domestic)	63
Chimpanzee	231
Chipmunk	31
Cow	284
Deer	201
Dog	61
Elk	250
Fox	52
Giraffe	425
Goat (dom.)	151
Goat (mtn.)	184
Gorilla	257
Guinea Pig	68
Horse	330
Kangaroo	42
Leopard	98
Lion	100
Monkey	165
Moose	240
Mouse (meadow)	21
Opossum	14–17
Pig	112
Puma	90
Rabbit	37
Rhinoceros	498
Sea Lion	350
Sheep	154
Squirrel	44
Tiger	105
Whale	365
Wolf	63
Zebra	365

Number of Days in Gestation Period	Frequency
0– 29	2
30– 59	6
60– 89	5
90–119	5
120–149	1
150–179	3
180–209	3
210–239	3
240–269	4
270–299	2
300–329	0
330–359	2
360–389	3
390–419	1
420–449	1
450–479	0
480–509	1

Measures of Central Tendency   Another approach to summarizing data is to represent all the values by one number, such as an average. In statistics there are three types of averages: the *mean,* the *median,* and the *mode.* These are called *measures of central tendency.* To illustrate these measures consider this table of major twentieth century earthquakes. Scanning the deaths column we see that the greatest life toll was 180,000 and the least was 65. Let's see how the three measures of central tendency can be used to "average" these extremes and to obtain "typical" values.

Major Twentieth Century Earthquakes

Year	Date	Location	Deaths
1906	Apr. 18–19	Cal., San Francisco	452
1906	Aug. 16	Chili, Valparaiso	1,500
1908	Dec. 28	Italy, Messina	75,000
1915	Jan. 13	Italy, Avezzano	29,970
1920	Dec. 16	China, Kansu	180,000
1923	Sep. 1	Japan, Tokyo	143,000
1932	Dec. 26	China, Kansu	70,000
1933	Mar. 10	Cal., Long Beach	115
1935	May 31	India, Quetta	60,000
1939	Jan. 24	Chili, Chillan	30,000
1939	Dec. 27	Turkey, Erzincan	23,000
1946	May 31	Eastern Turkey	1,300
1946	Dec. 21	Japan, Honshu	2,000
1948	June 28	Japan, Fukui	5,131
1949	Aug. 5	Ecuador, Pelileo	6,000
1950	Aug. 15	India, Assam	1,500
1953	Mar. 18	Northwestern Turkey	1,200
1954	Sep. 9–12	Northern Algeria	1,657
1956	June 10–17	Northern Afghanistan	2,000
1957	July 2	Northern Iran	2,500
1957	Dec. 13	Western Iran	2,000
1960	Feb. 29	Morocco, Agadir	12,000
1960	May 21–30	Southern Chili	5,700
1962	Sep. 1	Northwestern Iran	10,000
1963	July 26	Yugoslavia, Skopje	1,100
1964	Mar. 27	Alaska	131
1966	Aug. 19	Eastern Turkey	2,529
1968	Aug. 31	Northeastern Iran	11,588
1970	Mar. 28	Western Turkey	1,086
1970	May 31	Northern Peru	66,794
1971	Feb. 9	Southern California	65
1972	Apr. 10	Southern Iran	5,057
1972	Dec. 23	Nicaragua	6,000
1974	Dec. 28	Pakistan (around 9 towns)	5,200
1975	Sep. 6	Turkey	2,312
1976	Feb. 4	Guatemala	22,778
1976	May 6	Italy	946
1976	July 28	China	100,000
1976	Nov. 26	Turkey	3,500

The *mean* is what we often refer to as "the average." This is the sum of all values divided by the number of values. The sum of the number of deaths from the 39 quakes listed here is 895,111. The mean (average) is 895,111/39 or approximately 22,952 deaths per quake.

The *median* is the middle value of a list when the numbers are listed in increasing order. Half of the values are greater than or equal to the median and half are less than or equal to it. The numbers of deaths in the 39 earthquakes are listed next in increasing order. The median or middle number in this list is 5057 (circled number). Nineteen numbers are less than 5057 and 19 numbers are greater.

65,	115,	131,	452,	946,	1086,	1100,	1200,	1300,
1500,	1500,	1657,	2000,	2000,	2000,	2312,	2500,	
2529,	3500,	(5057)	5131,	5200,	5700,	6000,	6000,	
10,000,	11,588,	12,000,	22,778,	23,000,	29,970,	30,000,		
60,000,	66,794,	70,000,	75,000,	100,000,	143,000,	180,000		

In an ordered list with an odd number of values, such as the earthquake list, the median is always the middle value. If there are an even number of values, the median is halfway between the two middle numbers. For example, consider the first four values in the earthquake list: 65, 115, 131, and 452. The median is 123, the number which is halfway between 115 and 131.

The *mode* is the value which occurs most frequently. For the earthquakes it is the number 2000, which occurs in the list three times. If there are two different values which occur the same number of times, the data have two modes and are called *bimodal*.

The mean, median, and mode each have their advantages, depending on the type of data and information desired. Which do you feel is the best measure of central tendency for describing the earthquakes: the mean, 22,952; the median, 5057; or the mode, 2000?

Sometimes the mode and median are better descriptions than the mean. In this example of the salaries of a small company, the mean, $16,312.50, is misleading. It is more than twice the salary of most of the employees. In this case both the median, 6000, and the mode, 5000, are more representative of the majority of salaries.

One president	$100,000
One vice-president	60,000
One salesperson	20,000
One supervisor	11,000
One machine operator . . .	10,000
Five mill workers (each earning)	6,000
Six apprentice workers (each earning)	5,000

Measures of Dispersion These two lists of numbers have the same mean (23), median (20), and mode (20). However, since list B is more spread out than list A, this example illustrates the need for ways of measuring dispersion. The *range* and the *standard deviation* are two such measures.

List A

30, 28, 26, 20, 20, 19, 18

List B

60, 50, 20, 20, 10, 1, 0

The *range* is the difference between the largest and smallest values. The range for list A is 12, but for list B it is 60.

The *standard deviation* measures the spread of values about the mean. It is computed by the following steps: (1) Determine the mean; (2) Find the difference between each value and the mean; (3) Square these differences; (4) Determine the mean of these squared differences; and (5) Compute the square root of this mean. These steps are used in the following tables to compute the standard deviations for lists A and B.

List A (mean = 23)				List B (mean = 23)		
Values	Difference from Mean	Squares of Differences		Values	Difference from Mean	Squares of Differences
30	7	49		60	37	1369
28	5	25		50	27	729
26	3	9		20	3	9
20	3	9		20	3	9
20	3	9		10	13	169
19	4	16		1	22	484
18	5	25		0	23	529
	Total	142			Total	3298

Standard deviation $\approx \sqrt{20.3} \approx 4.5$ Standard deviation $\approx \sqrt{471.1} \approx 21.7$

Let's review these steps by computing the standard deviation for list A: (1) The mean for list A is 23; (2) The differences between 23 and the numbers in the first column of the table are the numbers in the second column (the difference is computed so that the numbers in the second column are positive); (3) The numbers in the third column are the squares of the corresponding numbers in the second column; (4) The sum of the numbers in the third column is 142 and the mean of these numbers is $142 \div 7$, which is approximately 20.3; (5) The square root of the mean, 20.3, is approximately 4.5. Carry out these five steps to obtain the numbers in the table for list B. The average of the numbers in the third column of this table is approximately 471.1. The square root of 471.1 gives a standard deviation of about 21.7 for list B.

The standard deviation for list A is about 4.5, compared to a standard deviation of approximately 21.7 for list B. The greater standard deviation is caused by the numbers being more dispersed or spread out about the mean.

Standard deviaitons determine intervals about the mean. For list A, 1 standard deviation above the mean is 23 + 4.5, or 27.5, and 1 standard deviation below the mean is 23 − 4.5, or 18.5. The interval within plus and minus 1 standard deviation of the mean is the interval from 18.5 to 27.5. The interval within plus and minus 2 standard deviations of the mean is from 14 to 32. What is the interval within plus and minus 1 standard deviation of the mean for list B?

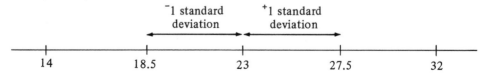

Inferential Statistics   Making predictions from samples is an important part of statistics. Because of the possibilities for errors, strict procedures must be followed in gathering a sample. The sample must be large enough and it must be a representative cross section of the whole.

The need for scientific sampling techniques was dramatically illustrated in the 1936 presidential election. The *Literary Digest,* which had been conducting surveys of elections since 1920, sent questionnaires to 10 million voters. Their sample was obtained from telephone directories and lists of automobile owners. Rather than sample voters from different income levels, they had selected people with above-average incomes. The Digest predicted that Alfred Landon would win, and instead the election was a landslide victory for Franklin D. Roosevelt. The Digest went out of publication the following year.

Topics of the day

LANDON, 1,293,669; ROOSEVELT, 972,897

Final Returns in The Digest's Poll of Ten Million Voters

Well, the great battle of the ballots in the Polls of ten million voters, scattered throughout the forty eight capital states of the Union, is now finished, and in the table below we record the figures received up to the hour of going to press.

These figures are exactly from more than op polled in our c

tran National Committee purchased *THE LITERARY DIGEST?*" And all types and variables, including: "Have the Jews purchased *THE LITERARY DIGEST?*" "Is the Pope of Rome a stockholder of *THE LITERARY DIGEST?*" And so it goes— all equally absurd and amusing. We could add more   this list, and not all of t   'ions   rent days

Many of the statistics we read in news-papers and magazines are obtained from samples. The largest regular sampling pro-cedure in the world is the survey of 47,000 households which is conducted monthly by the U.S. government's Bureau of Census. A rotation plan provides for changing one-fourth of the sample each month. To ensure that these monthly surveys will represent the civilian population, a pro-portional sample is taken from several categories, such as urban, central city, rural non-farm, and rural farm. The tech-nique of sampling in categories is called *stratified sampling.*

"*That's the worst set of opinions I've heard in my entire life.*"

Part of the survey information from the 47,000 households samples unemploy-ment. When we read that six million people are out of work, this number is derived from the government's survey as well as the numbers of people who sign up for unemployment benefits. Occasionally, the statisticians change their formulas for compiling unemployment figures. For example, from March to April 1975, the number which was obtained by the government as a measure of unemployment in Oregon increased by 15,000 people. This was not because there were fewer working, but because of a revision in the formulas for making predictions.

SUPPLEMENT *(Activity Book)*

Activity Set 10.3 Applications of Statistics (Averages, standard deviation, and sampling)

Additional Sources

Huff, D. *How to Lie with Statistics.* New York: W.W. Norton, 1954.

Fielker, D.S. *Statistics.* London: Cambridge University Press, 1967.

National Council of Teachers of Math-ematics. *Organizing Data.* Reston, Virginia: NCTM, 1969.

Newman, J.R. *The World of Mathematics,* **3.** New York: Simon and Schuster, 1956. "Foundations of Vital Statistics," pp. 1421–35, and "Sampling and Standard Error," pp. 1459–86.

"It is truth very certain that, when it is not in one's power to determine what is true, we ought to follow what is more probable."

René Descartes

EXERCISE SET 10.3

★ 1. Find the numbers of incomes from the list of per capita incomes which fall in the given intervals in the frequency distribution table. For example, there is only one income in the interval from 3800 to 4099. Use the frequencies from the table to complete the bar graph.

Frequency Distribution Table

Interval	Frequency
3800–4099	1
4100–4399	
4400–4699	
4700–4999	
5000–5299	
5300–5599	
5600–5899	
5900–6199	
6200–6499	
6500–6799	
6800–7099	

Per Capita Income by States, 1974
Source: Bureau of Labor Statistics.

State and region	Income	State and region	Income	State and region	Income
New England		**Plains**		**Southwest**	
Connecticut	6455	Iowa	5279	Arizona	5127
Maine	4590	Kansas	5500	New Mexico	4137
Massachusetts	5757	Minnesota	5422	Oklahoma	4581
New Hampshire	4944	Missouri	5036	Texas	4952
Rhode Island	5343	Nebraska	5278		
Vermont	4534	North Dakota	5583	**Rocky Mountains**	
		South Dakota	4685	Colorado	5515
Mideast				Idaho	4918
Delaware	6306	**Southeast**		Montana	4956
Dist. of Columbia	7044	Alabama	4215	Utah	4473
Maryland	5943	Arkansas	4200	Wyoming	5404
New Jersey	6247	Florida	5416		
New York	6159	Georgia	4751	**Far West**	
Pennsylvania	5447	Kentucky	4442	California	6032
		Louisiana	4391	Nevada	6016
Great Lakes		Mississippi	3803	Oregon	5284
Illinois	6234	North Carolina	4665	Washington	5710
Indiana	5184	South Carolina	4311		
Michigan	5883	Tennessee	4551	Alaska	7062
Ohio	5518	Virginia	5339	Hawaii	6042
Wisconsin	5247	West Virginia	4372		

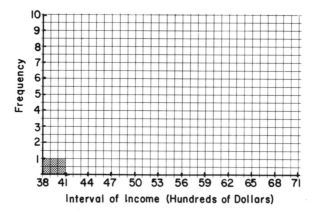

★ 2. Which interval in Exercise 1 has the most incomes (mode interval)? Which interval has the median income? The mean income for the United States in 1974 was 5486. Does this fall within the intervals containing the median or the mode?

3. Use the list at the right to count the numbers of states having different costs for gas taxes. Record these frequencies in the table and sketch the frequency polygon for this data.

Frequency Distribution

Tax	5	5.5	6	6.5	7	7.5	8	8.5	9	9.5	10
Frequency											

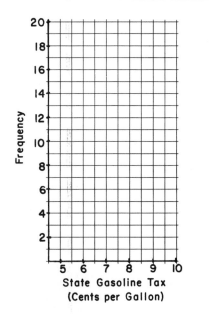

Frequency

State Gasoline Tax
(Cents per Gallon)

State	State Gas Tax (Cents per gal.)
Alabama	7
Alaska	8
Arizona	8
Arkansas	8.5
California	7
Colorado	7
Connecticut	10
Delaware	9
Florida	8
Georgia	7.5
Hawaii	5
Idaho	8.5
Illinois	7.5
Indiana	8
Iowa	7
Kansas	7
Kentucky	9
Louisiana	8
Maine	9
Maryland	9
Massachusetts	7.5
Michigan	9
Minnesota	7
Mississippi	9
Missouri	7
Montana	7
Nebraska	8.5
Nevada	6
New Hampshire	9
New Jersey	8
New Mexico	7
New York	8
North Carolina	9
North Dakota	7
Ohio	7
Oklahoma	6.5
Oregon	7
Pennsylvania	9
Rhode Island	8
South Carolina	8
South Dakota	7
Tennessee	7
Texas	5
Utah	7
Vermont	9
Virginia	9
Washington	9
West Virginia	8.5
Wisconsin	7
Wyoming	7
District of Columbia	8

4. Which of the "averages," mean, median, or mode, is the best number for describing the following instances?

★ a. The "average" size of hats sold in a store.

b. The heights of players on a basketball team.

★ c. The "average" age of seven people in a family if six of them are under 40 and one is 96 years old.

d. The "average" size of bicycles (by tire size) which are sold by a bike shop.

5. According to this description from Norris and McWhirter's 10 best oddities, a woman golfer took 166 strokes for the 130-yard 16th hole in a Ladies Invitational Golf Tournament. Hypothetical scores for the remaining 17 holes are given in the table.

> The *The Worst Woman Golfer.* "A woman player in the qualifying round of the Shawnee Invitational for Ladies at Shawnee-on-Delaware, Pa., in c. 1912, took 166 strokes for the 130-yard 16th hole. Her tee shot went into the Binniekill River and the ball floated. She put out in a boat with her exemplary, but statistically minded, husband at the oars. She eventually beached the ball 1½ mi. downstream, but was not yet out of the woods. She had to play through one on the home stretch."

Number of Hole	1	2	3	4	5	6	7	8	9	10	11	12	13	14	15	16	17	18
Score	4	5	4	5	4	6	4	5	5	4	7	3	7	5	5	166	7	6

★ a. What was her average (mean) score for the 18 holes?

b. What were her median and mode scores?

★ c. Which of these "averages" is most appropriate for summarizing her performance?

d. If it took 165 strokes to the green from 1.5 miles (7920 feet) downstream, what was the average (mean) length in feet of each return shot?

6. *Calculator Exercise:* Whenever anyone mentions the good old days, they were always "good." How good were they? Here are some surprising facts which were reported by the National Institute of Health Federal Credit Union.

★ a. Divide each 1925 time by the corresponding 1976 time to find out how many times longer you would have worked for each item in 1925. Compute your answer to 2 decimal places.

b. Compute the average (mean) of your answers in part **a**. On the average, how many times longer would you have had to work for these items in 1925 compared to 1976?

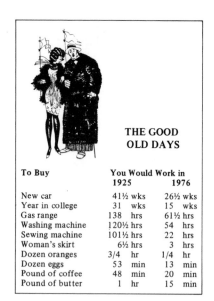

THE GOOD OLD DAYS

To Buy	You Would Work in	
	1925	1976
New car	41½ wks	26½ wks
Year in college	31 wks	15 wks
Gas range	138 hrs	61½ hrs
Washing machine	120½ hrs	54 hrs
Sewing machine	101½ hrs	22 hrs
Woman's skirt	6½ hrs	3 hrs
Dozen oranges	3/4 hr	1/4 hr
Dozen eggs	53 min	13 min
Pound of coffee	48 min	20 min
Pound of butter	1 hr	15 min

7. The home run leaders in the National and American Leagues from 1960 to 1977 are given at the top of the next page. Find the mean, median, and mode for each list of home runs.

Home Run Leaders

Year	National League	H.R.	Year	American League	H.R.
1960	Ernie Banks, Chicago	41	1960	Mickey Mantle, New York	40
1961	Orlando Cepeda, San Francisco	46	1961	Roger Maris, New York	61
1962	Willie Mays, San Francisco	49	1962	Harmon Killebrew, Minnesota	48
1963	Hank Aaron, Milwaukee		1963	Harmon Killebrew, Minnesota	45
	Willie McCovey, San Francisco	44			
1964	Willie Mays, San Francisco	47	1964	Harmon Killebrew, Minnesota	49
1965	Willie Mays, San Francisco	52	1965	Tony Conigliaro, Boston	32
1966	Hank Aaron, Atlanta		1966	Frank Robinson, Baltimore	49
	Willie McCovey, San Francisco	44			
1967	Hank Aaron, Atlanta	39	1967	Carl Yastrzemski, Boston	
				Harmon Killebrew, Minnesota	44
1968	Willie McCovey, San Francisco	36	1968	Frank Howard, Washington	44
1969	Willie McCovey, San Francisco	45	1969	Harmon Killebrew, Minnesota	49
1970	Johnny Bench, Cincinnati	45	1970	Frank Howard, Washington	44
1971	Willie Stargell, Pittsburgh	48	1971	Bill Melton, Chicago	33
1972	Johnny Bench, Cincinnati	40	1972	Dick Allen, Chicago	37
1973	Willie Stargell, Pittsburgh	44	1973	Reggie Jackson, Oakland	32
1974	Mike Schmidt, Philadelphia	36	1974	Dick Allen, Chicago	32
1975	Mike Schmidt, Philadelphia	38	1975	George Scott, Milwaukee	
				Reggie Jackson, Oakland	36
1976	Mike Schmidt, Philadelphia	38	1976	Greg Nettles, New York	32
1977	George Foster, Cincinnati	52	1977	Jim Rice, Boston	39

★ a. Which league's home run leaders have the greater mean (average) of home runs from 1960 to 1977?

b. Which league has the greater median?

★ c. Which league has the greater total?

d. In how many different years did the National League's home run leaders hit more home runs than the American League's?

e. Which league's home run leaders have the better record? Support your conclusion.

8. The grades of eight students on a 10-point test were 1, 3, 5, 5, 7, 8, 9, and 10.

★ a. Compute the mean and record it above the table.

b. Use this table to compute the standard deviation of these scores.

★ c. Another class took the same test and had the same mean. What can be said about the two sets of test scores if the second class had a standard deviation of 2?

Mean = [ ]

Score	Difference between Mean and Score	Squares of Differences
1		
3		
5		
5		
7		
8		
9		
10		
	Total	

Standard Deviation = _____

9. *Calculator Exercise:* In late 1975 the U.S. had 56 operating nuclear power plants. Their capacity of 37,995 megawatts represented 8% of the power industry's total. This table shows the top 10 nuclear power utilities.

The Top 10 in Nuclear Power

Utility	Total capacity	Nuclear capacity
	Megawatts	
Commonwealth Edison	15,337	5,100
Duke Power	11,228	2,613
TVA	26,726	2,304
Northern States Power	5,048	1,578
Florida Power & Light	9,015	1,332
Indiana & Michigan Electric	2,790	1,060
Wisconsin Electric	3,618	994
Philadelphia Electric	7,214	886
Con Edison	10,082	864
Niagara Mohawk Power	4,400	610

★ a. What is the average nuclear capacity in megawatts of these 10 companies?

b. Compute the total output in megawatts of electricity which is produced with non-nuclear power by these companies. (*Note:* Each nuclear capacity must be subtracted from the utilities' total capacity.) What is the average of these amounts?

★ c. About how many times greater is the average in part **b** than the average in part **a**?

10. Here are the pulse rates of 55 people, taken at rest. The mean of these rates is 72, and the standard deviation is approximately 9.2.

51 56 56 57 57	61 62 62 62 63	64 65 65 65 66	67 67 68 68 69
69 70 70 70 70	70 70 72 73 73	74 74 74 74 75	75 76 76 76 77
78 79 79 80 80	80 81 82 84 84	86 86 89 91 92	

a. Circle the pulse rates that are within 1 standard deviation above and 1 standard deviation below the mean. What percent of the total are these rates?

★ b. What percent of the rates are within 2 standard deviations above and 2 below the mean?

c. What percent of the rates are more than 2 standard deviations above or 2 standard deviations below the mean?

11. In the book *How to Take a Chance,** Darrell Huff relates the following study and its conclusion:

A large metropolitan police department made a check of the clothing worn by pedestrians killed in traffic at night. About four-fifths of the victims were wearing dark clothes and one-fifth light colored garments. This study points up the rule that pedestrians are less likely to encounter traffic mishaps at night if they wear or carry something white.

Explain why this conclusion does not necessarily follow from the study.

*D. Huff, *How to Take a Chance* (New York: W. W. Norton, 1959), p. 164.

12. Wildlife biologists have devised a sampling procedure for computing the numbers and types of fish in a lake. Sample catches are made by electronic shocking, and the fish are marked and returned to the lake. Sometime later, another sample is taken, and the ratio of marked fish to the number of fish in the sample is used to compute the total number of fish in the lake. In one survey 232 pickerel were caught and marked. About 2.5 months later a second sample of 329 was caught and 16 were found to be marked.

★ a. In the second sample, what is the ratio of the number of marked fish to the number of fish in the sample?

★ b. Assuming that the marked fish intermingled freely with the unmarked fish during the 2.5-month period, the ratio in part a should be equal to the ratio of 232 to the total pickerel population of the lake. Use a proportion to estimate the number of pickerel in the lake.

13. Television programming relies heavily on sampling techniques. The numbers and types of people watching television at given times determine the programs which are scheduled and the rates which are charged for advertising. The region on this map is surveyed regularly by station WOAY in West Virginia. The viewers in this region are called the Oak Hill Audience. The numbers of viewers in the following table were estimated from a sample of 704 participating households which each kept a diary for a 30-day period.

the OAK HILL television audience

	Adults 18+	Teens 12–17	Children 2–11	Total Number of Viewers
Tom Jones	25,700	6,800	8,100	40,600
Generation Gap	17,900	4,800	3,900	26,600
Let's Make A Deal	30,900	3,300	8,700	42,900

a. If the population of the Oak Hill Audience was 840,000, what percent watched Let's Make A Deal?

b. What percent of the people who watched Tom Jones were children?

c. If WOAY charged $500 for each minute of advertising on the Generation Gap program, what should they charge for each minute of advertising on the Tom Jones program?

Alexander Hamilton

James Madison

10.4 DISTRIBUTIONS

Studies of frequency distributions have a wide range of application. In some cases such studies have been used to determine authorship. Consider the solution to the authorship controversy of the *Federalist Papers.* * This is a collection of 85 political papers which were written in the eighteenth century by Alexander Hamilton, John Jay, and James Madison. There is general agreement on the authorship of all but 12 of these papers. To determine who should be given credit for these papers, frequency distributions have been compiled for certain filler words such as "by," "to," "of," "on," and several others. The graphs on the next page show the frequencies of usage of the words "by" and "to" from samples of Hamilton's and Madison's writings and the disputed papers. In both cases the graphs for the disputed papers (bottom graphs) are closer to the graphs for Madison than the graphs for Hamilton.

Let's look at the graphs for the rate of occurrence of "by." The top graph was obtained by counting all the words in each of Hamilton's 48 papers. It shows that his most frequent use of "by" was an average of 7 times (see horizontal axis) for every 1000 words (see tallest column of graph). This rate occurred in 18 out of his 48 (18/48) papers (see vertical axis). Madison's most frequent use of "by" was an average of 11 times for every 1000 words and this rate occurred in 16 out of his 50 papers (16/50). The bottom graph shows that the most frequent use of "by" in the 12 disputed papers was also 11 times per 1000. In general, the graphs for "by" show that the disputed papers more closely match Madison's papers than Hamilton's papers. The remaining three graphs show a comparison of papers for the rate of occurrence of "to." Similar distributions for other filler words also support Madison as the author of the disputed papers.

*F. Mosteller et al., *Statistics: A Guide to the Unknown* (New York: Holden-Day, 1972), pp. 164–75.

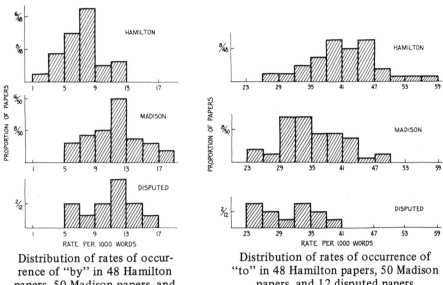

Distribution of rates of occurrence of "by" in 48 Hamilton papers, 50 Madison papers, and 12 disputed papers

Distribution of rates of occurrence of "to" in 48 Hamilton papers, 50 Madison papers, and 12 disputed papers

Randomness and Uniform Distributions

Alexander Hamilton and James Madison tended to favor certain words over others. That is, knowingly or unknowingly, their selection of words was biased. Opposite to the notion of bias is that of *randomness*. When events occur at random, there is no way to predict their outcome. For example, when selecting a digit (0 through 9) at random, any one of them is equally likely. Yet, if 1000 digits are selected at random, we can predict that each digit will occur about 100 times. Another way of saying this is that the distribution of digits should be a uniform distribution. In a *uniform distribution* each event occurs about the same number of times.

Randomness is difficult to achieve. Repeated tosses of a coin may appear to be a random method of making "yes or no" type decisions, but imbalances in the coin's weight and tossing it to approximately the same height each time are two ways of causing a biased result. Similarly, dice and spinners produce fairly random results, but these also have slight biases due to their physical imperfections.

Randomness is so important in sampling techniques and experiments, and so difficult to attain without very special efforts, that tables and books of random digits are published. One of these sources contains a million random digits.* There are hundreds of ways that computer programs can be written to generate random digits. The list of 250 digits on the next page is from such a program. The digits are printed in pairs and groups of ten for ease in reading and counting.

*Rand Corporation. *A Million Random Digits with 100,000 Normal Deviates.* (Glencoe, Illinois: Free Press, 1955).

```
40  09  18  94  06      62  89  97  10  02      58  63  02  91  44      79  03  55  47  69      14  11  42  33  99
33  19  98  40  42      13  73  63  72  59      26  06  08  92  65      63  08  82  45  85      14  45  81  65  21
69  49  02  58  44      45  45  19  69  33      51  68  97  99  05      77  54  22  70  97      59  06  64  21  68
17  49  43  65  45      04  95  82  76  31      85  53  15  21  70      59  17  27  54  67      07  76  13  95  00
43  13  78  80  55      90  80  88  19  13      13  89  11  00  60      41  86  23  07  60      22  77  93  30  83
```

Let's consider an example of how a table of random digits can be used. Suppose that there are 650 different items which are numbered with whole numbers from 1 to 650 and that you would like to take a random sample of 65. One way to accomplish this is to start at any digit in the table and list consecutive groups of three digits until you have found 65 numbers between 1 and 650. The numerals 001, 002, etc., represent 1, 2, etc., and any triples of numbers from the table which are greater than 650 will be discarded. If this method is applied to the previous table, beginning with the first line, 400 is the first number. Then 918, 940, 662, 899, and 710 are discarded because they are greater than 650. The next acceptable number is 025 which represents 25. Continuing this process will produce a random sample.

Normal Distributions   One of the most important distributions in statistics is called the *normal distribution*. This distribution has a symmetric bell-shaped curve, called the *normal curve*. The majority of values are clustered around the mean, with a gradual tapering off as the extremes are approached. The shape of these curves can vary somewhat as shown in the accompanying figures. The word "normal" is used to indicate that this type of curve is very common in nature. About 1833 the Belgian scientist L.A.J. Quetelet collected large amounts of data on human measurements: height, weight, length of limbs, intelligence, etc. He found that all measurements of mental and physical characteristics of human beings tended to be normally distributed. That is, the majority of people have measurements that are close to the mean (average), and moving away from the mean, the measurements occur less frequently. Quetelet was convinced that Nature's creation of people aims at the perfect person but misses the mark and thus creates deviations on both sides of the ideal.

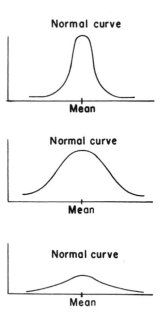

Normal distributions have the following important properties: About 68% of the values are within 1 standard deviation above and below the mean; about 96% are within 2 standard deviations of the mean; and about 99.8% fall within 3 standard deviations of the mean. These percentages hold for all normal distributions regardless of the mean or size of the standard deviation.

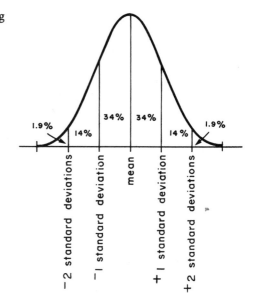

The two normal curves shown next are approximations to nearly normal distributions of measurements. On the left the distribution of college entrance examination scores has a mean of 500 and a standard deviation of 100. We know from this that 68% of the people taking these tests score between 400 and 600 and that 96% score between 300 and 700. The distribution on the right shows the frequency of heights of 8585 men. The mean is approximately 67 inches or 5 feet 7 inches. A standard deviation of 3 implies that 68% of these men are between 64 inches and 70 inches in height. Two standard deviations below the mean is 61, and 2 standard deviations above the mean is 73. The normal distribution tells us that 96% of these men have heights between 61 and 73 inches. We can also conclude that less than 2% of these men are taller than 73 inches (6 feet 1 inch).

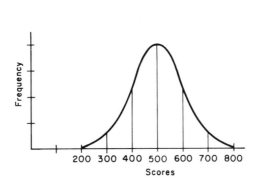

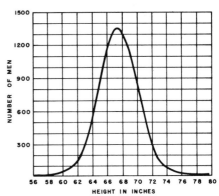

Skew Distributions   For a particular survey, this
graph shows that about 97 families have no
children, 110 families have one child, etc.
The most common number of children per
family is two. Graphs such as this, having
the data piled up at one end of the scale
and tapering off toward the other end,
are called *skewed*. The direction of skew-
ness is determined by the longer "tail" of
the distribution. This graph is *skewed to
the right* or *positively skewed*.

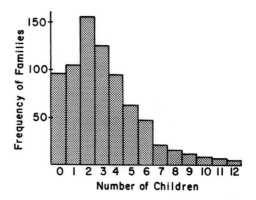

School test results sometimes produce
skewed graphs, especially if the test is too
difficult or too easy for the students. If a test which is designed for fifth graders is given
to a class of second graders, the majority of students will get low scores. In this case the
scores will pile up at the lower end of the scale and the distribution will be skewed to
the right, as shown in graph (a). If this test is given to a fifth grade class, the distribution
of scores will be more or less normal as illustrated by graph (b). Eighth graders taking
this test will obtain high scores and very few of them will score poorly. In this case the
distribution of scores will be skewed to the left, as shown in graph (c).

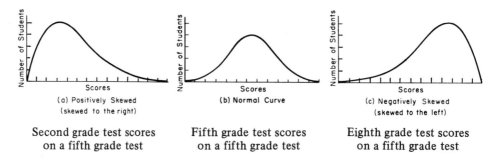

Second grade test scores   Fifth grade test scores   Eighth grade test scores
on a fifth grade test   on a fifth grade test   on a fifth grade test

Percentiles and Stanines   For a given set of values we often wish to compare one value with the
distribution of all values. This is especially important in analyzing test results. The mean
is one common method of comparison. If the mean test score is 70 and a student has a
score of 85, then we know the student has done better than "average." However, this
information does not tell us how many students scored higher than 85 or whether or not
85 was the highest score on the test.

Percentiles—One popular method of stating a person's relative performance on a test is to give the percentage of people who did not score as high. For example, a person who scores higher than 80 percent of the people taking a test is said to be in the 80th percentile. In this case only 19 percent of the people taking the test will have scored higher. Percentiles range from a low of 1 to a high of 99. The 50th percentile is the *median,* and the 25th and 75th percentiles are called the *lower quartile* and *upper quartile,* respectively.

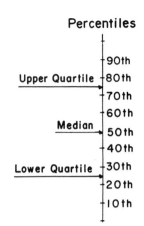

It is customary on standardized tests to establish percentiles for large samples of people. When you take such a test your score is compared to the sample. A percentile score of 65 means that you did better than 65 percent of the sample group. The following table and chart show a student's performance on a differential aptitude test. There are nine categories listed in the table at the top of this form: verbal reasoning, numerical ability, VR + NA (verbal reasoning and numerical ability), abstract reasoning, etc. Each raw score under these categories is the number of questions which the student answered correctly. The student's percentile score is obtained by comparing these raw scores with the scores from a sample of thousands of other students.

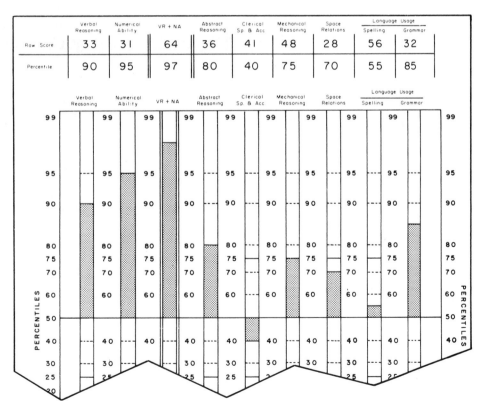

| | Verbal Reasoning | Numerical Ability | VR + NA | Abstract Reasoning | Clerical Sp. & Acc | Mechanical Reasoning | Space Relations | Language Usage | |
								Spelling	Grammar
Raw Score	33	31	64	36	41	48	28	56	32
Percentile	90	95	97	80	40	75	70	55	85

The verbal reasoning score is in the 90th percentile. This means that the student scored higher than 90 percent of the students in the sample. This student scored in the 95th percentile in numerical ability, etc. The chart below this table is a visual representation of these percentile scores. There is a horizontal line across the chart at the 50th percentile to make it easier to spot the scores above and below this level. We see that this student was below the 50th percentile in clerical speed and accuracy and just above the 50th percentile in spelling. Notice that the 75th and 25th percentiles are marked with dark horizontal line segments to show the upper quartile and the lower quartile. This student is in the upper quartile in six of these tests.

Stanines–A stanine is a value on a 9-point scale of a normal distribution. The word "stanine" is a contraction of "standard nine." Rather than dividing the scale into 100 parts, as in the case of percentiles, there are only 9 subdivisions for stanines. The stanines are numbered from a low of 1

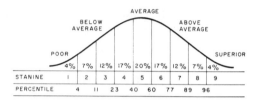

to a high of 9, with 5 representing average performance. The percentage of values for each of the stanine categories is shown under this normal curve. The first stanine has the lowest 4 percent; the second stanine has the next 7 percent; etc; to the 9th stanine which has the highest 4 percent. Thus, stanines are just subdivisions of the percentile range. If a person's test score is in the 8th stanine, we know this score is above the 88th percentile and below the 96th percentile. Stanines yield approximately the same information as percentiles, but there is the advantage of fewer subdivisions of the scale.

SUPPLEMENT *(Activity Book)*

Activity Set 10.4 Statistical Experiments (Uniform, non-uniform, and normal distributions)
Just for Fun: Cryptanalysis

Additional Sources

Gardner, M. *Mathematical Carnival.* New York: Knopf, 1975. "Random Numbers," pp. 161–72.

Kahn, D. *The Codebreakers.* New York: Macmillan, 1967.

Kramer, E. *The Main Stream of Mathematics.* New York: Oxford University Press, 1951. "Science and the Sweepstakes," pp. 154–88.

Menninger, K.W. *Mathematics in Your World.* New York: The Viking Press, 1962. "Normal and Not Normal," pp. 135–45.

Selkirk, K. "Random Models in the Classroom." *Mathematics in Schools,* **3** No. 1 (January 1974), 5–7.

Weaver, W. *Lady Luck.* New York: Doubleday, 1963.

EXERCISE SET 10.4

1. This article appeared in the *Washington Post* on November 27, 1965. The student recorded 17,950 coin flips and got 464 more heads than tails. He concluded that the United States Mint produces tail-heavy coins. For many repeated experiments of 17,950 tosses of a fair coin we can expect an approximately normal distribution with a mean of 8975 heads and a standard deviation of 67.

★ a. The area under a normal curve within plus and minus 1 standard deviation of the mean is 68% of the total area under the curve. Therefore, 68% of the time the number of heads should be between what two numbers?

b. Numbers above 3 standard deviations from the mean will occur only .1% of the time. Therefore, 99.9% of the time the number of heads should be below what number?

★ c. Edward Kelsey got 9207 heads. Was he justified in concluding that the coins are tail-heavy?

2. Computers have calculated π to 100,000 places. Sketch a frequency polygon on the grid for the occurrence of the first 200 decimal digits in π (see listing that follows part **b**, on the next page).

a. The digits of π are randomly distributed. This means that in an arbitrary sample of 200 digits, each digit would be expected to occur about 20 times. Which digits in the first 200 occur exactly 20 times?

★ b. Another condition for randomness is that each pair of digits should occur about 10 times in an arbitrary sample of 1000 digits, or about twice in a sample of 200. How many times do the following ordered pairs of digits occur in the first 200 digits of π: 23, 66, 74, 09?

Student Flips, Finds Penny Is Tail-Heavy

by Martin Weil

Washington Post Staff Writer

Edward J. Kelsey, 16, turned his dining room into a penny pitching parlor one day last spring, all in the name of science and statistics.

In ten hours the Northwestern High School senior registered 17,950 coin flips and showed the world that you didn't get as many heads as tails. You get more.

Edward got 464 more, enough to make him study the coins' balance and so discover that the United States Mint produces tail-heavy pennies.

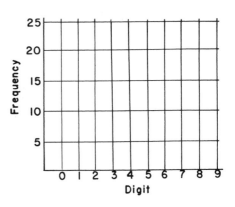

$$\pi \approx 3.14159 \ 26535 \ 89793 \ 23846 \ 26433 \ 83279 \ 50288 \ 41971 \ 69399 \ 37510$$
$$58209 \ 74944 \ 59230 \ 78164 \ 06286 \ 20899 \ 86280 \ 34825 \ 34211 \ 70679$$
$$82148 \ 08651 \ 32823 \ 06647 \ 09384 \ 46095 \ 50582 \ 23172 \ 53594 \ 08128$$
$$48111 \ 74502 \ 84102 \ 70193 \ 85211 \ 05559 \ 64462 \ 29489 \ 54930 \ 38196$$

3. The Montagnais-Naskapi, an Eastern Indian tribe, bake the shoulder blade of the caribou to help them make decisions concerning the well-being of their tribe. One bit of information they receive is determining the direction of the next hunt from the direction of the cracks which appear in the bone. This method of determining direction is a fairly random device which avoids human bias. It suggests that some practices in magic need to be reassessed.*

    a. Explain how a table of random numbers can be used to determine random directions of 0° to 360° for hunting.

★   b. How can a table of random numbers be used to determine both random directions and random distances to be travelled for the hunt?

    c. Use your method in part **b** and the list of random numbers in Exercise 4 to determine the direction and distance of your first hunt.

4. Computer programs generate random numbers for experiments and games. For example, a computer can't flip coins and roll dice, but it can read numbers and simulate these activities.

★   a. Explain how a computer can be programmed to "flip a coin" (give heads or tails) by using the following numbers from a random number table. Use your method and record the first ten "coin tosses."

    b. Devise a way for the computer to use random numbers to simulate the rolling of a die. Use your method and record the first ten "rolls of the die."

61 44 34 03 09	05 64 20 54 24	65 69 66 39 80	13 97 76 73 34
41 17 26 81 06	85 19 76 44 59	08 60 20 66 68	42 99 28 71 47
73 73 97 24 18	38 25 89 37 20		

5. Objects that are manufactured to certain specifications tend to vary slightly above and below their designed measurements. Answer the following questions by assuming the measurements are normally distributed.

★   a. A certain type of bulb has a mean life of 2400 hours with a standard deviation of 200 hours. What percent of these bulbs can be expected to burn longer than 2600 hours?

    b. A brand of crock pots has an average (mean) high temperature of 260°F and a standard deviation of 3°F. If the high temperature of these pots is below 254°F or above 266°F, they are considered defective. What percent of these pots can be expected to be defective?

*O.K. Moore, "Divination—A New Perspective," *American Anthropologist,* **59** (1965), 121–28.

6. A certain university's Watts line can handle as many as 20 calls per minute. The average number of calls per minute during peak periods is 16 with a standard deviation of 4. What percent of the time will the Watts line be overloaded during peak periods? (Assume that the numbers of phone calls are normally distributed during peak periods.)

7. One method of grading which uses a normal curve gives students within 1 standard deviation above and below the mean a grade of C. Grades of A and B are given for intervals of 2 and 3 standard deviations above the mean, and grades of D and F are given for intervals of 2 and 3 standard deviations below the mean. Answer the following questions for a test which was given to 50 students. The mean score was 78 and the standard deviation was 6.

★ a. How many students received a C?

b. How many students received a grade below C?

★ c. How many students received an A?

8. For each of the following distributions, would its curve tend to be skewed to the right, normal, or skewed to the left?

★ a. Distances that a 5th grade class of boys can throw a football.

b. Weights of newborn babies.

★ c. Numbers of people whose cars were built in the following years: 1960, 1961, . . . 1978, 1979, 1980.

d. Test scores of 5th graders taking a pretest on decimals at the beginning of the school year.

★ e. Shoe sizes of a college class of nurses.

9. Graphs of distributions are sometimes intended to be misleading. By using two different scales on the vertical axes of the following graphs, the distribution of sales for the five-year period shown in this table will be skewed in one case and uniform in the other. Illustrate these sales with bar graphs. How do these two graphs differ in the impression that they give to the reader? Which graph best illustrates the true sales increase? Explain why.

Year	Sales
1973	$191,000,000
1974	191,500,000
1975	193,000,000
1976	195,000,000
1977	198,000,000
1978	200,500,000

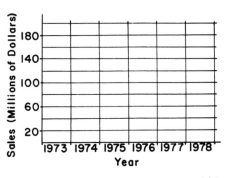

★ 10. This table and the corresponding bar graph contain the diameters to the nearest whole inch of 100 trees. The mean diameter is 12 inches and the standard deviation is approximately 2 inches. Answer the following questions to see how close this graph is to a normal distribution.

★   a. What percent of the diameters are within 1 standard deviation above and below the mean?

★   b. What percent of the diameters are within 2 standard deviations of the mean?

Frequency Table of Diameter Measures of 100 Trees of the Same Species

Diameters (inches)	Number of trees
7	2
8	5
9	8
10	10
11	13
12	26
13	12
14	9
15	8
16	4
17	3
Total	100

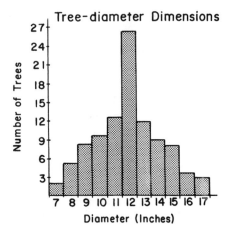

★ 11. These subtests are only a few of those on the Stanford Achievement Test, 1973 Edition. Convert each of the percentile scores to stanine scores.

NAME	SCORE TYPE	VOCAB.	ST. SKLS	MATH CONCEPTS	MATH COMP.	MATH APPL.	SPELL.	LANG.	SOCIAL SCIENCE	SCIENCE
GRADE TEST DATE AGE PUPIL NO. OTHER INFORMATION LVL FM NATL. NORMS *STANFORD* Achievement Test '73 Edition	G E NATL PR-S LOC PR-S SCALED SC NO RT	94	86	78	70	92	64	58	80	46

★ 12. A score greater than or equal to the 23rd percentile and less than or equal to the 39th is in the 4th stanine. For each stanine give the lower and upper percentiles.

Stanine	1	2	3	4	5	6	7	8	9
Lower Percentile				23					
Upper Percentile				39					

13. The device in the accompanying photo
is called a Quincunx and was described
by Sir Francis Galton in 1889. There
are 11 horizontal rows of pegs in the
top half of this Quincunx. As a ball
falls through the opening at the top
center, it strikes the center peg in the
top row and has an equal chance of
going right or left. At each lower row
the ball hits a peg and in each case has
a 50-50 chance of falling right or left.
The balls collect in compartments in
the lower half of the Quincunx. When
a large number of balls are dropped,
the distribution of the balls will be
approximately normal, as shown in
this picture.

The simplified Quincunx shown
next has only four rows of pegs. The
path of one ball is marked by arrows.
The ball has bounced left, right, left,
and left, into compartment B. By
tracing out the different ways the ball
can bounce, you will see why the
greatest number fall into the center
compartments.

In how many ways can the balls
bounce into:

★ a. compartment A?

b. compartment B?

★ c. compartment C?

d. compartment D?

★ e. compartment E?

f. How are your answers in parts a
through e related to Pascal's triangle?
(See page 7.)

★ g. If the four rows of pegs were increased
to five rows, there would be six compartments. Use Pascal's triangle to
obtain the numbers of ways the balls
can bounce into these compartments.

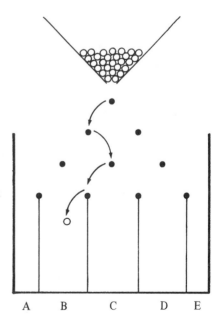

14. Cryptology is the science of making and breaking secret codes. One of the oldest recorded cryptograms dating from the fifteenth century B.C. has an unusual story surrounding it. A messenger journeyed from Persia to the house of Aristagoras, Greece, with a cryptogram tatooed on his scalp. When his head was shaved and the message was decoded, it was an order to Aristagoros from his father-in-law to start a revolt.

"I forgot the message!"

The frequencies of the letters which occur in our words is an important factor in enabling cryptographers to decode messages. Use this approach in the steps below to decode the following statement by the nineteenth century mathematician Pierre Laplace.

CA CW GZJHGBHYRZ AOHA H WVCZQVZ MOCVO YZFHQ MCAO AOZ VLQWCXZGHACLQ LK FHJZW LK VOHQVZ WOLERX YZ ZRZSHAZX AL AOZ GHQB LK AOZ JLWA CJILGAHQA WEYPZVAW LK OEJHQ BQLMRZXFZ.

a. Make a frequency distribution for the number of times each letter occurs. The four most often used letters in the English language are e, t, a, and o, in that order. This is also the order in which these letters are used in this "message." Substitute these letters into the four most frequently used letters of this code.

b. The letters, h, n, i and s, are four more which occur with high frequency in our language. Substitute these letters into the 5th, 6th, 7th and 8th most frequently used letters of the code. (*Note:* The letters which represent i and s both occur the same number of times in this message.)

c. Identify the short words. With these new letters you should be able to finish the decoding.

ANSWERS TO SELECTED EXERCISES

EXERCISE SET 1.1

1. **a.** The sum divided by 3 is the middle number. **c.** *Hint:* How many times greater is the sum than the middle number of the array?

2. **a.,b.** 3rd power 1 8 7 4 5 6 3 2 9
       4th power 1 6 1 6 5 6 1 6 1
       5th power 1 2 3 4 5 6 7 8 9

3. Each sum is one less than a Fibonacci number. This Fibonacci number is two Fibonacci numbers beyond the last number in the sum.

5. **a.** $1^2 + 1^2 + 2^2 + 3^2 + 5^2 = 5 \times 8$. **b.** In general, the sum is equal to the product of the last Fibonacci number in the sum times the next Fibonacci number.

6. **a.** 10, 14, 24, 38, 62, 100, 162, 262, 424, 686

9. 1, 2, 4, 8, 16, ... **a.** Geometric sequence **11. a.** 187

12. **a.** 1, 2, 4, 8, 16 Geometric sequence
    b. 1, 2, 4, 8, 15 Finite differences

13. **a.** $22\frac{1}{2}, 27, 31\frac{1}{2}$ Arithmetic sequence **c.** 8, 4, 0 Arithmetic sequence

15. **b.** 3, 5, 7, 9, 11, 13, 15, 17, 19, 21 Odd numbers

16. **c.** 35, 51, 70

17. **a.** 75 **b.** 75 It does not matter which of the three Wednesday dates are chosen.

EXERCISE SET 1.2

1. **a.** $4^2 + 5^2 + 20^2 = 21^2$ **3.** Try a two-digit number. **4. a.** 16

6. $6 = 1 + 5 = 2 + 4 = 3 + 3$ In general, the number of sums is the
    $7 = 1 + 6 = 2 + 5 = 3 + 4$ whole number of times 2 will divide
    $8 = 1 + 7 = 2 + 6 = 3 + 5 = 4 + 4$ into the number.
    $9 = 1 + 8 = 2 + 7 = 3 + 6 = 4 + 5$
    $10 = 1 + 9 = 2 + 8 = 3 + 7 = 4 + 6 = 5 + 5$

9. **a.** $\triangle + 1, \triangle + 2, \triangle + 3, \triangle + 4$ **b.** $(5\triangle + 10) \div 5 = \triangle + 2$. The sum of any five consecutive whole numbers divided by 5 is the middle number.

12. **a.** 48 miles per hour **c.** 11 seconds **e.** There is the same amount of liquid B in the first glass as liquid A in the second glass.

13. a. False   **c.** True   **e.** False

14. □, 5□, 5□ + 6, 20□ + 24, 20□ + 33, 100□ + 165, 100□ + 165 + △, 100□ + △

15. .4   .7   1.   1.6   2.8   5.2   10.   **a.** 2.8 astronomical units

17. Smith, English; Brown, French; Jones, logic; Robinson, mathematics

EXERCISE SET 1.3

1. *Hint:* Start both timers at the same time.

3.
Men	Women	Children	
11	15	74	There are other solutions.
14	10	76	
20	0	80	

5. *Hint:* Continue this table and look for a pattern.

Number of people	2	3	4	5	6	7
Number of handshakes	1	3	6			

6. b.

0	1	2	3	4	5	6	7	8	9
6	2	1	0	0	0	1	0	0	0

8. 1200 miles

10. a. 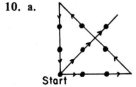   **b.**   **c.**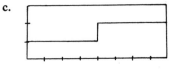

11. a. You would get 1 hour of sleep, unless the clock could be set for more than 12 hours.   **d.** The other one is a nickel and one is a quarter.

12. Smith, engineer; Robinson, fireman; Jones, brakeman

15. a. 6 marbles   **d.** Replace the toy top on the top scale by 1 block and 8 marbles (see middle scale). Then remove the same amounts from both sides of the scale.

EXERCISE SET 2.1

2. a. Column 2   **c.** These numbers are just before and just after 10 and 20.

3. a. Cardinal   **b.** Ordinal and naming use

4. a. Well-defined   **c.** Not well-defined

5. a. *SW* and *SB*, *W* and *L*   **c.** *SW* and *SB*, *SB* and *W*, *SW* and *L*, *SB* and *L*
   e. *lwt, lwr, lwh*   **6. a.** *lbt, lbr, lbh, sbt, sbr, lwt, lwr, lwh, swt, swr*

7. a. *lbt, lbr, lbh*   **8. a.** No   **c.** Yes   **9. a.** 4

10. a.

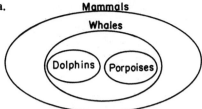

11. a.

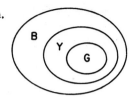

12. a. $x = 5, y = 3, z = 15$ **d.** 7   **13. a.**

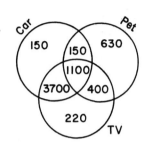

c. 150

15. a. True   **b.** False   **16.** 18 days. Five fine days and 13 days with rain

18. 10

EXERCISE SET 2.2

1. a. 1241   **c.**

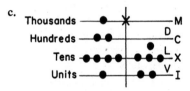

3. a. Four hands and two
b. Hand of hands, two hands and two

4. (8) Three on the other hand; (16) One on the other foot; (25) Hand on the next man.

6. b. DCIII   **7.** □□○○○ ○ − − − − − − − − ||||

8. b. Simple grouping   **9. a.** Ten   **c.** No

10. a. Five million, four hundred thirty-eight thousand, one hundred forty-six
c. Eight hundred sixteen billion, four hundred forty-seven million, two hundred ten thousand, three hundred sixty-one

11. a. $10^{23} - 1$ or 99,999,999,999,999,999,999,999

12. a.

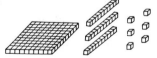

14. a. Last hands in the second and fourth columns
c. First and third columns; second and fourth columns

15. a.

16. a.

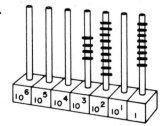

17. a. Add 800,000
19. a. 9#   **c.** 29#

EXERCISE SET 2.3

1. b. 21,600 **2. b.** $2^{16} - 1 = 65,535$ **3. a.** $2^{20} = 1,048,576$

4. a. $1101_{two} = 13$ **c.** $101101_{two} = 45$

5. a. $32 + 16 + 8 + 2 = 111010_{two}$ **6. a.** 9

7. a.

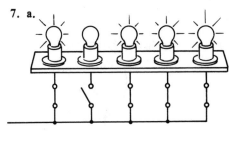

8. a.

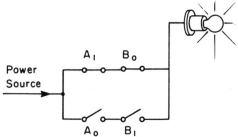

10. a. 111_{two}; Eight **c.** 32

12. a. Raise little finger, ring finger, index finger, and thumb. **c.** 1023

13. a. $2^{32} - 1$ or 4,294,967,295 **c.** I THINK. THEREFORE I AM.

15. a. Place the object on one side of the balance scale with the 1-g and 3-g weights. Place the 9-g weight on the other side.

EXERCISE SET 3.1

1. a. Units wheel and hundreds wheel **c.** Yes

2. a. The 10 was not regrouped ("1" was not carried) to the tens column.
  c. The "14" from $6 + 8$ and the "12" from $5 + 7$ were recorded side by side.

3. a and d. **a.** 4 **b.** 4 **c.** 4 **d.** 5 **5. a.** Associative **c.** Commutative

6. b.

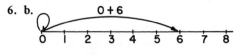

7. a.

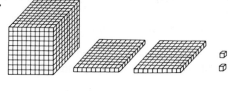

8. a.

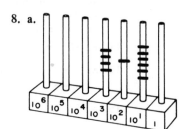

10. a.

Sum		Rounded off
238	→	200
709	→	700
+ 873	→	900
1820		1800

11. a. $2290 **c.** $2411

13. a.
```
   482
  6731
+ 2064
  8277
    92
```

16. Yes   **17. a.**   7 $\boxed{7}$ 2 $\boxed{9}$ **20. c.**   4510
                                                                          10
               + $\boxed{5}$ 4 $\boxed{6}$ 2   + 2786
               _____   _____
                   1 3 1 9 1   7306

There are other answers to **20.c.**

21. There is no missing money. The desk clerk has $25, the bellboy has $2, and each man has $1. The confusion arises when the $27 (which includes the $2 "tip") is added to the $2. The situation is similar to the sum of numbers on the Hillsville sign in #2.

EXERCISE SET 3.2

1. a. 6,297,100   **2. a.** $219 + \boxed{1} = 220$

$220 + \boxed{20} = 240$ $1 + 20 + 7 = 28$

$240 + \boxed{7} = 247$

5. a. Computed $6 - 4$ **b.** After regrouping a 10 to the units column, the "5" was not changed to a "4"   **c.** The numbers were added.   **d.** Computed $9 - 7 = 2$ and $4 - 3 = 1$

6. a.

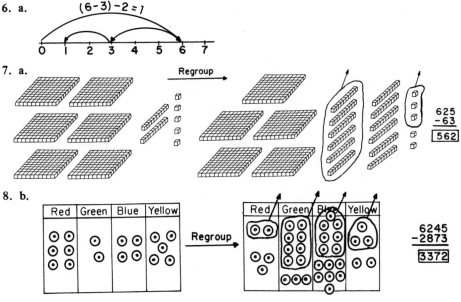

7. a.

8. b.

10. b. Add 4 to 3, add 2 to 2, add 5 to 0, and subtract 1 from 8.

13. The tens digit is always 9, and the sum of the units and hundreds digits is 9.

14. a. $\boxed{4}$ $\boxed{1}$ 2
             $-$ 3 $\boxed{7}$
             _____
                3 7 5

EXERCISE SET 3.3

1. b. 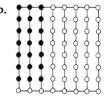 **f.**

The bulbs in the top face of the cube are all lighted.

2. a. 4 minutes and 9 seconds   **3. a.** $90

4.

```
   1   0   1
2 /8 /0 /5
 /   /   /
   2   0   2
0 /4 /0 /0
 /   /   /
   4   0   3
9 /2 /0 /5
   9   3   5
```

All partial products are recorded. The adding and carrying (if necessary) are done after the partial products are computed.

6. a. Commutative property for multiplication   **c.** Distributive property

7. a. $35 \times 19 = 35 \times (20 - 1) = 700 - 35 = 665$

8. a. 2500   **c.** Four hundred million

10. The product 42×30 is replaced by the equal product, 21×60, which can be written as $20 \times 60 + 60$. When 20×60 is replaced by 10×120, the 60 which is not used must be added later. Similarly, $5 \times 240 = 4 \times 240 + 240$. Therefore, when 4×240 is replaced by 2×480, the 240 which is not used must be added later. The last product, 1×960, plus the 60 and 240 which were not used, equals the product of 42×30. $(960 + 60 + 240 = 1260.)$

12. a.

Times 10   Times 4

Red	Green	Blue	Yellow

Red	Green	Blue	Yellow

Red	Green	Blue	Yellow

16. a. Yes   **c.** Multiply 8 times 7 to get the units digit in the product.

18. Yes   **20. a.**

```
   7   0   2
7 /2 /0 /4
   2   2   4
```

22. a. 51
          $\times$ 61

24. 775
        $\times$ 33

EXERCISE SET 3.4

1. a. Black   **c.** Black (396th), white (404th)   **3. a.** Not commutative

4. a. Measurement

6. a. The number of groups of markers on each column is a digit in the quotient. For example, there are 4 groups of two markers on the 10^3 column, and the digit in the thousands column of the algorithm is 4.

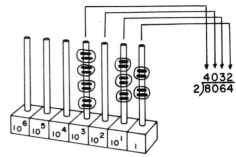

$$\begin{array}{r} 4032 \\ 2\overline{)8064} \end{array}$$

7. a. The sum of the numbers that correspond to the binary numbers 1, 2, and 8, equals 1232. Therefore, the quotient $1232 \div 112$ equals $1 + 2 + 8$, or 11.

8. a. 14,348,907   **c.** 268,435,456   **9. a.** 10^7 **c.** 10^{53}

10. a. Remainder of 28

12. $1 = (4 + 4) \div (4 + 4)$ $3 = (4 \times 4 - 4) \div 4$
$\quad\ 5 = (4 \times 4 + 4) \div 4$ $10 = (44 - 4) \div 4$

13. a. 5^{34} **c.** 10^{17} **15. a.** $([(22 - 19) + 2] \times 14) \div 10$ **16.** 2^{20}

17. a. 10^{19} **c.** 10^{13}

18. a., b. The steps in part c may produce the correct answer, depending on the type of calculator.

19. $354,393 \div 39$

EXERCISE SET 3.5

2. 47, 59, 73, 89, 107, 127, 149, 173, 199, 227, 257, $\boxed{289}$

4. 2, 3, 5, 7, 11, 13, 17, and 19   **5.** b and d   **6. a.** $924 = 2 \times 2 \times 3 \times 7 \times 11$

7. 1,000,000,000 has a unique factorization containing only 2s and 5s. Since there is no other factorization, 7 is not a factor.

8. a. 3 yellow rods or 15 white rods   **c.** 1, 2, 3, 6, 7, 14, 21, and 42

9. a. $n = 4, n = 6$

10. a. $21 = 3 + 7 + 11; 27 = 3 + 11 + 13; 31 = 7 + 11 + 13$ **c.** True

12. a. 5, 13, 17, 29, 37, and 41

13. a. 2 is a factor of 3628800 and a factor of itself. Therefore, 2 is a factor of 3628802. Three is a factor of 3628800 and also a factor of itself. Therefore, 3 is a factor of 3628803, etc.   **c.** Let p = the product of the whole numbers from 2 through 100. The numbers $p + 2, p + 3, p + 4, ..., p + 100$ form a sequence of 99 consecutive composite whole numbers.

14. a. K is greater than any of the primes $p_1, p_2, ..., p_n$.   **b.** A composite number has at least two prime factors.

EXERCISE SET 3.6

1. b. If the greatest number of the floors which must be delivered to is odd, deliver to all the odd-numbered floors first. Then walk down one flight of stairs and use the elevator for the even-numbered floors. A similar plan can be used if the greatest number of the floors which must be delivered to is even.

2. a. 900 seconds or 15 minutes   **c.** 3

3. a. White rods, green rods, or dark green rods   **c.** Relatively prime

4. a. 7 purple rods and 4 black rods

5. a. Prime numbers   **c.** Square numbers have an odd number of factors.

6. a. Prime numbers

7. a. 2, 3, 4, 5, 7, 8, 9, 10, 11, 13, 14, 15, 16, 17, 19   **c.** 6   **9. a.** Not necessarily

10. 512,112 and 4,328,104,292 are divisible by 4.   **11. d.** Yes

13. a.

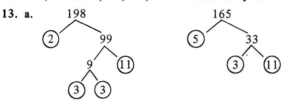

$$\text{g.c.f. } (198,165) = 3 \times 11 = 33$$

14. a.

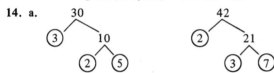

$$\text{l.c.m. } (30,42) = 2 \times 3 \times 5 \times 7 = 210$$

15. a. 3, 15, 21, 33, and in general numbers of the form $3n$ where 2 and 3 are not factors of n

16. 2520   **18. a.**

64796	Excess	The nines excess for	**d.** The remainder
$\times$ 1560	5	101181760 is 7, and	is 7.
101181760	$\times$ 3	for 15 it is 6. This	
	15	indicates an error.	

21. a. 1, 4, 9, and 16

23. Lockers whose numbers have an odd number of factors will be open (1, 4, 9, 16, 25, ...).

EXERCISE SET 4.1

1. b. 120°   **2. b.** 60°

3. a. Four new line segments and a total of 10 line segments   **b.** Five more line segments and a total of 15 line segments

4.

Number of sides	3	4	5	6	7	8	9	10	15
Number of diagonals	0	2	5	9	14	20	27	35	90

The number of diagonals of a polygon is the number of line segments connecting the vertices, minus the number of sides.

5. a. 105   **6. a.** 0, 1, 3, 4, 5, 6

7. a. Three new points of intersection and a total of 6 points of intersection

   b. Four new points of intersection and a total of 10 points of intersection

8. a. Three new regions and a total of 7 regions   **b.** Four new regions and a total of 11 regions

10. a. Angles G, I, J **c.** Angles B, E

11. The number of regions plus the number of vertices minus 1 equals the number of sides.

12.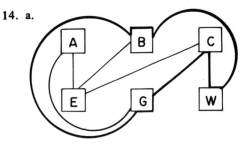

14. a.

EXERCISE SET 4.2

1. a. Hexagons; no   **2. a.** $72°$ **b.** Hexagon $60°$, octagon $45°$

3. Triangle $180°, 60°$; pentagon $540°, 108°$ **6. a.** Approximately $51.4°$

7. a. Pentagon   **8. a.** False   **c.** False   **e.** True

9. c. No. Each vertex angle has a measure of $135°$, and 135 is not a factor of 360.

11. a, c, e   **13. c.**

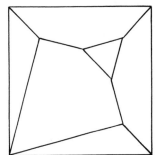

14. a. $72°$
   b. Triangle $120°$;
   Pentagon $72°$

EXERCISE SET 4.3

2. a. No   **b.** Angle a Angles b and c are right angles.

3. $CFGH, AEFH, ABCF, ACDH$ **5.** $v + f = e + 2$

7.

	Vertices	Faces	Edges
Tetrahedron	4	4	6
Cube	8	6	12
Octahedron	6	8	12
Dodecahedron	20	12	30
Icosahedron	12	20	30

9. Tokyo $(35°N, 140°E)$
San Francisco $(38°N, 120°W)$
Melbourne $(38°S, 145°E)$
Glasgow $(56°N, 4°W)$
Capetown $(35°S, 20°E)$

10. a. 1 through 7   **c.** 26   **11. a.** $(20°S, 60°E)$ **12.** Buenos Aires

13. a. Conic   **b.** Plane   **c.** Cylindrical   **14. a.** September 15: $(32.5°N, 48°W)$

15. *Hint:* Build in three dimensions.

16. White; one location is the North Pole. For other locations start 15 miles north of a 15-mile circle around the South Pole.

EXERCISE SET 4.4

1. c. Rectangular windows   **2. a.** Eight

3. a. Two lines of reflection and two rotational symmetries   **c.** Six lines of reflection and six rotational symmetries

4. b, g   a. 4   **b.** 2   **c.** 5   **d.** 1   **e.** 2   **f.** 7   **g.** 3   **5. a, c, e, f**

6. *Hint:* There are four letters with two lines of reflection and three letters with two rotational symmetries but no lines of reflection.

7. a. A, H, I, M, O, T, U, V, W, X, Y   Each of these letters has a vertical line of reflection.

8. a. Two, $(180°, 360°)$ **c.** Three, $(120°, 240°, 360°)$ **e.** Four, $(90°, 180°, 270°, 360°)$

9. a. 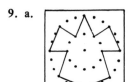   **10. a.** 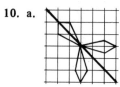   **11. b.**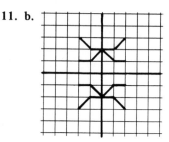

12. b. 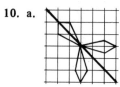   **14. a.** Cylinder, prism, sphere, cube   **c.** Cone (infinite), cylinder (infinite), prism (3), sphere (infinite), cube (4), pyramid (4)

16. a. Eight rotational symmetries and no lines of reflection

17. a. Sixteen   **c.** Six   **18.** Highest rating f; lowest rating b and d

EXERCISE SET 5.1

1. a. 92   **c.** 60   **3.** *Across* 1. 165   4. 3150   5. 120
                                        8. 92   9. 7920   10. 5550

4. Potatoes, 4.536 kg; motor oil, 946 ml

5. a. From 111.5 kg to 112.5 kg   **c.** From 48.25 cm to 48.35 cm

6. a. 11.6 kg   **c.** $35.64   **e.** 320 days; 5¢ per day   **7. a.** 12 ℓ   **8. a.** 7.2

9. a. 56   **c.** $108.48   **10.** 2

11. **b.** Length 158 m, breadth 26 m, height 16 m

12. **a.** Longer   **c.** Shorter   13. **a.** 57.8°C   **c.** ⁻14.8°F

14. **b.** Place a 100-g weight on one side of the scale with the 900-g item, and place the 1-kg weight on the other side.

15. **a.** 1651   **b.** 2   17. 5.8 cm

EXERCISE SET 5.2

1. **a.** Approximately 1088 m²   **b.** 106 m by 40 m; Area = 4240 m²

3. **b.** Dimensions (or perimeter) and area; the sum of the dimensions of the 3 by 5 rectangle, 3 + 5, is equal to the sum of the dimensions of the 4 by 4 square, 4 + 4.

4. Areas of rectangles in square centimeters: 13, 24, 33, 40, 45, 48, 49.

5. **a.** Type A   **c.** $8.28   **e.** Approximately 1428.6 m²

6. **a.** 1740 cm²   **c.** 1696.25 cm²

7. **a.** 1020 mm²   **c.** Approximately 1146 mm²   9. **a.** 1600 cm²

10. **a.** 5.76 cm²   **c.** 15 cm²

11. **a.** 1375 mm² or 13.75 cm²   **c.** 870 mm² or 8.7 cm²

12. **a.** 982 m² (to the nearest square meter)   **c.** 111 m   **e.** 3220 m²

13. **a.** 90,675 cm²   14. **a.** 56 cm²

15. **a.** $20,790   **c.** $1081.08 for the house in part **a**

16. The sectors of the circle should cover approximately 3.1 squares. This approximation will improve if the circle is cut into smaller sectors. Theoretically, the area of the circle is equal to π times the area of the square or approximately 3.14 squares.

17. **a.** 152 mm²   **c.** Approximately 22%   20. **a**

EXERCISE SET 5.3

1. **b.** Roof of the tower   **f.** Hands of the clock   **j.** Edges of the roof of the tower

2. **a.** 946 cm³   3. **a.** 166,779 cm³

4. **a.** 21,000 Btu unit   **c.** Type B   5. **a.** 37.5

6. **a.** They counted the surface area that can be seen.   7. **a.** 2.68 m   **c.** No

8. **b.** 1570 kg   9. **a.** 25,658 m³ (to nearest m³)

11. **a.** 509 m³ (to nearest m³)   **c.** 410 kg

12. **a.** About 25 times greater   14. **a.** Two times greater   **c.** 2^{20}

15. **a.** 72 cm³   **c.** 26.25 cm³   **e.** 9.42 cm³

16. **a.** 351.68 mm³   **b.** 7,033,600 cm³ or 7.0336 m³

EXERCISE SET 6.1

1. **a.** More   3. **a.** Dow Ch: $44.38   **b.** RCA: up 13 cents

4. **a.** Alaska Airlines   **c.** Less than   5. **a.** $1\frac{2}{3}$ **c.** $4\frac{1}{6}$ **e.** $3\frac{2}{5}$

6. a. $\frac{7}{4}$ **c.** $\frac{14}{3}$ **8. a.** Dark green rod   **c.** Brown rod

9. a.
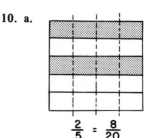
$$\frac{7}{10} = \frac{14}{20}$$

10. a.
$$\frac{2}{5} = \frac{8}{20}$$

11. a. $\frac{28}{32}$ **c.** $\frac{12}{18}$ **12. a.** $\frac{2}{9}$ **c.** $\frac{3}{9}$

13. a. $\frac{2}{3} = \frac{10}{15}$ **c.** $\frac{3}{15} = \frac{6}{30}$

$\frac{4}{5} = \frac{12}{15}$ $\frac{5}{6} = \frac{25}{30}$

14. a. $\frac{2}{3} > \frac{4}{7}$ $\frac{8}{11} < \frac{3}{4}$ **b.**

$\frac{2}{3} = \frac{14}{21}$		$\frac{8}{11} = \frac{32}{44}$	
$\frac{4}{7} = \frac{12}{21}$		$\frac{3}{4} = \frac{33}{44}$	

15. a. $\frac{6}{32} < \frac{5}{24}$ because $.1875 < .208\overline{3}$ **17.** Iron

19. a. 1.5 liters   **b.** $\frac{40}{1040} = \frac{1}{26}$ **20. a.** $\frac{1}{3}$ **c.** $\frac{2}{5}$

EXERCISE SET 6.2

1. a. $3\frac{5}{8}$ **c.** Drew National $1\frac{1}{4}$ MEM Company $\frac{5}{8}$ Old Town $1\frac{3}{8}$

2. b.
$$\frac{4}{5} = \frac{32}{40}$$
$$+ 1\frac{3}{8} = 1\frac{15}{40}$$
$$\overline{1\frac{47}{40} = 2\frac{7}{40}}$$

3. b.
$$1\frac{1}{8} = 1\frac{5}{40} = \frac{45}{40}$$
$$- \frac{7}{10} = \frac{28}{40} = \frac{28}{40}$$
$$\overline{\frac{17}{40}}$$

4. a. 5.4 meters

5. a.

$1\frac{1}{3}$	$\frac{1}{2}$	$\frac{2}{3}$
$\frac{1}{6}$	$\frac{5}{6}$	$1\frac{1}{2}$
1	$1\frac{1}{6}$	$\frac{1}{3}$

Magic number $= 2\frac{1}{2}$

6. a.
$$\frac{1}{2} \quad \frac{1}{3} \quad \frac{1}{4} \quad \frac{1}{5} \quad \frac{1}{n}$$
$$-\frac{1}{3} \quad -\frac{1}{4} \quad -\frac{1}{5} \quad -\frac{1}{6} \quad -\frac{1}{n+1}$$
$$\overline{\frac{1}{6}} \quad \overline{\frac{1}{12}} \quad \overline{\frac{1}{20}} \quad \overline{\frac{1}{30}} \quad \overline{\frac{1}{n(n+1)}}$$

The denominator of the difference is the product of the denominators of the two fractions, and the numerator is 1.

8. a. Denominators were added.   **c.** Denominators were not multiplied.

9. a. Commutative property for addition   **c.** Associative property for addition
e. Distributive property

11. 1. $7333.33   **13. a.** D, E, G, A, B   **c.** $\frac{1}{128}$ as long
    2. $2400.00
    3. $9733.33

14. a. $\frac{2}{7}$ **17.** 84

EXERCISE SET 6.3

1. a. ⁻29°C   **2. a.** 16   **c.** Sweden ⁺517 million kilograms
                                        West Germany ⁻1507 million kilograms

3. a. 17 days and 17 hours

4. a. 6 + ⁻5 = 1

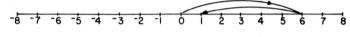

5. b. 4 − 7 = ⁻3

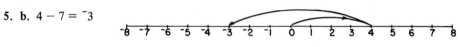

7. a. $\frac{1}{4}$ **c.** 2.7%   **8.** ⁻1   **11. a.** ⁻14   **c.** ⁻4

12. a. Commutative property for addition   **c.** Associative property for multiplication

14. The negative of $\frac{7}{8}$ is $\frac{⁻7}{8}$ and the reciprocal of $\frac{7}{8}$ is $\frac{8}{7}$.

15. a. ⁻12   **c.** ⁻4   **e.** ⁻5   **16. a.** 2   **c.** 5   **17. a.** 1

18. a. The reciprocals of nonzero numbers in the interval from ⁻1 to 1 will be outside this interval.

19. a. Yes   **c.** No   **e.** No   **21. a.** ⁻3, ⁻6, ⁻9, ⁻12, ⁻15, ⁻18, ⁻21, ⁻24

EXERCISE SET 7.1

1. a. 9,193,000,000   **3. a.** ⁻11.9° Celsius   **c.** ⁻13.6° Celsius

5. a. Three hundred sixty and two thousand eight hundred seventy-six ten-thousandths
    e. Three hundred forty-seven and ninety-six hundredths

7. $\frac{7}{12} = .58\overline{3}$ $\frac{5}{6} = .8\overline{3}$ $\frac{11}{20} = .55$ **a.** Finite   **8. a.** $\frac{837}{1000}$ **c.** $\frac{64}{99}$

9. a. .44   **c.** 3.72   **10. a.** .8236   **12. a.** .4266   **c.** .43365   **e.** .436

13. a. .41$\overline{6}$ **14. a.** 0.1, 0.65, 1.0, 1.24   **15. a.** $\frac{9}{9}$ **b.** Yes

16. M. Sweeney 6.46875 feet   W. G. George 4.21$\overline{3}$ minutes

EXERCISE SET 7.2

1. **a.** Airlines vs. Railroads, 2.01   Life-insurance companies vs. automobile dealers, .78

2. **a.**

3. **a.** $1210.90   **c.** $3.76   **e.** $158.55

4. **a.** $38.50   **c.** $13.31   **e.** $161.81

5. **a.** 852.6 kilowatt-hours   **c.** $34.10

6. **a.**

$$\begin{array}{r} 1 \\ 4.821 \\ +61.73 \\ \hline 66.551 \end{array}$$

$$\frac{8}{10} + \frac{7}{10} = \frac{15}{10} = \frac{10}{10} + \frac{5}{10} = 1 + \frac{5}{10}$$

7. **a.**

$$\begin{array}{r} 5 \\ 6\!\!\!/6.43 \\ -41.72 \\ \hline 24.71 \end{array}$$

$$6 = 5 + 1 = 5 + \frac{10}{10}$$

8. **b.** 37.4°C

9.

161.001

11. **a.**

$$\begin{array}{r} 300 \\ 30 \\ +700 \\ \hline 1030 \end{array}$$

c.

$$\begin{array}{r} 600 \\ \times\ 4 \\ \hline 2400 \end{array}$$

These computations do not indicate an error.

13. **a.** .8874106   **c.** 33

14. **a.** The numbers will become so large the calculator display will be exceeded.
    c. The numbers will become so small the calculator display will show 0.

15. **b.** 18   **16. a.** 3206.88 feet   **c.** 5245 new wells and 16.8 million feet of drilling

17. **a.** 4 minutes and 11.76 seconds   **18. a.** 1.3 seconds

19. **a.** $118.3 billion   **c.** Hospital bills

EXERCISE SET 7.3

1. **a.** Approximately 5.55 hours or 5 hours and 33 minutes

3. Mercury 36,002,000   **4. a.** 3.2×10^{-2} **5. a.** .000000003048
   Venus 6.7273×10^7 **c.** 635,000,000,000
   Earth   93,003,000

6. **a.** 3.9952×10^4 kph   **c.** 2.551176×10^{18} km

7. The distributive property   **a.** $178.08   **8. a.** $50.54

9. **a.** 2,822,200   **c.** 1,421,400   **10. a.** Favored: 638. Opposed: 623.

11. **a.** Large size   **12. a.** $5742   **c.** 2.8   **13. a.** 1 to 45

15. **a.** Approximately .2%   **c.** 27.168 tons   **16. a.** $1060   **c.** 12 years

17. Alabama 21.4 to 1   **c.** Oregon and Florida
    Florida   23.8 to 1
    Hawaii   20.9 to 1

19. a. New Jersey   **20. a.** 12,114,714,773.7   **c.** 480,742,649.75

21. a. 64% dismissed or dropped; 26% solved by pleas of guilty; 10% brought to trial

EXERCISE SET 7.4

1. $\frac{1}{8}$, rational number and real number; $\sqrt{3}$, real number.

2. a. Rational   **c.** Irrational   **4. a.** $3\sqrt{5}$ **c.** $2\sqrt{15}$ **5. a.** 10   **c.** 14

6. a. 4.5   **c.** 2.3   **7. a.** 5.7   **c.** 4.1

9. a. A line segment of length $\sqrt{5}$ can be constructed as the hypotenuse of a right triangle whose legs have lengths of 1 and 2.

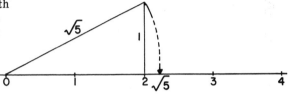

10. a. $\dfrac{3\sqrt{7}}{7}$ **c.** $\sqrt{5}$

11. $\sqrt{.2} \approx .4$ $\sqrt{.5} \approx .7$ $\sqrt{.6} \approx .8$ $\sqrt{.8} \approx .9$ The square root of a number between 0 and 1 is also between 0 and 1 but is greater than the given number.

12. a. 1   **b.** 1   **13. a.** No; $3 - 5 = {}^-2$ **d.** Yes

14. a. Associative property for multiplication   **c.** Commutative property for addition

15. a. Inverses for addition, inverses for multiplication, and the completeness property
    c. Completeness property

16. a. ${}^-7; (Z, Q, R)$ **d.** $\sqrt{3}$ and ${}^-\sqrt{3}; (R)$ **18. b.** Yes   **c.** $\sqrt{7.5} \approx 2.7$ Yes

19. a. 3.140845   **c.** 3.1415929   **e.** 3.1416

20. b. Area $= \dfrac{1}{2^{17}}$ square units; dimensions: $\dfrac{1}{\sqrt{2^{17}}}$ by $\dfrac{1}{\sqrt{2^{17}}}$

EXERCISE SET 8.1

1. a. $(5,420^\circ)$ **c.** $(6, {}^-80^\circ)$ **2. a.** 7 seconds   **b.** .12 second

3. a. £1.50 $\approx$ \$2.50   **b.** \$6 $\approx$ £3.50   **4. a.** 10   **c.** 12.7 meters

5. a. 3.2 ounces cost 54 cents.   **b.** 15, 28, 41, 54, etc.   **d.** No

7. a. Slope of line I is 3. Slope of line II is ${}^-3$.   **b.** II and III.

8. a. Slope of L_1 is 10. Slope of L_3 is $\frac{1}{2}$.   **b.** No

9. a.

x	${}^-2$	${}^-1\frac{1}{2}$	${}^-1$	${}^-\frac{1}{2}$	0	$\frac{1}{2}$	1	$1\frac{1}{2}$	2
y	${}^-8$	${}^-3\frac{3}{8}$	${}^-1$	${}^-\frac{1}{8}$	0	$\frac{1}{8}$	1	$3\frac{3}{8}$	8

10. b. \$5

11. a. L_2 is for oil heat and L_1 is for solar heat.   **c.** 10; this is the x-coordinate of the point of intersection of the two lines.   **e.** \$1200

12. a. Speed factor for force equals the square of the speed. **b.** 4

13. a. Tangent $30° \approx .57735$; tangent $60° \approx 1.7321$; tangent $85° \approx 11.430$. The tangents become greater and greater. **b.** Tangent $23° \approx .424$; tangent $13° \approx .231$; tangent $4° \approx .070$. The angles of inclination get closer and closer to 0.

14. a. Cloud x $(70,75°)$; Cloud y $(50,115°)$; Cloud z $(80,250°)$

16. a. $(12,9)$; $(14,10)$; $(16,11)$

EXERCISE SET 8.2

1. a. Paraboloid

3. The greatest distance is obtained from a $45°$ angle of elevation. The same distance is obtained for $35°$ and $55°$ angles of elevation.

4. a. Parabola **c.** 6 seconds **5. a.** 10 and 10 **6. a.** III

7. a.

x	0	0	5	⁻5
y	4	⁻4	0	0

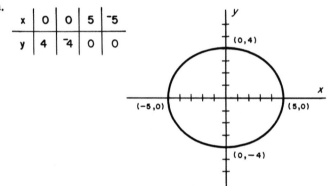

8. a. 15π or approximately 47.1 square units

9. a. The eccentricity approaches zero, and the ellipse becomes more circular.

10. a. y increases **c.** Hyperbola **11. a.** y approaches 0 **c.** No

EXERCISE SET 8.3

1. a. I **2.**

d	1	2	3	4	5	6
I	10	2.5	$\frac{10}{9}$	$\frac{5}{8}$	$\frac{2}{5}$	$\frac{5}{18}$

a. Intensity increases

3. $y = 2^x$

x	⁻3	⁻2	⁻1	0	1	2	3
y	$\frac{1}{8}$	$\frac{1}{4}$	$\frac{1}{2}$	1	2	4	8

4. $y = 3^x$

x	⁻3	⁻2	⁻1	0	1	2	3
y	$\frac{1}{27}$	$\frac{1}{9}$	$\frac{1}{3}$	1	3	9	27

6. a. 264 million **7. a.** 7 weeks (6th week to 13th week)

8. a. Two **b.** Each succeeding revolution takes longer **9. a.** E **c.** A

10. a. 4 hours **c.** No, not for relatively short distances

11. D has a wavelength of 1.14 m and E has a wavelength of 1.02 m.   **a.** Two

12. c. 1600   **13. a.** .5   **c.** .5   **e.** Close to ⁻1 (⁻.98 to 2 decimal places)

15. a. Sin $20°$ ≈ .34202; sin $60°$ ≈ .86603; sin $86°$ ≈ .99756
   b. Sin $40°$ = sin $140°$ ≈ .64279   **c.** Sin $10°$ = sin $370°$ ≈ .17365

16. a. This graph's domain is the years from 1900 to 2100. The years from 1970 to 2100 represent prediction of the future.   **c.** Population: 1950 to 2050; pollution: 1960 to 2025; food supply is decreasing.

18. a. 2.71828183   **b.** .049787068

EXERCISE SET 9.1

1. a. No   **c.** Translation   **2. b.** Reflections or $180°$ rotations

3. b. Infinite number   **d.** No   **5. a.** Yes   **c.** Their areas are equal

6. a. $90°$ **c.** All points on line L **7. a.** $90°$ **8. a.** Over 6 and up 1

9. a. Translation   **10. b.** It has twice as many degrees as $\angle POQ$.

12. a.

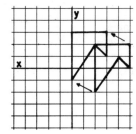

13. Reflect across the rows and down the columns.

14. b. A' (3,3) B' (0,⁻3) C' (5,⁻4)

16. a. Translation.

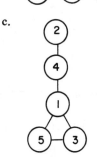

18. a. ③—①—④—②

   c.
   ②
   |
   ④
   |
   ①
   / \
   ⑤—③

EXERCISE SET 9.2

1. a. 10,000   **c.** 6 cm

5. c. $\dfrac{^-7}{40}$ **6. a.** Yes   **b.** Yes

7. a. $\triangle ABC$ is similar to $\triangle AEF$. Therefore, $\dfrac{AB}{AE} = \dfrac{BC}{EF} = \dfrac{AC}{AF}$. Furthermore, $\triangle ACD$ is similar to $\triangle AFG$. Therefore, $\dfrac{AD}{AG} = \dfrac{DC}{GF} = \dfrac{AC}{AF}$. All angles in the rectangles equal $90°$.

b. Rectangle (2) is similar to rectangle (4).

9. a. Grid A to Grid B: scale factor $= \dfrac{3}{2}$. Grid B to Grid C: scale factor $= \dfrac{1}{5}$.

c. $\dfrac{9}{4}$ or $2\dfrac{1}{4}$

4. a.

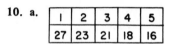

Scale factor = 2

10. a.

1	2	3	4	5
27	23	21	18	16

11.

1	2	3
3 x 2 x 1	6 x 4 x 2	9 x 6 x 3
22 cm^2	88 cm^2	198 cm^2
6 cm^3	48 cm^3	162 cm^3

12. a. 56 cubic units

13. a. 21 mm   **c.** 64

14. a. Height: 320 cm; width: 192 cm
c. 512 kg

15. b. 1728

EXERCISE SET 9.3

1. a. Figure 3   **c.** Figure 1   **e.** Figure 1

3. a.

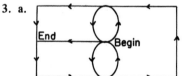

4. a.

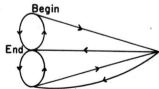

One odd vertex will be a beginning point and the other odd vertex will be an ending point.

5. a. Octahedron   **b.** Three, connecting six different vertices.

6. Yes

8. No

9. a.

e. There must be exactly zero, one, or two regions with an odd number of sides for a solution.

10. c. The point is inside if the line crosses the simple closed curve an odd number of times. Points of tangency should not be counted.

11. a. No   **12. a.** 4

13. e. If there are one or more odd vertices, at least 3 colors will be needed.

15. b.

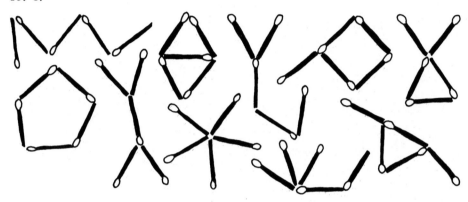

EXERCISE SET 10.1

1. a. $\frac{15}{36}$ **b.** $\frac{15}{36}$ **2. b.** $\frac{1}{9}$ **3. b.** $\frac{4}{9}$

	Tetrahedron	Cube	Octahedron
4. a. A number less than 3	$\frac{1}{2}$	$\frac{1}{3}$	$\frac{1}{4}$
b. An even number	$\frac{1}{2}$	$\frac{1}{2}$	$\frac{1}{2}$
c. The number 2	$\frac{1}{4}$	$\frac{1}{6}$	$\frac{1}{8}$

5. a. Probability is $\frac{1}{13}$. Odds are 1 to 12.   **c.** Probability is $\frac{1}{4}$. Odds are 1 to 3.

6. a. $\frac{12}{13}$ **c.** $\frac{3}{4}$ **7. a.** 7 to 3   **c.** 1 to 499   **8. a.** $\frac{4}{7}$

9. a. $\frac{9,929,200}{10,000,000} \approx .99$ **c.** $\frac{6,800,531}{9,396,358} \approx .72$ **e.** 14   **10. b.** $30

11. **a.** $1.58   12. **a.** $\dfrac{1}{38}$ **c.** Equal   13. **a.** $\dfrac{1}{9}$ **b.** $\dfrac{1}{81}$ **c.** $\dfrac{1}{9}$

14. **a.** $\dfrac{14}{1540} = \dfrac{1}{110}$ **c.** $6

EXERCISE SET 10.2

1. **a.** $\left(\dfrac{18}{38}\right)^2 \approx .224$ **c.** $\left(\dfrac{18}{38}\right)^3 \approx .106$ 2. $\dfrac{1}{2}$

3. **a.** Independent, $\dfrac{1}{8}$ **b.** Dependent, $\dfrac{1}{13} \times \dfrac{1}{17} \approx .005$

4. **a.** $\left(\dfrac{1}{12}\right)^2 = \dfrac{1}{144}$ **c.** $\left(\dfrac{11}{12}\right)^3 \approx .77$ 5. **b.** $\dfrac{3}{8}$

6. **b.** The numbers in the tables are the fourth and fifth rows of Pascal's triangle.
   d. $\dfrac{5}{32}$

7. **a.** $\dfrac{10}{20} \times \dfrac{9}{19} \approx .24$ **c.** 1   9. **a.** (1) $\dfrac{1}{10}$ (2) $\dfrac{1}{5} \times \dfrac{1}{5} = \dfrac{1}{25}$

10. **a.** $1 - \left(\dfrac{5}{6}\right)^4 \approx .52$ **c.** $1 - \left(\dfrac{28}{36}\right)^3 \approx .53$ 12. **b.** $\left(\dfrac{1}{365}\right)^2 \approx .0000075$

13. **a.** $\dfrac{1}{20}$ **c.** $\dfrac{1}{20} \times \dfrac{3}{20} \times \dfrac{1}{20} = .000375$ 14. **a.** $\dfrac{1}{144}$

EXERCISE SET 10.3

1.

Interval	Frequency
3800–4099	I
4100–4399	6
4400–4699	8
4700–4999	5

2. Mode: 5300 to 5599
   Median: 5000 to 5299
   The mean income in the United States in 1974 falls in the mode interval.

4. **a.** Mode   **c.** Median   5. **a.** 14   **c.** Mode or median

6. **a.** About 1.57 times longer for a new car; about 4.61 times longer for a sewing machine; 4 times longer for a pound of butter.

7. **a.** National   **c.** National

8. **a.** 6   **c.** The test scores for the second class were closer to the mean of 6 and not as spread out as the test scores for the first class.

9. **a.** 1734.1   **c.** 4.5   10. **b.** Approximately 94.5%

12. **a.** 16 to 329   **b.** 4770

1. **a.** 8908 and 9042   **c.** If the coins were tossed in a random manner and the outcomes were not affected by deformed or dirty coins, the conclusion was justified since 9207 is more than 3 standard deviations from the mean.

2. **b.** 23, four times; 66, once; 74, twice; 09, three times

3. **b.** Use groups of five numbers. The first three numbers determine a direction from 0 to 360 degrees. [If the number is greater than 360, subtract 360 (or 720) and use the remainder.] The second two digits determine the distance. (*Note:* There are other ways to answer this question.)

4. **a.** Use pairs of digits and add the two numbers. If the sum is even, the "toss" is heads, and if the sum is odd, the "toss" is tails.

5. **a.** 16%   **7. a.** 34   **c.** 1

8. **a.** Normal   **c.** Skewed to the left   **e.** Normal   **10. a.** 70%   **b.** 95%

11. The 94th percentile is in the 8th stanine. The 70th percentile is in the 6th stanine.

12.

Stanine	1	2	3
Lower Percentile	1	4	11
Upper Percentile	3	10	22

13. **a.** 1   **c.** 6   **e.** 1
   g. 1, 5, 10, 10, 5, 1 for the six compartments from left to right

rule for subtracting, 128–
129, 351
scientific notation, 350
Exponential equation, 412,
417
Exponential form, 128
Exponential function, 412,
423
Expressible, 364
Extrapolation, 392

Factor, 138
common, 149
greatest common, 149
models, 140–141, 151
proper, 155–156
Factor tree, 140, 150–151, 157
Faded document, 95, 104, 122,
136
Fahrenheit, Gabriel, 217
Fahrenheit scale, 217, 224, 306
Fair game, 477
Fathom, 212
Ferguson, D.F., 368
Fermat, Pierre, 144, 388, 474
Fibonacci, Leonardo, 7, 96
Fibonacci numbers, 6–7,
10–11, 277
Finger counting:
base ten, 68
binary, 81
Finger multiplication, 111,
121
Finite differences, 8, 11–12
Finite set, 47
Fixed decimal point display,
322
Fixed point, 427–428
Focus:
ellipse, 401
hyperbola, 403
parabola, 399–400, 405
Folio, 455
Folium of Descartes, 410
Foot, 212–213
Four-Color Problem, 32, 466,
471
Fourth dimension, 184
Fraction Bars, 266
addition, 279–280
common denominator,
268–269
division, 283–284
equality, 267–268, 274
multiplication, 282–283

subtraction, 280–281
Fractions, 264
addition, 279–280
Babylonian, 263
calculator, 271, 276, 285–
287, 292
common denominator,
268–269
converting to decimals, 271,
285–287, 320–321
denominator, 264, 268–269
density, 270
division, 283–284
division of whole numbers,
265, 271
Egyptian, 223, 263
equality, 267–270, 274–276
improper, 266
inequality, 269–270
inverses, 285
lowest terms, 268
mixed number, 266, 280–281
models, 266–269, 274–275
multiplication, 282–283
music, 293–294
number line, 267, 270, 279–
280, 284, 289
numerator, 264
subtraction, 280–281
unit, 263, 267, 274
Frame, M.R., 144
Fraser spiral, 423
Frend, William, 296
Frequency:
distribution, 496–498, 504–
505
polygon, 497
Frequency of sound, 414,
420–421
Fujii, John, 471
Function, 381–382
domain, 382
exponential, 412, 423
linear, 388–389, 394
many-to-one, 382
mapping, 425
one-to-one, 383
range, 382
Fundamental Theorem of
Arithmetic, 140

Galaxy:
in Andromeda, 2
Messier, 47
Milky Way, 129

Galilei, Galileo, 15, 21, 211,
398, 405
Galton, Sir Francis, 521
Game, 19, 104, 135–136, 158,
440, 459
Gardner, Martin, 36, 120, 200,
416, 441, 472
Gelosia algorithm:
decimals, 337
whole numbers, 107, 117
Gematria, 137–138
Generalized distributive prop-
erty, 108
Genus, 464
Geoboard:
circular, 206
rectangular, 397
Geometric number, 8, 13
Geometric sequence, 8, 12,
114, 130, 312, 335
common ratio, 8
Glide reflection, 429, 440
Gnomon, 12
Goff, G., 33
Goldbach, C., 18, 31, 144
Goldbach's conjecture, 18, 144
Golden Mean, The, 293
Golden rule, 347
Gram, 213, 216
Graph:
bar, 497, 511, 514, 520
equations, 387–389, 392–
393
frequency polygon, 497
*Graphic Works of M. C. Escher,
The,* 429
Grating method, 107
Graunt, J., 496
Greater than (>):
decimals, 319
fractions, 269–270
origin of symbol, 88
signed numbers, 298
whole numbers, 88
Great circle, 23, 189
Greatest common factor, 149–
150, 157
Greek:
letters and numbers, 137
numeration system, 67
*Guinness Book of World
Records,* 327
Gulliver's Travels, 458

Haeckel, Ernst, 179

Parabola, 399–401, 404
Paraboloid, 400
Parallel lines, 162, 393
Parallelogram, 167
  altitude, 229
  area, 229
Parallel postulate, 17, 23
Parallels of latitude, 189
Partial products, 107, 112–113
Partial sums, 87
Pascal, Blaise, 7, 91, 474
Pascal's triangle, 7, 11, 491,
  521
Patterns:
  nature, 2–3, 6–7, 173, 176,
    179, 184, 199, 204, 277,
    314, 411, 412–413
  numbers, 7–8, 79, 122, 168–
    170, 291, 327–328, 378
  ornaments, 4–5, 171, 177,
    208, 416, 423, 430, 435
Peano, Giuseppe, 16
Pendulum, 15, 21
Pentagon, 165, 174, 175
Pentagonal number, 13
Percent, 348
  applications, 349, 351–352,
    356–362
  calculator, 351–352
Percentile, 515–516
Perceptual blocks, 34
Perfect number, 156
Perimeter, 227–229
  simple closed curve, 232
Perpendicular, 163, 392
Pi (π):
  area of circle, 234, 243
  circumference of circle, 234
  distribution of digits, 517
  early values, 368, 378
  mnemonics, 368
Piaget, Jean, 468
Piele, D.T., 440
Pitch, 414
Placeholder, 18
Place value:
  decimals, 315–318
  whole numbers, 58, 61–64
Plane, 162
Plane of symmetry, 200–201,
  203, 207
Plane projection, 190
Plane region, 165
Planimeter, 232

Plateau, Joseph, 31
Plato, 2, 313
Platonic solid, 184–185, 193,
  469, 480
Point, 162
Polar coordinate system, 385,
  390, 396–397
Polya, George, 15, 26
Polygon, 165
  angle, 174–175, 180
  area, 226–244
  circumscribed, 176, 180
  convex, 165, 178
  diagonal, 166, 169
  inscribed, 179–180
  nonconvex, 165, 178
  perimeter-area relationship,
    15–16, 227–229
  regular, 173–174, 179–180
  tessellation, 176–178
Polygonal number:
  pentagonal, 13
  square, 8, 13
  triangular, 8, 11, 13
Polyhedron, 184
  area-volume relationship, 16,
    246
  convex, 185
  duals, 193
  nonconvex, 185
  prism, 187
  pyramid, 186
  regular, 184–185, 480
  semiregular, 185–186
  volume, 246–252
Pope, Alexander, 1
Positional numeration system,
  58–60, 69
  Hindu-Arabic, 59–60, 315
  Mayan, 59
Positive and negative numbers
  (see also Integers)
  addition, 299
  applications, 296–297
  decimals, 319
  division, 303
  inequality, 298, 311
  models, 298–299
  multiplication, 301–302
  number line, 298–299, 308,
    311
  subtraction, 300–301
Positively skewed, 514, 519
Positive slope, 387, 392–393

Postulate, 17
Pound, 213
Precision, 218–219, 235–236
Prime numbers, 138
  conjecture, 31, 144
  distribution, 138
  infinitude, 145
  largest known, 138
  relatively prime, 150
Prime number test, 138–139
Principles of Algebra, 296
Prism:
  altitude, 247
  crystal, 187
  oblique, 187, 247
  rectangular, 246
  right prism, 187
  surface area, 247
  volume, 246–247
Probability, 474–476
  binomial, 487–488
  complementary, 488–489
  compound, 485–494
  dependent events, 486–487,
    490
  empirical, 476
  fair game, 477
  independent events, 486, 490
  life expectancy, 476, 481–
    482
  mathematical expectation,
    477–478
  odds, 474, 476–477, 481–
    482
  theoretical, 474–475
Problem solving, 26–37
  drawing pictures, 26
  forming tables, 29–30
  guessing, 27
  simplifying, 29
  working backwards, 30–31
Projection point, 443
Proper factor, 155
Properties of operations (see
  Addition, Subtraction,
  Multiplication, and
  Division)
Proportion, 347
  applications, 347–348, 357
  rule of three, 347
Protractor:
  circular, 163
  semicircular, 175–176
Ptolemy, Cladius, 378